Algebra and Applications 2

SCIENCES

Mathematics, Field Director – Nikolaos Limnios

Algebra and Geometry, Subject Head – Abdenacer Makhlouf

Algebra and Applications 2

Combinatorial Algebra and Hopf Algebras

Coordinated by

Abdenacer Makhlouf

iSTE

WILEY

First published 2021 in Great Britain and the United States by ISTE Ltd and John Wiley & Sons, Inc.

ISTE Ltd
27-37 St George's Road
London SW19 4EU
UK

www.iste.co.uk

John Wiley & Sons, Inc.
111 River Street
Hoboken, NJ 07030
USA

www.wiley.com

Library of Congress Control Number: 2021942616

British Library Cataloguing-in-Publication Data
A CIP record for this book is available from the British Library
ISBN 978-1-78945-018-7

ERC code:
PE1 Mathematics
PE1_2 Algebra
PE1_5 Lie groups, Lie algebras
PE1_12 Mathematical physics

MIX
Paper from responsible sources
FSC www.fsc.org FSC® C013604

Contents

Preface

Abdenacer MAKHLOUF

University of Haute Alsace, Mulhouse, France

The aim of this series of books is to report on the new trends of research in algebra and related topics. We provide an insight into the fast development of new concepts and theories related to algebra and present self-contained chapters on various topics, with each chapter combining some of the features of both graduate-level textbooks and research-level surveys. Each chapter includes an introduction with motivations and historical remarks, as well as the basic concepts, main results and perspectives. Moreover, the authors have commented on the relevance of the results in relation to other results and applications.

In this volume, the chapters encompass surveys of basic theories on non-associative algebras like Lie theories, using modern tools and more recent algebraic structures like Hopf algebras, which are related to Quantum groups and Mathematical Physics. The algebraic background of pre-Lie algebras, other non-associative algebras (non-associative permutative, assosymmetric, dendriform, etc.) and algebraic operads is presented. This volume also deals with noncommutative symmetric functions, Lie series, descent algebras, chronological and Rota–Baxter algebras. We focus on the increasing role played by Combinatorial algebra and Hopf algebras, as well as some non-associative algebraic structures in iterated integrals, chronological calculus, differential equations, numerical methods and control theory. It turns out that the Hopf algebra of rooted trees is an adequate tool, not only for vector fields, but also for studying the numerical approximation of their integral curves. Runge–Kutta methods form a group (called the Butcher group), which is the character group of the Connes–Kreimer Hopf algebra. The algebraic theory of Runge–Kutta methods B-series and related formal expansions are considered. Algebraic structures underlying calculus with iterated integrals lead naturally to

Algebra and Applications 2,
coordinated by Abdenacer MAKHLOUF.

the notions of descent (Hopf) algebra, as well as permutation Hopf algebra. In this volume we discuss the Lie-theoretic perspective and advances of chronological calculus. Chronological algebras and time-ordered products appear in an (almost) uncountable number of places, especially in theoretical physics and control theory. Noncommutative symmetric functions are applied to the study of formal power series with coefficients in a noncommutative algebra, in particular to Lie series. Moreover, Lie idempotents, Eulerian idempotents and Magnus expansion are considered. In addition, the interaction of algebra and (infinite-dimensional) geometry in the guise of Hopf algebras and certain associated character groups is examined. It turns out that fundamental concepts in control theory are inherently linked to combinatorial and algebraic structures. It is shown how modern combinatorial algebraic tools provide deeper insight and facilitate analysis, computations and design. The emphasis is on exhibiting the algebraic structures that map combinatorial structures to geometric and dynamic objects.

I thank Kurusch Ebrahimi-Fard for suggesting these topics, presented at Benasque Intensive School, and express my deep gratitude to all of the contributors of this volume and to ISTE for their support.

August 2021

1

Algebraic Background for Numerical Methods, Control Theory and Renormalization

Dominique MANCHON

University of Clermont Auvergne, Clermont-Ferrand, France

1.1. Introduction

Since the pioneering work of Cayley in the 19th century (Cayley 1857), we have known that rooted trees and vector fields on the affine space are closely related. Surprisingly enough, rooted trees were also revealed to be a fundamental tool for studying not only the integral curves of vector fields, but also their Runge–Kutta numerical approximations (Butcher 1963).

The rich algebraic structure of the k-vector space $\mathcal{T}$ spanned by rooted trees (where k is some field of characteristic zero) can be, in a nutshell, described as follows: $\mathcal{T}$ is both the free pre-Lie algebra with one generator and the free non-associative permutative algebra with one generator (Chapoton and Livernet 2001; Dzhumadil'daev and Löfwall 2002), and moreover, there are two other pre-Lie structures on $\mathcal{T}$, of operadic nature, which show strong compatibility with the first pre-Lie (respectively the NAP) structure (Chapoton and Livernet 2001; Calaque *et al.* 2011; Manchon and Saïdi 2011). The Hopf algebra of coordinates on the Butcher group (Butcher 1963), that is, the graded dual of the enveloping algebra $\mathcal{U}(\mathcal{T})$ (with respect to the Lie bracket given by the first pre-Lie structure), was first investigated

Algebra and Applications 2,
coordinated by Abdenacer MAKHLOUF.

in Dür (1986), and intensively studied by Kreimer for renormalization purposes in Quantum Field Theory (Connes and Kreimer 1998; Kreimer 2002), see also Brouder (2000).

This chapter is organized as follows: the first section is devoted to general connected graded or filtered Hopf algebras, including the renormalization of their characters. The second section gives a short presentation of operads in the symmetric monoidal category of vector spaces, and the third section will treat pre-Lie algebras in some detail: in particular, we will give a "pedestrian" proof of the Chapoton–Livernet theorem on free pre-Lie algebras. In the last section, Rota–Baxter, dendriform and NAP algebras will be introduced.

1.2. Hopf algebras: general properties

We choose a base field k of characteristic zero. Most of the material here is borrowed from Manchon (2008), to which we can refer for more details.

1.2.1. *Algebras*

A k-algebra is by definition a k-vector space A together with a bilinear map $m : A \otimes A \to A$ which is *associative*. The associativity is expressed by the commutativity of the following diagram:

$$\begin{array}{ccc} A \otimes A \otimes A & \xrightarrow{m \otimes I} & A \otimes A \\ \big\downarrow {\scriptstyle I \otimes m} & & \big\downarrow {\scriptstyle m} \\ A \otimes A & \xrightarrow{\quad m \quad} & A \end{array}$$

The algebra A is *unital* if there is a unit $\mathbf{1}$ in it. This is expressed by the commutativity of the following diagram:

$$\begin{array}{ccccc} k \otimes A & \xrightarrow{u \otimes I} & A \otimes A & \xleftarrow{I \otimes u} & A \otimes k \\ & {\scriptstyle \sim}\searrow & \big\downarrow {\scriptstyle m} & \swarrow{\scriptstyle \sim} & \\ & & A & & \end{array}$$

where u is the map from k to A defined by $u(\lambda) = \lambda\mathbf{1}$. The algebra A is *commutative* if $m \circ \tau = m$, where $\tau : A \otimes A \to A \otimes A$ is the *flip*, defined by $\tau(a \otimes b) = b \otimes a$.

A subspace $J \subset A$ is called a *subalgebra* (respectively a *left ideal, right ideal and two-sided ideal*) of A if $m(J \otimes J)$ (respectively $m(A \otimes J)$, $m(J \otimes A)$, $m(J \otimes A + A \otimes J)$) is included in J.

With any vector space V, we can associate its *tensor algebra* $T(V)$. As a vector space, it is defined by:

$$T(V) = \bigoplus_{k \geq 0} V^{\otimes k},$$

with $V^{\otimes 0} = k$ and $V^{\otimes k+1} := V \otimes V^{\otimes k}$. The product is given by the *concatenation*:

$$m(v_1 \otimes \cdots \otimes v_p,\, v_{p+1} \otimes \cdots \otimes v_{p+q}) = v_1 \otimes \cdots \otimes v_{p+q}.$$

The embedding of $k = V^{\otimes 0}$ into $T(V)$ gives the unit map u. The tensor algebra $T(V)$ is also called the *free (unital) algebra generated by* V. This algebra is characterized by the following universal property: for any linear map φ from V to a unital algebra A, there is a unique unital algebra morphism $\widetilde{\varphi}$ from $T(V)$ to A extending φ.

Let A and B be the unital k-algebras. We put a unital algebra structure on $A \otimes B$ in the following way:

$$(a_1 \otimes b_1).(a_2 \otimes b_2) = a_1 a_2 \otimes b_1 b_2.$$

The unit element $\mathbf{1}_{A \otimes B}$ is given by $\mathbf{1}_A \otimes \mathbf{1}_B$, and the associativity is clear. This multiplication is thus given by:

$$m_{A \otimes B} = (m_A \otimes m_B) \circ \tau_{23},$$

where $\tau_{23} : A \otimes B \otimes A \otimes B \to A \otimes A \otimes B \otimes B$ is defined by the flip of the two middle factors:

$$\tau_{23}(a_1 \otimes b_1 \otimes a_2 \otimes b_2) = a_1 \otimes a_2 \otimes b_1 \otimes b_2.$$

1.2.2. *Coalgebras*

Coalgebras are the objects which are somehow dual to algebras: axioms for coalgebras are derived from axioms for algebras by reversing the arrows of the corresponding diagrams:

A k-coalgebra is by definition a k-vector space C together with a bilinear map $\Delta : C \to C \otimes C$, which is *coassociative*. The coassociativity is expressed by the commutativity of the following diagram:

$$\begin{array}{ccc} C\otimes C\otimes C & \xleftarrow{\Delta\otimes I} & C\otimes C \\ \uparrow{\scriptstyle I\otimes\Delta} & & \uparrow{\scriptstyle \Delta} \\ C\otimes C & \xleftarrow{\Delta} & C \end{array}$$

Coalgebra C is *counital* if there is a counit $\varepsilon : C \to k$, such that the following diagram commutes:

$$\begin{array}{ccccc} k\otimes C & \xleftarrow{\varepsilon\otimes I} & C\otimes C & \xrightarrow{I\otimes\varepsilon} & C\otimes k \\ & \nwarrow{\scriptstyle\sim} & \uparrow{\scriptstyle\Delta} & {\scriptstyle\sim}\nearrow & \\ & & C & & \end{array}$$

A subspace $J \subset C$ is called a *subcoalgebra* (respectively a *left coideal, right coideal and two-sided coideal*) of C if $\Delta(J)$ is contained in $J\otimes J$ (respectively $C\otimes J$, $J\otimes C$, $J\otimes C + C\otimes J$) is included in J. The duality alluded to above can be made more precise:

PROPOSITION 1.1.–

1) The linear dual C^* of a counital coalgebra C is a unital algebra, with product (respectively unit map) the transpose of the coproduct (respectively of the counit).

2) Let J be a linear subspace of C. Denote by $J^\perp$ the orthogonal of J in C^*. Then:

- *J is a two-sided coideal if and only if $J^\perp$ is a subalgebra of C^*.*
- *J is a left coideal if and only if $J^\perp$ is a left ideal of C^*.*
- *J is a right coideal if and only if $J^\perp$ is a right ideal of C^*.*
- *J is a subcoalgebra if and only if $J^\perp$ is a two-sided ideal of C^*.*

PROOF.– For any subspace K of C^*, we will denote by $K^\perp$ the subspace of those elements of C on which any element of K vanishes. It coincides with the intersection of the orthogonal of K with C, via the canonical embedding $C \hookrightarrow C^{**}$. Therefore, for any linear subspaces $J \subset C$ and $K \subset C^*$ we have:

$$J^{\perp\perp} = J, \qquad\qquad K^{\perp\perp} \supset K.$$

Suppose that J is a two-sided coideal. Take any ξ, η in $J^\perp$. For any $x \in J$, we have:

$$< \xi\eta, x > = < \xi\otimes\eta, \Delta x > = 0,$$

as $\Delta x \subset J \otimes C + C \otimes J$. Therefore, $J^{\perp}$ is a subalgebra of C^*. Conversely if $J^{\perp}$ is a subalgebra, then:

$$\Delta J \subset (J^{\perp} \otimes J^{\perp})^{\perp} = J \otimes C + C \otimes J,$$

which proves the first assertion. We leave the reader to prove the three other assertions along the same lines. □

Dually, we have the following:

PROPOSITION 1.2.– *Let K be a linear subspace of C^*. Then:*

– *$K^{\perp}$ is a two-sided coideal if and only if K is a subalgebra of C^*.*

– *$K^{\perp}$ is a left coideal if and only if K is a left ideal of C^*.*

– *$K^{\perp}$ is a right coideal if and only if K is a right ideal of C^*.*

– *$K^{\perp}$ is a subcoalgebra if and only if K is a two-sided ideal of C^*.*

PROOF.– The linear dual $(C \otimes C)^*$ naturally contains the tensor product $C^* \otimes C^*$. Take as a multiplication the restriction of ${}^t\Delta$ to $C^* \otimes C^*$:

$$m = {}^t\Delta : C^* \otimes C^* \longrightarrow C^*,$$

and put $u = {}^t\varepsilon : k \to C^*$. It is easily seen, by just reverting the arrows of the corresponding diagrams, that the coassociativity of Δ implies the associativity of m, and that the counit property for ε implies that u is a unit. □

Note that the duality property is not perfect: if the linear dual of a coalgebra is always an algebra, the linear dual of an algebra is generally not a coalgebra. However, the *restricted dual* A° of an algebra A is a coalgebra. It is defined as the space of linear forms on A vanishing on some finite-codimensional ideal (Sweedler 1969).

The coalgebra C is *cocommutative* if $\tau \circ \Delta = \Delta$, where $\tau : C \otimes C \to C \otimes C$ is the flip. It will be convenient to use *Sweedler's notation*:

$$\Delta x = \sum_{(x)} x_1 \otimes x_2$$

Cocommutativity then expresses as:

$$\sum_{(x)} x_1 \otimes x_2 = \sum_{(x)} x_2 \otimes x_1$$

In Sweedler's notation coassociativity reads as:

$$\begin{aligned}(\Delta \otimes I) \circ \Delta(x) &= \sum_{(x)} x_{1:1} \otimes x_{1:2} \otimes x_2 \\ &= \sum_{(x)} x_1 \otimes x_{2:1} \otimes x_{2:2} = (I \otimes \Delta) \circ \Delta(x)\end{aligned}$$

We will sometimes write the iterated coproduct as:

$$\sum_{(x)} x_1 \otimes x_2 \otimes x_3$$

Sometimes, we will even mix the two ways of using Sweedler's notation for the iterated coproduct, in the case where we want to partially keep track of how we have constructed it (Dăscălescu *et al.* 2001). For example,

$$\begin{aligned}\Delta_3(x) &= (I \otimes \Delta \otimes I) \circ (\Delta \otimes I) \circ \Delta(x) \\ &= (I \otimes \Delta \otimes I)(\sum_{(x)} x_1 \otimes x_2 \otimes x_3) \\ &= \sum_{(x)} x_1 \otimes x_{2:1} \otimes x_{2:2} \otimes x_3.\end{aligned}$$

With any vector space V, we can associate its *tensor coalgebra* $T^c(V)$. It is isomorphic to $T(V)$ as a vector space. The coproduct is given by the *deconcatenation*:

$$\Delta(v_1 \otimes \cdots \otimes v_n) = \sum_{p=0}^{n} (v_1 \otimes \cdots \otimes v_p) \bigotimes (v_{p+1} \otimes \cdots \otimes v_n).$$

The counit is given by the natural projection of $T^c(V)$ onto k.

Let C and D be the unital k-coalgebras. We put a counital coalgebra structure on $C \otimes D$ in the following way: the comultiplication is given by:

$$\Delta_{C \otimes D} = \tau_{23} \circ (\Delta_C \otimes \Delta_D),$$

where τ_{23} is again the flip of the two middle factors, and the counity is given by $\varepsilon_{C \otimes D} = \varepsilon_C \otimes \varepsilon_D$.

1.2.3. *Convolution product*

Let A be an algebra and C be a coalgebra (over the same field k). Then, there is an associative product on the space $\mathcal{L}(C, A)$ of linear maps from C to A, called the *convolution product*. It is given by:

$$\varphi * \psi = m_A \circ (\varphi \otimes \psi) \circ \Delta_C.$$

In Sweedler's notation, it reads:

$$\varphi * \psi(x) = \sum_{(x)} \varphi(x_1)\psi(x_2).$$

The associativity is a direct consequence of both the associativity of A and coassociativity of C.

1.2.4. *Bialgebras and Hopf algebras*

A (unital and counital) *bialgebra* is a vector space $\mathcal{H}$ endowed with a structure of unital algebra (m, u) and a structure of counital coalgebra (Δ, ε), which are compatible. The compatibility requirement is that Δ is an algebra morphism (or equivalently that m is a coalgebra morphism), ε is an algebra morphism and u is a coalgebra morphism. It is expressed by the commutativity of the three following diagrams:

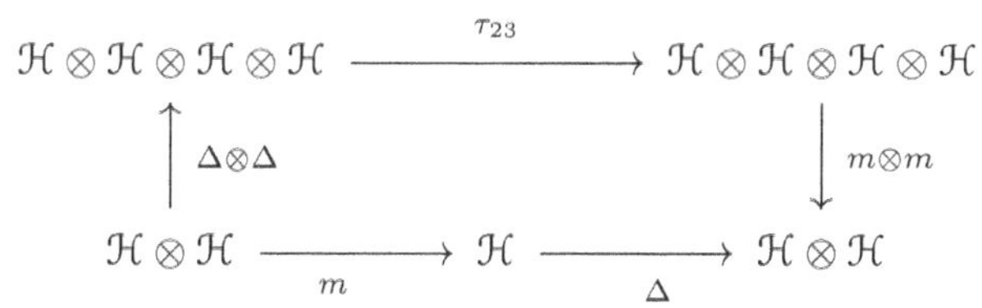

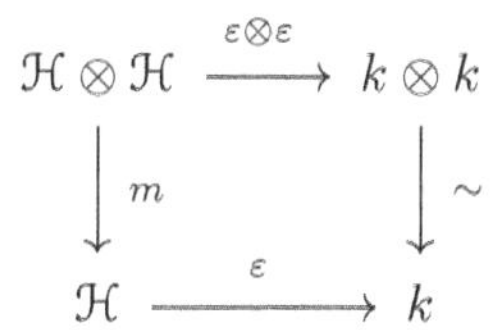

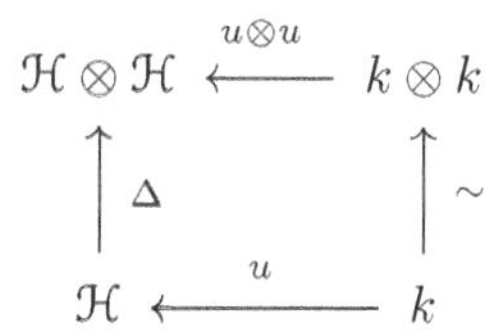

A *Hopf algebra* is a bialgebra $\mathcal{H}$ together with a linear map $S : \mathcal{H} \to \mathcal{H}$, called the *antipode*, such that the following diagram commutes:

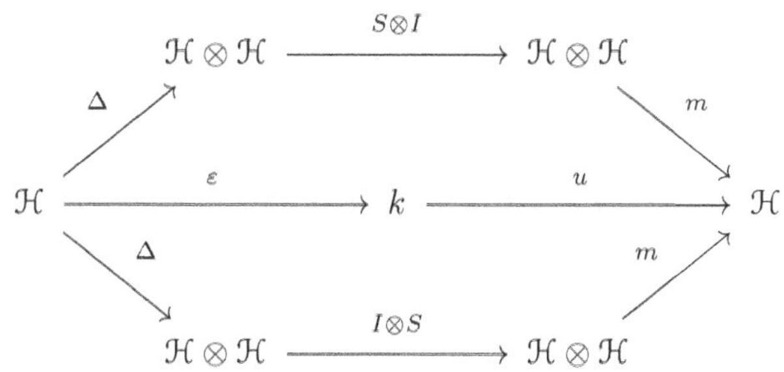

In Sweedler's notation, it reads:

$$\sum_{(x)} S(x_1)x_2 = \sum_{(x)} x_1 S(x_2) = (u \circ \varepsilon)(x).$$

In other words, the antipode is an inverse of the identity I for the convolution product on $\mathcal{L}(\mathcal{H},\mathcal{H})$. The unit for the convolution is the map $u \circ \varepsilon$.

A *primitive element* in a bialgebra $\mathcal{H}$ is an element x, such that $\Delta x = x \otimes 1 + 1 \otimes x$. A *grouplike element* is a nonzero element x, such that $\Delta x = x \otimes x$. Note that grouplike elements make sense in any coalgebra.

A *bi-ideal* in a bialgebra $\mathcal{H}$ is a two-sided ideal, which is also a two-sided coideal. A *Hopf ideal* in a Hopf algebra $\mathcal{H}$ is a bi-ideal J, such that $S(J) \subset J$.

1.2.5. *Some simple examples of Hopf algebras*

1.2.5.1. *The Hopf algebra of a group*

Let G be a group, and let kG be the group algebra (over the field k). It is by definition the vector space freely generated by the elements of G: the product of G extends uniquely to a bilinear map from $kG \times kG$ into kG, hence, a multiplication $m : kG \otimes kG \to kG$, which is associative. The neutral element of G gives the unit for m. The space kG is also endowed with a counital coalgebra structure, given by:

$$\Delta(\sum \lambda_i g_i) = \sum \lambda_i . g_i \otimes g_i$$

and:

$$\varepsilon(\sum \lambda_i g_i) = \sum \lambda_i .$$

This defines the *coalgebra of the set* G: it does not take into account the extra group structure on G, as the algebra structure does.

PROPOSITION 1.3.– *The vector space kG endowed with the algebra and coalgebra structures defined above is a Hopf algebra. The antipode is given by:*

$$S(g) = g^{-1}, g \in G.$$

PROOF.– The compatibility of the product and the coproduct is an immediate consequence of the following computation: for any $g, h \in G$, we have:

$$\Delta(gh) = gh \otimes gh = (g \otimes g)(h \otimes h) = \Delta g . \Delta h.$$

Now, $m(S \otimes I)\Delta(g) = g^{-1}g = e$ and similarly for $m(I \otimes S)\Delta(g)$. But, $e = u \circ \varepsilon(g)$ for any $g \in G$, so the map S is indeed the antipode. □

REMARK 1.1.– *If G were only a semigroup, the same construction would lead to a bialgebra structure on kG: the Hopf algebra structure (i.e. the existence of an* *antipode) reflects the group structure (the existence of the inverse). We have $S^2 = I$ in this case; however, the involutivity of the antipode is not true for general Hopf algebras.*

1.2.5.2. *Tensor algebras*

There is a natural structure of cocommutative Hopf algebra on the tensor algebra $T(V)$ of any vector space V. Namely, we define the coproduct Δ as the unique algebra morphism from $T(V)$ into $T(V) \otimes T(V)$, such that:

$$\Delta(1) = 1 \otimes 1, \qquad \Delta(x) = x \otimes 1 + 1 \otimes x, \; x \in V.$$

We define the counit as the algebra morphism, such that $\varepsilon(1) = 1$ and $\varepsilon_{|_V} = 0$. This endows $T(V)$ with a cocommutative bialgebra structure. We claim that the principal anti-automorphism:

$$S(x_1 \otimes \cdots \otimes x_n) = (-1)^n x_n \otimes \cdots \otimes x_1$$

verifies the axioms of an antipode, so that $T(V)$ is indeed a Hopf algebra. For $x \in V$, we have $S(x) = -x$; hence, $S * I(x) = I * S(x) = 0$. As V generates $T(V)$ as an algebra, it is easy to conclude.

1.2.5.3. *Enveloping algebras*

Let $\mathfrak{g}$ be a Lie algebra. The universal enveloping algebra is the quotient of the tensor algebra $T(\mathfrak{g})$ by the ideal J generated by $x \otimes y - y \otimes x - [x, y]$, $x, y \in \mathfrak{g}$.

LEMMA 1.1.– *J is a Hopf ideal, that is, $\Delta(J) \subset \mathcal{H} \otimes J + J \otimes \mathcal{H}$ and $S(J) \subset J$.*

PROOF.– The ideal J is generated by primitive elements (according to Proposition 1.5 below), and any ideal generated by primitive elements is a Hopf ideal (very easy and left to the reader). □

The quotient of a Hopf algebra by a Hopf ideal is a Hopf algebra. Hence, the universal enveloping algebra $\mathcal{U}(\mathfrak{g})$ is a cocommutative Hopf algebra.

1.2.6. ***Some basic properties of Hopf algebras***

In the proposition below we summarize the main properties of the antipode in a Hopf algebra:

PROPOSITION 1.4.– *(see Sweedler (1969, Proposition 4.0.1)). Let $\mathcal{H}$ be a Hopf algebra with multiplication m, comultiplication Δ, unit $u : 1 \mapsto \mathbf{1}$, counit ε and antipode S. Then:*

1) $S \circ u = u$ and $\varepsilon \circ S = \varepsilon$.

2) S is an algebra antimorphism and a coalgebra antimorphism, that is, if τ denotes the flip, we have:

$$m \circ (S \otimes S) \circ \tau = S \circ m, \qquad \tau \circ (S \otimes S) \circ \Delta = \Delta \circ S.$$

3) If $\mathcal{H}$ is commutative or cocommutative, then $S^2 = I$.

For a detailed proof, see Kassel (1995).

PROPOSITION 1.5.–

1) If x is a primitive element, then $S(x) = -x$.

2) The linear subspace Prim $\mathcal{H}$ *of primitive elements in $\mathcal{H}$ is a Lie algebra.*

PROOF.– If x is primitive, then $(\varepsilon \otimes \varepsilon) \circ \Delta(x) = 2\varepsilon(x)$. On the contrary, $(\varepsilon \otimes \varepsilon) \circ \Delta(x) = \varepsilon(x)$, so $\varepsilon(x) = 0$. Then:

$$0 = (u \circ \varepsilon)(x) = m(S \otimes I)\Delta(x) = S(x) + x.$$

Now let x and y be the primitive elements of $\mathcal{H}$. Then, we can easily compute:

$$\begin{aligned}\Delta(xy - yx) &= (x \otimes \mathbf{1} + \mathbf{1} \otimes x)(y \otimes \mathbf{1} + \mathbf{1} \otimes y) - (y \otimes \mathbf{1} + \mathbf{1} \otimes y)(x \otimes \mathbf{1} + \mathbf{1} \otimes x)\\ &= (xy - yx) \otimes \mathbf{1} + \mathbf{1} \otimes (xy + yx) + x \otimes y + y \otimes x - y \otimes x - x \otimes y\\ &= (xy - yx) \otimes \mathbf{1} + \mathbf{1} \otimes (xy - yx).\end{aligned}$$

□

1.3. Connected Hopf algebras

We introduce the crucial property of connectedness for bialgebras. The main interest resides in the possibility of implementing recursive procedures in connected bialgebras, the induction taking place with respect to a filtration or a grading. An important example of these techniques is the recursive construction of the antipode, which then "comes for free", showing that any connected bialgebra is in fact a connected Hopf algebra.

1.3.1. *Connected graded bialgebras*

A *graded Hopf algebra* on k is a graded k-vector space:

$$\mathcal{H} = \bigoplus_{n \geq 0} \mathcal{H}_n$$

endowed with a product $m : \mathcal{H} \otimes \mathcal{H} \to \mathcal{H}$, a coproduct $\Delta : \mathcal{H} \to \mathcal{H} \otimes \mathcal{H}$, a unit $u : k \to \mathcal{H}$, a counit $\varepsilon : \mathcal{H} \to k$ and an antipode $S : \mathcal{H} \to \mathcal{H}$, fulfilling the usual axioms of a Hopf algebra, and such that:

$$\mathcal{H}_p.\mathcal{H}_q \subset \mathcal{H}_{p+q} \qquad [1.1]$$

$$\Delta(\mathcal{H}_n) \subset \bigoplus_{p+q=n} \mathcal{H}_p \otimes \mathcal{H}_q. \qquad [1.2]$$

If we do not ask for the existence of an antipode $\mathcal{H}$, we get the definition of a *graded bialgebra*. In a graded bialgebra $\mathcal{H}$, we will consider the increasing filtration:

$$\mathcal{H}^n = \bigoplus_{p=0}^{n} \mathcal{H}_p.$$

It is an easy exercise (left to the reader) to prove that the unit u and the counit ε are degree zero maps, that is, $\mathbf{1} \in \mathcal{H}_0$ and $\varepsilon(\mathcal{H}_n) = \{0\}$ for $n \geq 1$. We can also show that the antipode S, when it exists, is also of degree zero, that is, $S(\mathcal{H}_n) \subset \mathcal{H}_n$. It can be proven as follows: let $S' : \mathcal{H} \to \mathcal{H}$ be defined, so that $S'(x)$ is the nth homogeneous component of $S(x)$ when x is homogeneous of degree n. We can write down the coproduct $\Delta(x)$ with Sweedler's notation:

$$\Delta(x) = \sum_{(x)} x_1 \otimes x_2,$$

where x_1 and x_2 are the homogeneous of degree, say, k and $n - k$. We then have:

$$\begin{aligned} m \circ (S' \otimes \mathrm{Id}) \circ \Delta(x) &= \sum_{(x)} S'(x_1)x_2 \\ &= n^{\text{th}}\text{component of} \sum_{(x)} S(x_1)x_2 = \varepsilon(x)\mathbf{1}. \end{aligned} \qquad [1.3]$$

Similarly, $m \circ (\mathrm{Id} \otimes S') \circ \Delta(x) = \varepsilon(x)\mathbf{1}$. By uniqueness of the antipode, we then get $S' = S$.

Suppose that $\mathcal{H}$ is *connected*, that is, $\mathcal{H}_0$ is one-dimensional. Then, we have:

$$\mathrm{Ker}\, \varepsilon = \bigoplus_{n \geq 1} \mathcal{H}_n.$$

PROPOSITION 1.6.– *For any $x \in \mathcal{H}^n, n \geq 1$, we can write:*

$$\Delta x = x \otimes \mathbf{1} + \mathbf{1} \otimes x + \widetilde{\Delta} x, \qquad \widetilde{\Delta} x \in \bigoplus_{p+q=n,\, p\neq 0,\, q\neq 0} \mathcal{H}_p \otimes \mathcal{H}_q.$$

The map $\widetilde{\Delta}$ is coassociative on Ker ε *and $\widetilde{\Delta}_k = (I^{\otimes k-1} \otimes \widetilde{\Delta})(I^{\otimes k-2} \otimes \widetilde{\Delta})...\widetilde{\Delta}$ sends $\mathcal{H}^n$ into $(\mathcal{H}^{n-k})^{\otimes k+1}$.*

PROOF.– Thanks to connectedness we can clearly write:

$$\Delta x = a(x \otimes 1) + b(1 \otimes x) + \widetilde{\Delta} x$$

with $a, b \in k$ and $\widetilde{\Delta} x \in \operatorname{Ker} \varepsilon \otimes \operatorname{Ker} \varepsilon$. The counit property then tells us that, with $k \otimes \mathcal{H}$ and $\mathcal{H} \otimes k$ canonically identified with $\mathcal{H}$:

$$x = (\varepsilon \otimes I)(\Delta x) = bx, \qquad x = (I \otimes \varepsilon)(\Delta x) = ax, \qquad [1.4]$$

hence, $a = b = 1$. We will use the following two variants of Sweedler's notation:

$$\Delta x = \sum_{(x)} x_1 \otimes x_2, \qquad [1.5]$$

$$\widetilde{\Delta} x = \sum_{(x)} x' \otimes x'', \qquad [1.6]$$

the second being relevant only for $x \in \operatorname{Ker} \varepsilon$. If x is homogeneous of degree n, we can suppose that the components x_1, x_2, x', x'' in the expressions above are homogeneous as well, and we then have $|x_1| + |x_2| = n$ and $|x'| + |x''| = n$. We easily compute:

$$\begin{aligned}(\Delta \otimes I)\Delta(x) &= x \otimes 1 \otimes 1 + 1 \otimes x \otimes 1 + 1 \otimes 1 \otimes x \\ &\quad + \sum_{(x)} x' \otimes x'' \otimes 1 + x' \otimes 1 \otimes x'' + 1 \otimes x' \otimes x'' \\ &\quad + (\widetilde{\Delta} \otimes I)\widetilde{\Delta}(x)\end{aligned}$$

and

$$\begin{aligned}(I \otimes \Delta)\Delta(x) &= x \otimes 1 \otimes 1 + 1 \otimes x \otimes 1 + 1 \otimes 1 \otimes x \\ &\quad + \sum_{(x)} x' \otimes x'' \otimes 1 + x' \otimes 1 \otimes x'' + 1 \otimes x' \otimes x'' \\ &\quad + (I \otimes \widetilde{\Delta})\widetilde{\Delta}(x),\end{aligned}$$

hence, the coassociativity of $\widetilde{\Delta}$ comes from the one of Δ. Finally, it is easily seen by induction on k that for any $x \in \mathcal{H}^n$, we can write:

$$\widetilde{\Delta}_k(x) = \sum_{x} x^{(1)} \otimes \cdots \otimes x^{(k+1)}, \qquad [1.7]$$

with $|x^{(j)}| \geq 1$. The grading imposes:

$$\sum_{j=1}^{k+1} |x^{(j)}| = n,$$

so the maximum possible for any degree $|x^{(j)}|$ is $n - k$. □

1.3.2. *An example: the Hopf algebra of decorated rooted trees*

A *rooted tree* is an oriented graph with a finite number of vertices, one among them being distinguished as the *root*, such that any vertex admits exactly one outgoing edge, except the root which has no outgoing edges. Here is the list of rooted trees up to five vertices, where the edges are tacitly oriented from top to bottom:

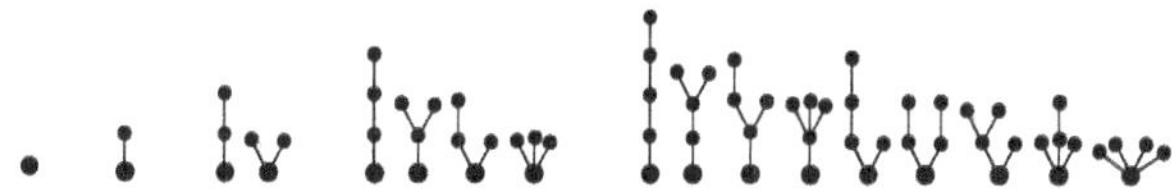

A *rooted forest* is a finite collection of rooted trees. The Connes–Kreimer Hopf algebra $\mathcal{H}_{\text{CK}} = \bigoplus_{n\geq 0} \mathcal{H}_{\text{CK}}^{(n)}$ is the Hopf algebra of rooted forests over k, graded by the number of vertices. It is the free commutative algebra on the linear space $\mathcal{T}$ spanned by nonempty rooted trees. The coproduct on a rooted forest u (i.e. a product of rooted trees) is described as follows: the set U of vertices of a forest u is endowed with a partial order defined by $x \leq y$ if and only if there is a path from a root to y passing through x. Any subset W of the set of vertices U of u defines a *subforest* w of u in an obvious manner, that is, by keeping the edges of u which link two elements of W. The coproduct is then defined by:

$$\Delta_{\text{CK}}(u) = \sum_{\substack{V \amalg W = U \\ W < V}} v \otimes w. \qquad [1.8]$$

Here, the notation $W < V$ means that $y < x$ for any vertex x of v and any vertex y of w, such that x and y are comparable. Such a couple (V, W) is also called an *admissible cut*, with crown (or pruning) v and trunk w. We have, for example:

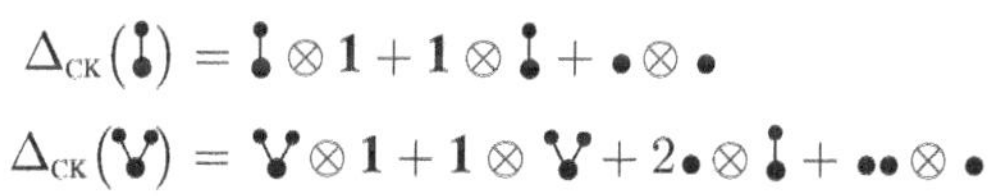

With the restriction that V and W will be nonempty (i.e. if V and W give rise to an ordered partition of U into two blocks), we get the restricted coproduct:

$$\widetilde{\Delta}_{\text{CK}}(u) = \Delta_{\text{CK}}(u) - u \otimes \mathbf{1} - \mathbf{1} \otimes u = \sum_{\substack{V \amalg W = U \\ W < V,\ V, W \neq \emptyset}} v \otimes w, \qquad [1.9]$$

which is often displayed as $\sum_{(u)} u' \otimes u''$ in Sweedler's notation. The iterated restricted coproduct, in terms of ordered partitions of U into n blocks, writes:

$$\widetilde{\Delta}_{\text{CK}}^{n-1}(u) = \sum_{\substack{V_1 \amalg \cdots \amalg V_n = U \\ V_n < \cdots < V_1,\ V_j \neq \emptyset}} v_1 \otimes \cdots \otimes v_n, \qquad [1.10]$$

and we get the full-iterated coproduct $\Delta_{\text{CK}}^{n-1}(u)$ by allowing empty blocks in the formula above. The coassociativity of the coproduct follows immediately.

Note, however, that the relation $<$ on the subsets of vertices is not transitive. The notation $V_n < \cdots < V_1$ is to be understood as $V_i < V_j$ for any $i > j$, $i, j \in \{1, \ldots, n\}$.

1.3.3. *Connected filtered bialgebras*

A *filtered bialgebra* on k is a k-vector space together with an increasing $\mathbb{N}$-indexed filtration:

$$\mathcal{H}^0 \subset \mathcal{H}^1 \subset \cdots \subset \mathcal{H}^n \subset \cdots, \bigcup_n \mathcal{H}^n = \mathcal{H}$$

endowed with a product $m : \mathcal{H} \otimes \mathcal{H} \to \mathcal{H}$, a coproduct $\Delta : \mathcal{H} \to \mathcal{H} \otimes \mathcal{H}$, a unit $u : k \to \mathcal{H}$, a counit $\varepsilon : \mathcal{H} \to k$ and an antipode $S : \mathcal{H} \to \mathcal{H}$, fulfilling the usual axioms of a bialgebra, and such that:

$$\mathcal{H}^p.\mathcal{H}^q \subset \mathcal{H}^{p+q} \qquad [1.11]$$

$$\Delta(\mathcal{H}^n) \subset \sum_{p+q=n} \mathcal{H}^p \otimes \mathcal{H}^q. \qquad [1.12]$$

It is easy (and left to the reader) to show that the unit u and the counit ε are of degree zero, if we consider the filtration on the base field k given by $k^0 = \{0\}$ and $k^n = k$ for any $n \geq 1$. Namely, $u(k^n) \subset \mathcal{H}^n$ and $\varepsilon(\mathcal{H}^n) \subset k^n$ for any $n \geq 0$.

If we ask for the existence of an antipode S, we get the definition of a *filtered Hopf algebra* if the antipode is of degree zero, that is, if:

$$S(\mathcal{H}^n) \subset \mathcal{H}^n \qquad [1.13]$$

for any $n \geq 0$. Contrarily to the graded case, it is not likely that a filtered bialgebra with antipode is automatically a filtered Hopf algebra (see, for example, Montgomery (1993, Lemma 5.2.8), Andruskiewitsch and Schneider (2002) and Andruskiewitsch and Cuadra (2013)). The antipode is, however, of degree zero in the connected case: for any $x \in \mathcal{H}$, we set

$$|x| := \min\{n \in \mathbb{N},\ x \in \mathcal{H}^n\}. \qquad [1.14]$$

We say that the filtered bialgebra $\mathcal{H}$ is connected if $\mathcal{H}^0$ is one-dimensional. There is an analogue of Proposition 1.6 in the connected filtered case, the proof of which is very similar:

PROPOSITION 1.7.– *For any $x \in \mathcal{H}^n, n \geq 1$, we can write:*

$$\Delta x = x \otimes \mathbf{1} + \mathbf{1} \otimes x + \widetilde{\Delta} x, \qquad \widetilde{\Delta} x \in \sum_{p+q=n,\, p\neq 0,\, q\neq 0} \mathcal{H}^p \otimes \mathcal{H}^q. \qquad [1.15]$$

The map $\widetilde{\Delta}$ is coassociative on $\operatorname{Ker} \varepsilon$ *and $\widetilde{\Delta}_k = (I^{\otimes k-1} \otimes \widetilde{\Delta})(I^{\otimes k-2} \otimes \widetilde{\Delta})...\widetilde{\Delta}$ sends $\mathcal{H}^n$ into $(\mathcal{H}^{n-k})^{\otimes k+1}$.*

As an easy corollary, the degree of the antipode is also zero in the connected case, that is, $S(\mathcal{H}^n) \subseteq \mathcal{H}^n$ for any n. This is an immediate consequence of the recursive formulae [1.19] and [1.20] below.

Any graded bialgebra, or the Hopf algebra, is obviously filtered by the canonical filtration associated with the grading:

$$\mathcal{H}^n := \bigoplus_{i=0}^{n} \mathcal{H}_i, \qquad [1.16]$$

and in that case, if x is a homogeneous element, x is of degree n if and only if $|x| = n$.

1.3.4. *The convolution product*

An important result is that any connected filtered bialgebra is indeed a filtered Hopf algebra, in the sense that the antipode comes for free. We give a proof of this fact as well as a recursive formula for the antipode using the *convolution product*: let $\mathcal{H}$ be a (connected filtered) bialgebra, and let $\mathcal{A}$ be any k-algebra (which will be called the *target algebra*): the convolution product on $\mathcal{L}(\mathcal{H}, \mathcal{A})$ is given by:

$$\begin{aligned} \varphi * \psi(x) &= m_{\mathcal{A}}(\varphi \otimes \psi)\Delta(x) \\ &= \sum_{(x)} \varphi(x_1)\psi(x_2). \end{aligned}$$

PROPOSITION 1.8.– *The map $e = u_{\mathcal{A}} \circ \varepsilon$, given by $e(\mathbf{1}) = \mathbf{1}_{\mathcal{A}}$ and $e(x) = 0$ for any $x \in \operatorname{Ker} \varepsilon$, is a unit for the convolution product. Moreover, the set $G(\mathcal{A}) := \{\varphi \in \mathcal{L}(\mathcal{H}, \mathcal{A}),\ \varphi(\mathbf{1}) = \mathbf{1}_{\mathcal{A}}\}$ endowed with the convolution product is a group.*

PROOF.– The first statement is straightforward. To prove the second, let us consider the formal series:

$$\begin{aligned} \varphi^{*-1}(x) &= \big(e - (e - \varphi)\big)^{*-1}(x) \\ &= \sum_{k \geq 0} (e - \varphi)^{*k}(x). \end{aligned}$$

Using $(e - \varphi)(\mathbf{1}) = 0$, we have $(e - \varphi)^{*k}(\mathbf{1}) = 0$ immediately, and for any $x \in \operatorname{Ker} \varepsilon$:

$$(e - \varphi)^{*k}(x) = (-1)^k m_{\mathcal{A},k-1}(\underbrace{\varphi \otimes \cdots \otimes \varphi}_{k\text{times}})\widetilde{\Delta}_{k-1}(x). \qquad [1.17]$$

When $x \in \mathcal{H}^n$, this expression vanishes and then for $k \geq n+1$. The formal series then ends up with a finite number of terms for any x, which proves the result. □

COROLLARY 1.1.– *Any connected filtered bialgebra $\mathcal{H}$ is a filtered Hopf algebra. The antipode is defined by:*

$$S(x) = \sum_{k\geq 0}(u\varepsilon - I)^{*k}(x). \qquad [1.18]$$

It is given by $S(\mathbf{1}) = \mathbf{1}$ and recursively by any of the two formulae for $x \in \operatorname{Ker}\varepsilon$:

$$S(x) = -x - \sum_{(x)} S(x')x'' \qquad [1.19]$$

$$S(x) = -x - \sum_{(x)} x'S(x''). \qquad [1.20]$$

PROOF.– The antipode, when it exists, is the inverse of the identity for the convolution product on $\mathcal{L}(\mathcal{H}, \mathcal{H})$. We then just need to apply Proposition 1.8 with $\mathcal{A} = \mathcal{H}$. The two recursive formulas come directly from the two equalities:

$$m(S \otimes I)\Delta(x) = m(I \otimes S)\Delta(x) = 0$$

fulfilled by any $x \in \operatorname{Ker}\varepsilon$. □

Let $\mathfrak{g}(\mathcal{A})$ be the subspace of $\mathcal{L}(\mathcal{H}, \mathcal{A})$ formed by the elements α, such that $\alpha(\mathbf{1}) = 0$. It is clearly a subalgebra of $\mathcal{L}(\mathcal{H}, \mathcal{A})$ for the convolution product. We have:

$$G(\mathcal{A}) = e + \mathfrak{g}(\mathcal{A}).$$

From now on, we will suppose that the ground field k is of characteristic zero. For any $x \in \mathcal{H}^n$, the exponential:

$$e^{*\alpha}(x) = \sum_{k\geq 0} \frac{\alpha^{*k}(x)}{k!} \qquad [1.21]$$

is a finite sum (ending up at $k = n$). It is a bijection from $\mathfrak{g}(\mathcal{A})$ onto $G(\mathcal{A})$. Its inverse is given by:

$$\operatorname{Log}(1 + \alpha)(x) = \sum_{k\geq 1} \frac{(-1)^{k-1}}{k}\alpha^{*k}(x). \qquad [1.22]$$

This sum again ends up at $k = n$ for any $x \in \mathcal{H}^n$. Let us introduce a decreasing filtration on $\mathcal{L} = \mathcal{L}(\mathcal{H}, \mathcal{A})$:

$$\mathcal{L}^n := \{\alpha \in \mathcal{L},\ \alpha_{|\mathcal{H}^{n-1}} = 0\}. \qquad [1.23]$$

Clearly, $\mathcal{L}_0 = \mathcal{L}$ and $\mathcal{L}_1 = \mathfrak{g}(\mathcal{A})$. We define the valuation val φ of an element φ of $\mathcal{L}$ as the greatest integer k, such that φ is in $\mathcal{L}_k$. In the sequel we will consider the ultrametric distance on $\mathcal{L}$ induced by the filtration:

$$d(\varphi, \psi) = 2^{-\operatorname{val}(\varphi-\psi)}. \qquad [1.24]$$

For any $\alpha, \beta \in \mathfrak{g}(\mathcal{A})$ let $[\alpha, \beta] = \alpha * \beta - \beta * \alpha$.

PROPOSITION 1.9.– *We have the inclusion:*

$$\mathcal{L}^p * \mathcal{L}^q \subset \mathcal{L}^{p+q}, \qquad [1.25]$$

and moreover, the metric space $\mathcal{L}$ *endowed with the distance defined by* [1.24] *is complete.*

PROOF.– Take any $x \in \mathcal{H}^{p+q-1}$, and any $\alpha \in \mathcal{L}_p$ and $\beta \in \mathcal{L}_q$. We have:

$$(\alpha * \beta)(x) = \sum_{(x)} \alpha(x_1)\beta(x_2).$$

Recall that we denote by $|x|$ the minimal n, such that $x \in \mathcal{H}^n$. Since $|x_1| + |x_2| = |x| \leq p + q - 1$, either $|x_1| \leq p - 1$ or $|x_2| \leq q - 1$, so the expression vanishes. Now, if (ψ_n) is a Cauchy sequence in $\mathcal{L}$, it is seen immediately that this sequence is *locally stationary*, that is, for any $x \in \mathcal{H}$ there exists $N(x) \in \mathbb{N}$, such that $\psi_n(x) = \psi_{N(x)}(x)$ for any $n \geq N(x)$. Then, the limit of (ψ_n) exists and is clearly defined by:

$$\psi(x) = \psi_{N(x)}(x).$$

□

As a corollary, the Lie algebra $\mathcal{L}_1 = \mathfrak{g}(\mathcal{A})$ is *pro-nilpotent*, in a sense that it is the projective limit of the Lie algebras $\mathfrak{g}(\mathcal{A})/\mathcal{L}^n$, which are nilpotent.

1.3.5. *Characters*

Let $\mathcal{H}$ be a connected filtered Hopf algebra over k, and let $\mathcal{A}$ be a k-algebra. We will consider unital algebra morphisms from $\mathcal{H}$ to the target algebra $\mathcal{A}$. When the algebra $\mathcal{A}$ is commutative, we will call them, slightly abusively, *characters*. We recover, of course, the usual notion of character when the algebra $\mathcal{A}$ is the ground field k.

The notion of character involves only the algebra structure of $\mathcal{H}$. On the contrary, the convolution product on $\mathcal{L}(\mathcal{H}, \mathcal{A})$ involves only the *coalgebra* structure on $\mathcal{H}$. Let us now consider the full Hopf algebra structure on $\mathcal{H}$ and see what happens to algebra morphisms with the convolution product:

PROPOSITION 1.10.– *Let $\mathcal{H}$ be any Hopf algebra over k, and let $\mathcal{A}$ be a commutative k-algebra. Then, the characters from $\mathcal{H}$ to $\mathcal{A}$ form a group $G_1(\mathcal{A})$ under the convolution product, and for any $\varphi \in G_1(\mathcal{A})$, the inverse is given by:*

$$\varphi^{*-1} = \varphi \circ S. \qquad [1.26]$$

PROOF.– Using the fact that Δ is an algebra morphism, we have for any $x, y \in \mathcal{H}$:

$$f * g(xy) = \sum_{(x)(y)} f(x_1y_1)g(x_2y_2).$$

If $\mathcal{A}$ is commutative and if f and g are characters, we get:

$$\begin{aligned} f * g(xy) &= \sum_{(x)(y)} f(x_1)f(y_1)g(x_2)g(y_2) \\ &= \sum_{(x)(y)} f(x_1)g(x_2)f(y_1)g(y_2) \\ &= (f * g)(x)(f * g)(y). \end{aligned}$$

The unit $e = u_{\mathcal{A}} \circ \varepsilon$ is an algebra morphism. The formula for the inverse of a character comes easily from the commutativity of the following diagram:

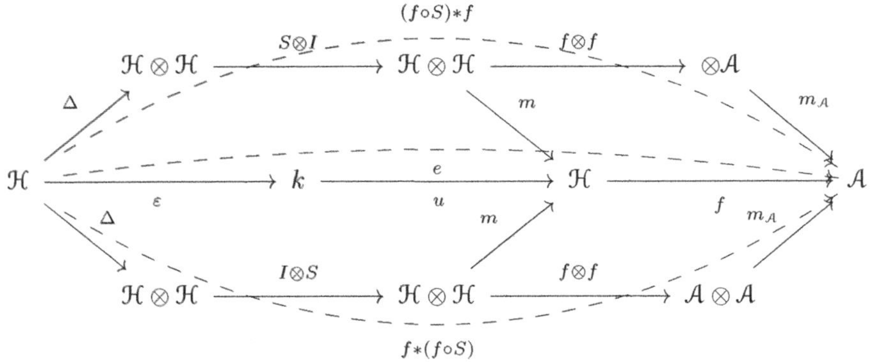

□

We call *infinitesimal characters with values in the algebra* $\mathcal{A}$ those elements α of $\mathcal{L}(\mathcal{H}, \mathcal{A})$, such that:

$$\alpha(xy) = e(x)\alpha(y) + \alpha(x)e(y).$$

PROPOSITION 1.11.– *Let $\mathcal{H}$ be a connected filtered Hopf algebra, and suppose that $\mathcal{A}$ is a commutative algebra. Let $G_1(\mathcal{A})$ (respectively $\mathfrak{g}_1(\mathcal{A})$) be the set of characters of $\mathcal{H}$ with values in $\mathcal{A}$ (respectively the set of infinitesimal characters of $\mathcal{H}$ with values in $\mathcal{A}$). Then, $G_1(\mathcal{A})$ is a subgroup of G, the exponential restricts to a bijection from $\mathfrak{g}_1(\mathcal{A})$ onto $G_1(\mathcal{A})$, and $\mathfrak{g}_1(\mathcal{A})$ is a Lie subalgebra of $\mathfrak{g}(\mathcal{A})$.*

PROOF.– Take two infinitesimal characters α and β with values in $\mathcal{A}$ and compute:

$$\begin{aligned}(\alpha * \beta)(xy) &= \sum_{(x)(y)} \alpha(x_1y_1)\beta(x_2y_2) \\ &= \sum_{(x)(y)} \big(\alpha(x_1)e(y_1) + e(x_1)\alpha(y_1)\big).\big(\beta(x_2)e(y_2) + e(x_2)\alpha(y_2)\big) \\ &= (\alpha * \beta)(x)e(y) + \alpha(x)\beta(y) + \beta(x)\alpha(y) + e(x)(\alpha * \beta)(y).\end{aligned}$$

Using the commutativity of $\mathcal{A}$, we immediately get:

$$[\alpha, \beta](xy) = [\alpha, \beta](x)e(y) + e(x)[\alpha, \beta](y),$$

which shows that $\mathfrak{g}_1(\mathcal{A})$ is a Lie algebra. Now, for $\alpha \in \mathfrak{g}_1(\mathcal{A})$, we have:

$$\alpha^{*n}(xy) = \sum_{k=0}^{n} \binom{n}{k} \alpha^{*k}(x)\alpha^{*(n-k)}(y),$$

as easily seen by induction on n. A straightforward computation then yields:

$$\exp(\alpha)(xy) = \exp(\alpha)(x)\exp(\alpha)(y),$$

with

$$\exp \alpha := \sum_{k \geq 0} \frac{\alpha^{*k}}{k!} = e + \alpha + \frac{\alpha^{*2}}{2} + \cdots.$$

The series above makes sense thanks to connectedness, as explained in section 1.3.4. Now let $\varphi = e + \gamma \in G_1(\mathcal{A})$, and let $\log(\varphi) = \sum_{j\geq 1} \frac{(-1)^{j-1}\gamma^{*j}}{j}$. Set $\varphi^{*t} := \exp(t \log \varphi)$ for $t \in k$. It coincides with the nth convolution power of φ for any integer n. Hence, φ^{*t} is an $\mathcal{A}$-valued character of $\mathcal{H}$ for any $t \in k$. Indeed, for any $x, y \in \mathcal{H}$, the expression $\varphi^{*t}(xy) - \varphi^{*t}(x)\varphi^{*t}(y)$ is polynomial in t and vanishes on all integers, and hence, vanishes identically. Differentiating with respect to t at $t = 0$, we immediately find that $\log \varphi$ is an infinitesimal character. □

1.3.6. *Group schemes and the Cartier–Milnor–Moore–Quillen theorem*

THEOREM 1.1 (Cartier, Milnor, Moore, Quillen).– *Let $\mathcal{U}$ be a cocommutative connected filtered Hopf algebra and let $\mathfrak{g}$ be the Lie algebra of its primitive elements, endowed with the filtration induced by the one of $\mathcal{U}$, which in turns induces a filtration on the enveloping algebra $\mathcal{U}(\mathfrak{g})$. Then, $\mathcal{U}$ and $\mathcal{U}(\mathfrak{g})$ are the isomorphic as filtered Hopf algebras. If $\mathcal{U}$ is graded, then the two Hopf algebras are isomorphic as graded Hopf algebras.*

PROOF.– The following proof is borrowed from Foissy's thesis. The embedding $\iota : \mathfrak{g} \to \mathcal{U}$ obviously induces an algebra morphism

$$\varphi : \mathcal{U}(\mathfrak{g}) \longrightarrow \mathcal{U}. \qquad [1.27]$$

It is easy to show that φ is also a coalgebra morphism. It remains to show that φ is surjective, injective and respects the filtrations. Let us first prove the surjectivity by induction on the coradical filtration degree:

$$d(x) := \min\{n \in \mathbb{N},\ \widetilde{\Delta}^n(x) = 0\}. \qquad [1.28]$$

Set $\mathcal{U}^n := \{x \in \mathcal{U},\ d(x) \leq n\}$, and similarly for $\mathcal{U}(\mathfrak{g})$. We can limit ourselves to the kernel of the counit. Any $x \in \mathcal{U}^1 \cap \operatorname{Ker}\varepsilon$ is primitive, hence $\varphi : \mathcal{U}(gg)^1 \to \mathcal{U}^1$ is obviously a linear isomorphism. Now, for $x \in \mathcal{U}^n \cap \operatorname{Ker}\varepsilon$ (for some integer $n \geq 2$), we can write, using cocommutativity:

$$\begin{aligned}\widetilde{\Delta}^{n-1}(x) &= \sum_{(x)} x^{(1)} \otimes \cdots \otimes x^{(n)} \\ &= \frac{1}{n!} \sum_{\sigma \in S_n} \sum_{(x)} x^{(\sigma_1)} \otimes \cdots \otimes x^{(\sigma_n)}\end{aligned}$$

where the $x^{(j)}$s are of coradical filtration degree 1, hence primitive. But, we also have:

$$\widetilde{\Delta}^{n-1}(x^{(1)} \cdots x^{(n)}) = \sum_{\sigma \in S_n} x^{(\sigma_1)} \otimes \cdots \otimes x^{(\sigma_n)}. \qquad [1.29]$$

Hence, the element $y = x - \frac{1}{n!} \sum_{(x)} x^{(1)} \otimes \cdots \otimes x^{(n)}$ belongs to $\mathcal{U}^{n-1}$. It is a linear combination of products of primitive elements by induction hypothesis, hence so is x. We have thus proven that $\mathcal{U}$ is generated by $\mathfrak{g}$, which amounts to the surjectivity of φ.

Now consider a nonzero element $u \in \mathcal{U}(\mathfrak{g})$, such that $\varphi(u) = 0$, and such that $d(u)$ is minimal. We have already proven $d(u) \geq 2$. We now compute:

$$\begin{aligned}0 = \Delta\big(\varphi(u)\big) &= (\varphi \otimes \varphi)\Delta(u) \\ &= (\varphi \otimes \varphi)\left(u \otimes 1 + 1 \otimes u + \sum_{(x)} u' \otimes u''\right) \\ &= \sum_{(u)} \varphi(u') \otimes \varphi(u'').\end{aligned}$$

By minimality hypothesis on $d(u)$, we then get $\sum_{(u)} u' \otimes u'' = 0$. Hence, u is primitive, that is, $d(u) = 1$, a contradiction. Hence, φ is injective. The compatibility with the original filtration or graduation is obvious. □

Now, let $\mathcal{H} : \bigcup_{n\geq 0} \mathcal{H}^n$ be a connected filtered Hopf algebra and let $\mathcal{A}$ be a commutative unital algebra. We suppose that the components of the filtration are finite-dimensional. The group $G_1(\mathcal{A})$ defined in the previous section depends functorially on the target algebra $\mathcal{A}$: in particular, when the Hopf algebra $\mathcal{H}$ itself is commutative, the correspondence $\mathcal{A} \mapsto G_1(\mathcal{A})$ is a *group scheme*. In the graded case with finite-dimensional components, it is possible to reconstruct the Hopf algebra $\mathcal{H}$ from the group scheme. We have indeed:

PROPOSITION 1.12.–

$$\mathcal{H} = \Big(\mathcal{U}\big(\mathfrak{g}_1(k)\big)\Big)^{\circ},$$

where $\mathfrak{g}_1(k)$ is the Lie algebra of infinitesimal characters with values in the base field k, where $\mathcal{U}\big(\mathfrak{g}_1(k)\big)$ stands for its enveloping algebra, and $(-)^{\circ}$ stands for the graded dual.

In the case when the Hopf algebra $\mathcal{H}$ is not commutative, it is no longer possible to reconstruct it from $G_1(k)$.

1.3.7. *Renormalization in connected filtered Hopf algebras*

In this section we describe the renormalization *à la* Connes–Kreimer (Connes and Kreimer 1998; Kreimer 2002) in the abstract context of connected filtered Hopf algebras: the objects to be renormalized are characters with values in a commutative unital target algebra $\mathcal{A}$ endowed with a *renormalization scheme*, that is, a splitting $\mathcal{A} = \mathcal{A}_- \oplus \mathcal{A}_+$ into two subalgebras. An important example is given by the *minimal subtraction scheme* of the algebra $\mathcal{A}$ of meromorphic functions of one variable z, where $\mathcal{A}_+$ is the algebra of meromorphic functions which are holomorphic at $z = 0$, and $\mathcal{A}_- = z^{-1}\mathbb{C}[z^{-1}]$ stands for the "polar parts". Any $\mathcal{A}$-valued character φ admits a unique *Birkhoff decomposition*:

$$\varphi = \varphi_-^{*-1} * \varphi_+,$$

where φ_+ is an $\mathcal{A}_+$-valued character, and $\varphi(\operatorname{Ker}\varepsilon) \subset \mathcal{A}_-$. In the MS scheme case described above, the renormalized character is the scalar-valued character given by the evaluation of φ_+ at $z = 0$ (whereas the evaluation of φ at $z = 0$ does not necessarily make sense).

Here, we consider the situation where the algebra $\mathcal{A}$ admits a *renormalization scheme*, that is, a splitting into two subalgebras:

$$\mathcal{A} = \mathcal{A}_- \oplus \mathcal{A}_+$$

with $\mathbf{1} \in \mathcal{A}_+$. Let $\pi : \mathcal{A} \twoheadrightarrow \mathcal{A}_-$ be the projection on $\mathcal{A}_-$ parallel to $\mathcal{A}_+$.

THEOREM 1.2.–

1) Let $\mathcal{H}$ be a connected filtered Hopf algebra. Let $G(\mathcal{A})$ be the group of those $\varphi \in \mathcal{L}(\mathcal{H}, \mathcal{A})$, such that $\varphi(\mathbf{1}) = \mathbf{1}_{\mathcal{A}}$ endowed with the convolution product. Any $\varphi \in G(\mathcal{A})$ admits a unique Birkhoff decomposition:

$$\varphi = \varphi_-^{*-1} * \varphi_+, \qquad [1.30]$$

where φ_- sends $\mathbf{1}$ to $\mathbf{1}_{\mathcal{A}}$ and $\operatorname{Ker}\varepsilon$ into $\mathcal{A}_-$, and φ_+ sends $\mathcal{H}$ into $\mathcal{A}_+$. The maps φ_- and φ_+ are given on $\operatorname{Ker}\varepsilon$ by the following recursive formulae:

$$\varphi_-(x) = -\pi\Big(\varphi(x) + \sum_{(x)} \varphi_-(x')\varphi(x'')\Big) \qquad [1.31]$$

$$\varphi_+(x) = (I - \pi)\Big(\varphi(x) + \sum_{(x)} \varphi_-(x')\varphi(x'')\Big). \qquad [1.32]$$

2) If the algebra $\mathcal{A}$ is commutative and if φ is a character, the components φ_- and φ_+ occurring in the Birkhoff decomposition of χ are characters as well.

PROOF.– The proof goes along the same lines as the proof of Theorem 4 from Connes and Kreimer (1998): for the first assertion, it is immediate from the definition of π that φ_- sends $\operatorname{Ker}\varepsilon$ into $\mathcal{A}_-$, and that φ_+ sends $\operatorname{Ker}\varepsilon$ into $\mathcal{A}_+$. It only remains to check equality $\varphi_+ = \varphi_- * \varphi$, which is an easy computation:

$$\begin{aligned}\varphi_+(x) &= (I - \pi)\Big(\varphi(x) + \sum_{(x)} \varphi_-(x')\varphi(x'')\Big).\\ &= \varphi(x) + \varphi_-(x) + \sum_{(x)} \varphi_-(x')\varphi(x'')\\ &= (\varphi_- * \varphi)(x).\end{aligned}$$

The proof of assertion 2 goes exactly as in Connes and Kreimer (1998) and relies on the following *Rota–Baxter relation* in $\mathcal{A}$:

$$\pi(a)\pi(b) = \pi\big(\pi(a)b + a\pi(b)\big) - \pi(ab), \qquad [1.33]$$

which is easily verified by decomposing a and b into their $\mathcal{A}_\pm$-parts. Let φ be a character of $\mathcal{H}$ with values in $\mathcal{A}$. Suppose that we have $\varphi_-(xy) = \varphi_-(x)\varphi_-(y)$ for any $x, y \in \mathcal{H}$, such that $|x| + |y| \leq d - 1$, and compute for x, y, such that $|x| + |y| = d$:

$$\varphi_-(x)\varphi_-(y) = \pi(X)\pi(Y),$$

with $X = \varphi(x) - \sum_{(x)} \varphi_-(x')\varphi(x'')$ and $Y = \varphi(y) - \sum_{(y)} \varphi_-(y')\varphi(y'')$. Using the formula:

$$\pi(X) = -\varphi_-(x),$$

we get:

$$\varphi_-(x)\varphi_-(y) = -\pi\big(XY + \varphi_-(x)Y + X\varphi_-(y)\big),$$

hence:

$$\begin{aligned}\varphi_-(x)\varphi_-(y) = -\pi\Big(&\varphi(x)\varphi(y) + \varphi_-(x)\varphi(y) + \varphi(x)\varphi_-(y)\\ &+ \sum_{(x)} \varphi_-(x')\varphi(x'')\big(\varphi(y) + \varphi_-(y)\big)\\ &+ \sum_{(y)} \big(\varphi(x) + \varphi_-(x)\big)\varphi_-(y')\varphi(y'')\\ &+ \sum_{(x)(y)} \varphi_-(x')\varphi(x'')\varphi_-(y')\varphi(y'')\Big).\end{aligned}$$

We have to compare this expression with:

$$\begin{aligned}\varphi_-(xy) = -\pi\Big(&\varphi(xy) + \varphi_-(x)\varphi(y) + \varphi_-(y)\varphi(x)\\ &+ \sum_{(x)} \big(\varphi_-(x'y)\varphi(x'') + \varphi_-(x')\varphi(x''y)\big)\\ &+ \sum_{(y)} \big(\varphi_-(xy')\varphi(y'') + \varphi_-(y')\varphi(xy'')\big)\\ &+ \sum_{(x)(y)} \varphi_-(x'y')\varphi(x''y'')\Big).\end{aligned}$$

These two expressions are easily seen to be equal using the commutativity of the algebra $\mathcal{A}$, the character property for φ and the induction hypothesis. □

REMARK 1.2.– *Assertion 2 admits a more conceptual proof (see the notes by Ebrahimi-Fard in the present volume), which is based on the following recursive expressions for the components of the Birkhoff decomposition: define the* ***Bogoliubov preparation map*** *as the map* $b : G(\mathcal{A}) \to \mathcal{L}(\mathcal{H}, \mathcal{A})$*, recursively given by:*

$$b(\varphi)(x) = \varphi(x) + \sum_{(x)} \varphi_-(x')\varphi(x''). \qquad [1.34]$$

Then, the components of φ *in the Birkhoff decomposition read:*

$$\varphi_- = -\pi \circ b(\varphi), \qquad \varphi_+ = (I - \pi) \circ b(\varphi). \qquad [1.35]$$

The Bogoliubov preparation map also writes in a more concise form:

$$b(\varphi) = \varphi_- * (\varphi - e). \qquad [1.36]$$

Plugging equation [1.36] into equation [1.35] and setting $\alpha := e - \varphi$, we get the following expression for φ_-:

$$\varphi_- = e + P(\varphi_- * \alpha) \quad [1.37]$$

$$= e + P(\alpha) + P\big(P(\alpha) * \alpha\big) + \cdots + \underbrace{P\Big(P(...P}_{n \text{ times}}(\alpha) * \alpha\big) \cdots * \alpha\Big) + \cdots \quad [1.38]$$

where $P : \mathcal{L}(\mathcal{H}, \mathcal{A}) \to \mathcal{L}(\mathcal{H}, \mathcal{A})$ is the projection defined by $P(\alpha) = \pi \circ \alpha$. The renormalized part φ_+ satisfies an analogous recursive expression:

$$\varphi_+ = e + \tilde{P}(\varphi_- * \alpha) \quad [1.39]$$

$$= e + \tilde{P}(\varphi_+ * \beta) \quad [1.40]$$

$$= e + \tilde{P}(\beta) + \tilde{P}\big(\tilde{P}(\beta) * \beta\big) + \cdots + \underbrace{\tilde{P}\Big(\tilde{P}(\ldots \tilde{P}}_{n \text{ times}}(\beta) * \beta\big) \cdots * \beta\Big) + \cdots$$

with $\beta := \varphi^{-1} * \alpha = e - \varphi^{-1}$, and where $\tilde{P} = I - P$ is the projection on $\mathcal{L}(\mathcal{H}, \mathcal{A})$ defined by $\tilde{P}(\alpha) = (I - \pi) \circ \alpha$.

1.4. Pre-Lie algebras

Pre-Lie algebras are sometimes called *Vinberg algebras*, as they appear in the work of Vinberg (1963) under the name "left-symmetric algebras" on the classification of homogeneous cones. They appear independently at the same time in the work of Gerstenhaber (1963) on Hochschild cohomology and deformations of algebras, under the name "pre-Lie algebras", which is now the standard terminology. The term "chronological algebras" has also been used sometimes, for example, in the fundamental work of Agrachev and Gamkrelidze (1981). The notion itself can, however, be traced back to the work of Cayley (1857) which, in modern language, describes the pre-Lie algebra morphism F_a from the pre-Lie algebra of rooted trees into the pre-Lie algebra of vector fields on $\mathbb{R}^n$, sending the one-vertex tree to a given vector field a. For a survey emphasizing geometric aspects, see Burde (2006).

1.4.1. *Definition and general properties*

A *left pre-Lie algebra* over a field k is a k-vector space A with a bilinear binary composition $\rhd$ that satisfies the left pre-Lie identity:

$$a \rhd (b \rhd c) - (a \rhd b) \rhd c = b \rhd (a \rhd c) - (b \rhd a) \rhd c, \quad [1.41]$$

for $a, b, c \in A$. Analogously, a *right pre-Lie algebra* is a k-vector space A with a binary composition $\triangleleft$ that satisfies the right pre-Lie identity:

$$(a \triangleleft b) \triangleleft c - a \triangleleft (b \triangleleft c) = (a \triangleleft c) \triangleleft b - a \triangleleft (c \triangleleft b). \qquad [1.42]$$

The left pre-Lie identity is rewritten as:

$$L_{[a,b]} = [L_a, L_b], \qquad [1.43]$$

where $L_a : A \to A$ is defined by $L_a b = a \triangleright b$, and the bracket on the left-hand side is defined by $[a, b] := a \triangleright b - b \triangleright a$. As an easy consequence, this bracket satisfies the Jacobi identity: If A is unital (i.e. there exists $\mathbf{1} \in A$, such that $\mathbf{1} \triangleright a = a \triangleright \mathbf{1} = \mathbf{1}$ for any $a \in A$), it is immediate thanks to the fact that $L : A \to \operatorname{End} A$ is injective. If not, we can add a unit by considering $\overline{A} := A \oplus k.\mathbf{1}$ and extend $L : \overline{A} \to \operatorname{End} \overline{A}$ accordingly. As any right pre-Lie algebra $(A, \triangleleft)$ is also a left pre-Lie algebra with product $a \triangleright b := b \triangleleft a$, we can stick to left pre-Lie algebras, which we will do unless specifically indicated.

1.4.2. *The group of formal flows*

The following is taken from the paper by Agrachev and Gamkrelidze (1981). Suppose that A is a left pre-Lie algebra endowed with a compatible decreasing filtration, namely, $A = A_1 \supset A_2 \subset A_3 \supset \cdots$, such that the intersection of the A_j's reduces to $\{0\}$, and such that $A_p \triangleright A_q \subset A_{p+q}$. Suppose, moreover, that A is complete with respect to this filtration. The Baker–Campbell–Hausdorff formula:

$$C(a, b) = a + b + \frac{1}{2}[a, b] + \frac{1}{12}([a, [a, b]] + [b, [b, a]]) + \cdots \qquad [1.44]$$

then endows A with a structure of a pro-unipotent group. An example of this situation is given by $A = hB[[h]]$, where B is any pre-Lie algebra, and $A_j = h^j B[[h]]$. This group admits a more transparent presentation as follows: introduce a fictitious unit $\mathbf{1}$, such that $\mathbf{1} \triangleright a = a \triangleright \mathbf{1} = a$ for any $a \in A$, and define $W : A \to A$ by:

$$W(a) := e^{L_a}\mathbf{1} - \mathbf{1} = a + \frac{1}{2} a \triangleright a + \frac{1}{6} a \triangleright (a \triangleright a) + \cdots . \qquad [1.45]$$

The application W is clearly a bijection. The inverse, denoted by Ω, also appears under the name "pre-Lie Magnus expansion" in Ebrahimi-Fard and Manchon (2009b). It verifies the equation:

$$\Omega(a) = \frac{L_{\Omega(a)}}{e^{L_{\Omega(a)}} - Id} a = \sum_{i \geq 0} B_i L^i_{\Omega(a)} a, \qquad [1.46]$$

where the B_is are the Bernoulli numbers. The first few terms are:

$$\Omega(a) = a - \frac{1}{2} a \triangleright a + \frac{1}{4}(a \triangleright a) \triangleright a + \frac{1}{12} a \triangleright (a \triangleright a) + \cdots \qquad [1.47]$$

By transferring the BCH product by means of the map W, namely:

$$a\#b = W\Big(C\big(\Omega(a),\Omega(b)\big)\Big), \tag{1.48}$$

we have $W(a)\#W(b) = W\big(C(a,b)\big) = e^{L_a}e^{L_b}\mathbf{1} - \mathbf{1}$, hence $W(a)\#W(b) = W(a) + e^{L_a}W(b)$. The product $\#$ is thus given by the simple formula:

$$a\#b = a + e^{L_{\Omega(a)}}b. \tag{1.49}$$

The inverse is given by $a^{\#-1} = W\big(-\Omega(a)\big) = e^{-L_{\Omega(a)}}\mathbf{1}-\mathbf{1}$. If $(A,\rhd)$ and $(B,\rhd)$ are two such pre-Lie algebras and $\psi : A \to B$ is a filtration-preserving pre-Lie algebra morphism, we should immediately check that for any $a, b \in A$ we have:

$$\psi(a\#b) = \psi(a)\#\psi(b). \tag{1.50}$$

In other words, the group of formal flows is a functor from the category of complete filtered pre-Lie algebras to the category of groups.

When the pre-Lie product $\rhd$ is associative, all of this simplifies to:

$$a\#b = a \rhd b + a + b \tag{1.51}$$

and

$$a^{\#-1} = \frac{1}{1+a} - 1 = \sum_{n\geq 1}(-1)^n a^n. \tag{1.52}$$

1.4.3. *The pre-Lie Poincaré–Birkhoff–Witt theorem*

This section exposes a result by Guin and Oudom (2005).

THEOREM 1.3.– *Let A be any left pre-Lie algebra, and let $S(A)$ be its symmetric algebra, that is, the free commutative algebra on A. Let A_{Lie} be the underlying Lie algebra of A, that is, the vector space A endowed with the Lie bracket given by $[a,b] = a\rhd b - b\rhd a$ for any $a, b \in A$, and let $\mathcal{U}(A)$ be the enveloping algebra of A_{Lie}, endowed with its usual increasing filtration. Let us consider the associative algebra $\mathcal{U}(A)$ as a left module over itself.*

There exists a left $\mathcal{U}(A)$-module structure on $S(A)$ and a canonical left $\mathcal{U}(A)$-module isomorphism $\eta : \mathcal{U}(A) \to S(A)$, such that the associated graded linear map $\mathrm{Gr}\,\eta : \mathrm{Gr}\,\mathcal{U}(A) \to S(A)$ is an isomorphism of commutative graded algebras.

PROOF.– The Lie algebra morphism

$$\begin{aligned} L : A &\longrightarrow \mathrm{End}\,A \\ a &\longmapsto (L_a : b \mapsto a \rhd b) \end{aligned}$$

extends by the Leibniz rule to a unique Lie algebra morphism $L : A \to \operatorname{Der} S(A)$. Now we claim that the map $M : A \to \operatorname{End} S(A)$ defined by:

$$M_a u = au + L_a u \qquad [1.53]$$

is a Lie algebra morphism. Indeed, for any $a, b \in A$ and $u \in S(A)$ we have:

$$\begin{aligned} M_a M_b u &= M_a(bu + L_b u) \\ &= abu + aL_b u + L_a(bu) + L_a L_b u \\ &= abu + aL_b u + bL_a u + (a \rhd b)u + L_a L_b u. \end{aligned}$$

Hence

$$\begin{aligned} [M_a, M_b]u &= (a \rhd b - b \rhd a)u + [L_a, L_b]u \\ &= M_{[a,b]}u, \end{aligned}$$

which proves the claim. Now M extends, by universal property of the enveloping algebra, to a unique algebra morphism $M : \mathcal{U}(A) \to \operatorname{End} S(A)$. The linear map:

$$\begin{aligned} \eta : \mathcal{U}(A) &\longrightarrow S(A) \\ u &\longmapsto M_u.1 \end{aligned}$$

is clearly a morphism of left $\mathcal{U}(A)$-modules. It is immediately seen by induction that for any $a_1, \ldots, a_n \in A$, we have $\eta(a_1 \cdots a_n) = a_1 \cdots a_n + v$, where v is a sum of terms of degree $\leq n - 1$. This proves the theorem. □

REMARK 1.3.– *Let us recall that the symmetrization map $\sigma : \mathcal{U}(A) \to S(A)$, uniquely determined by $\sigma(a^n) = a^n$ for any $a \in A$ and any integer n, is an isomorphism for the two A_{Lie}-module structures given by the adjoint action. This is* **not** *the case for the map η defined above. The fact that it is possible to replace the adjoint action of $\mathcal{U}(A)$ on itself by the simple left multiplication is a remarkable property of pre-Lie algebras, and makes Theorem 1.3 different from the usual Lie algebra PBW theorem.*

Let us finally note that, if p stands for the projection from $S(A)$ onto A, for any $a_1, \ldots, a_k \in A$, we easily get:

$$p \circ \eta(a_1 \cdots a_k) = L_{a_1} \cdots L_{a_k} \mathbf{1} = a_1 \rhd \Big(a_2 \rhd \big(\cdots (a_{k-1} \rhd a_k) \ldots\big)\Big) \qquad [1.54]$$

by a simple induction on k. The linear isomorphism η transfers the product of the enveloping algebra $\mathcal{U}(A)$ into a noncommutative product $*$ on $\mathcal{S}(A)$ defined by:

$$s * t = \eta\big(\eta^{-1}(s)\eta^{-1}(t)\big). \qquad [1.55]$$

Suppose now that A is endowed with a complete decreasing compatible filtration as in section 1.4.2. This filtration induces a complete decreasing filtration $S(A) = S(A)_0 \supset S(A)_1 \supset S(A)_2 \supset \cdots$, and the product $*$ readily extends to the completion $\widehat{S}(A)$. For any $a \in A$, the application of equation [1.54] gives:

$$p(e^{*a}) = W(a) \qquad [1.56]$$

as an equality in the completed symmetric algebra $\widehat{S}(A)$.

According to equation [1.48], we can identify the pro-unipotent group $\{e^{*a}, a \in A\} \subset \widehat{S}(A)$ and the group of formal flows of the pre-Lie algebra A by means of the projection p, namely:

$$p(e^{*a})\#p(e^{*b}) = p(e^{*a} * e^{*b}) \qquad [1.57]$$

for any $a, b \in A$.

1.5. Algebraic operads

An operad is a combinatorial device which appeared in algebraic topology (May 1972), coined for coding "types of algebras". Hence, for example, a Lie algebra is an algebra over some operad denoted by LIE, an associative algebra is an algebra over some operad denoted by ASSOC, a commutative algebra is an algebra over some operad denoted by COM and so on.

1.5.1. *Manipulating algebraic operations*

Algebra starts, in most cases, with some set E and some binary operation $* : E \times E \to E$. The set E shows some extra structure most of the time. Here, we will stick to the linear setting, where E is replaced by a vector space V (over some base field k), and $*$ is bilinear, that is, a linear map from $V \otimes V$ into V. A second bilinear map is deduced from the first by permuting the entries:

$$a *^{\mathrm{op}} b := b * a. \qquad [1.58]$$

It also makes sense to look at tri-, quadri- and multi-linear operations, that is, linear maps from $V^{\otimes n}$ to V for any V. For example, it is very easy to produce 12 tri-linear maps starting with the bilinear map $*$ by considering:

$$(a, b, c) \mapsto (a * b) * c,$$
$$(a, b, c) \mapsto a * (b * c),$$

and the others deduced by permuting the three entries a, b and c. We could also introduce some tri- or multi-linear operations from scratch, that is, without deriving them from the bilinear operation $*$. We can even consider 1-ary and 0-ary operations, the latter being just distinguished elements of V. Note that there is a canonical 1-ary operation, namely, the identity map $e : V \to V$. At this stage note that the symmetric group S_n obviously acts on the n-ary operations *on the right* by permuting the entries before composing them.

The bilinear operation $*$ is not arbitrary in general: its properties determine the "type of algebra" considered. For example, V will be an associative or a Lie algebra if for any $a, b, c \in V$, we have respectively:

$$(a * b) * c = a * (b * c), \qquad [1.59]$$

$$(a * b) * c + (b * c) * a + (c * a) * b = 0, \qquad a * b - b * a = 0. \qquad [1.60]$$

The concept of operad emerges when we try to rewrite such relations in terms of the operation $*$ only, discarding the entries a, b, c. For example, the associativity axiom equation [1.59] informally expresses itself as follows: *composing the operation $*$ twice in two different ways gives the same result*. Otherwise said:

$$* \circ_1 * = * \circ_2 *. \qquad [1.61]$$

The Lie algebra axioms (equation [1.60]), involving flip and circular permutations, are clearly rewritten as:

$$(* \circ_1 *) + (* \circ_1 *) \circ \sigma + (* \circ_1 *) \circ \sigma^2 = 0, \qquad * + * \circ \tau = 0, \qquad [1.62]$$

where τ is the flip (21) and σ is the circular permutation (231). The next section will give a precise meaning to these "partial compositions", and we will end up giving the axioms of an operad, which is the natural framework in which equations like [1.61] and [1.62] make sense.

1.5.2. *The operad of multi-linear operations*

Let us now look at the prototype of algebraic operads: for any vector space V, the operad Endop(V) is given by:

$$\text{Endop}(V)_n = \mathcal{L}(V^{\otimes n}, V). \qquad [1.63]$$

The right action of the symmetric group S_n on Endop$(V)_n$ is induced by the left action of S_n on $V^{\otimes n}$ given by:

$$\sigma.(v_1 \otimes \cdots \otimes v_n) := v_{\sigma^{-1}(1)} \otimes \cdots \otimes v_{\sigma^{-1}(n)}. \qquad [1.64]$$

Elements of $\mathrm{Endop}(V)_n$ are conveniently represented as boxes with n inputs and one output: as illustrated by the graphical representation below, the partial composition $a \circ_i b$ is given by:

$$a \circ_i b(v_1 \otimes \cdots \otimes v_{k+l-1}) :=$$

$$a\big(v_1 \otimes \cdots \otimes v_{i-1} \otimes b(v_i \otimes \cdots \otimes v_{i+l-1}) \otimes v_{i+l} \otimes \cdots \otimes v_{k+l-1}\big). \qquad [1.65]$$

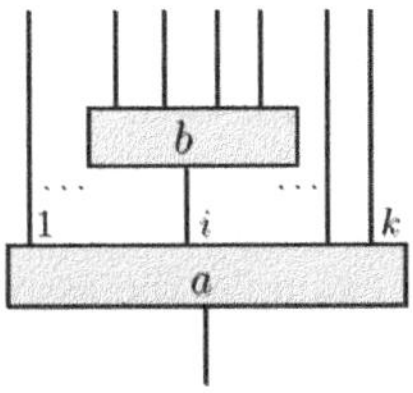

Partial composition $a \circ_i b$

The following result is straightforward:

PROPOSITION 1.13.– *For any* $a \in \mathrm{Endop}(V)_k$, $b \in \mathrm{Endop}(V)_l$ *and* $c \in \mathrm{Endop}(V)_m$, *we have:*

$$(a \circ_i b) \circ_{i+j-1} c = a \circ_i (b \circ_j c), \qquad i \in \{1,\ldots,k\},\ j \in \{1,\ldots,l\}$$

(nested associativity property),

$$(a \circ_i b) \circ_{l+j-1} c = (a \circ_j c) \circ_i b, \qquad i,j \in \{1,\ldots,k\},\ i < j$$

(disjoint associativity property).

The identity $e : V \to V$ *satisfies the following unit property:*

$$e \circ a = a \qquad [1.66]$$

$$a \circ_i e = a, \qquad i = 1,\ldots,k, \qquad [1.67]$$

and finally, the following equivariance property is satisfied:

$$a.\sigma \circ_i b.\tau = (a \circ_{\sigma_i} b).\iota_i(\sigma,\tau) \qquad [1.68]$$

where $\iota_i(\sigma,\tau) \in S_{k+l-1}$ *is defined by letting* τ *permute the set* $E_i = \{i, i+1,\ldots,i+l-1\}$ *of cardinality* l, *and then by letting* σ *permute the set* $\{1,\ldots,i-1,E_i,i+l,\ldots,k+l-1\}$ *of cardinality* k.

The two associativity properties are graphically represented as follows:

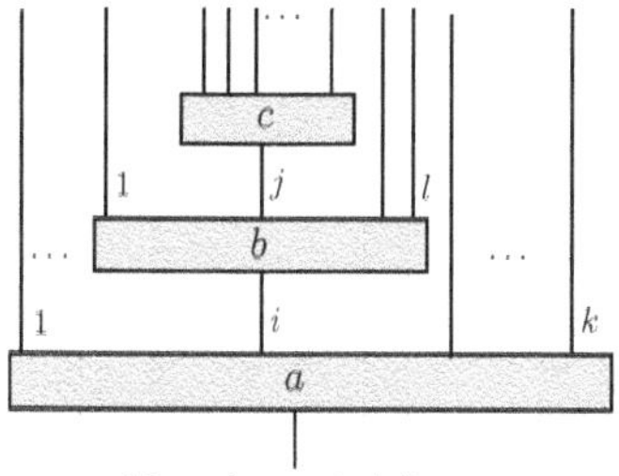

Nested associativity

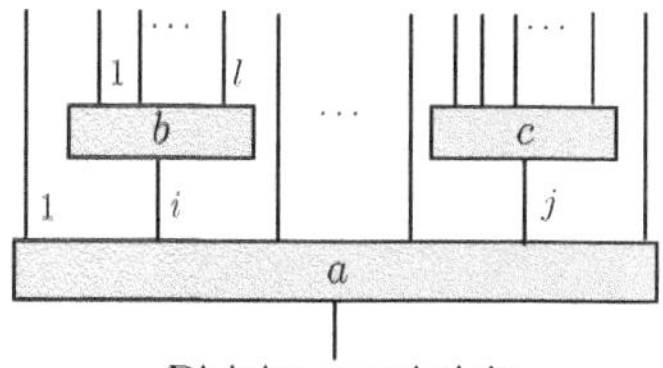

Disjoint associativity

1.5.3. *A definition for linear operads*

We are now ready to give the precise definition of a linear operad:

DEFINITION 1.1.– *An operad $\mathcal{P}$ (in the symmetric monoidal category of k-vector spaces) is given by a collection of vector spaces $(\mathcal{P}_n)_{n\geq 0}$, a right action of the symmetric group S_n on $\mathcal{P}_n$, a distinguished element $e \in \mathcal{P}_1$ and a collection of partial compositions:*

$$\begin{aligned} \circ_i : \mathcal{P}_k \otimes \mathcal{P}_l &\longrightarrow \mathcal{P}_{k+l-1}, \qquad i = 1, \ldots, k \\ (a, b) &\longmapsto a \circ_i b \end{aligned}$$

subject to the associativity, unit and equivariance axioms of Proposition 1.13.

The *global composition* is defined by:

$$\begin{aligned} \gamma : \mathcal{P}_n \otimes \mathcal{P}_{k_1} \otimes \cdots \otimes \mathcal{P}_{k_n} &\longrightarrow \mathcal{P}_{k_1+\cdots+k_n} \\ (a, b_1, \ldots, b_n) &\longmapsto \Big(\ldots\big((a \circ_n b_n) \circ_{n-1} b_{n-1}\big) \cdots\Big) \circ_1 b_1 \end{aligned}$$

and is graphically represented as follows:

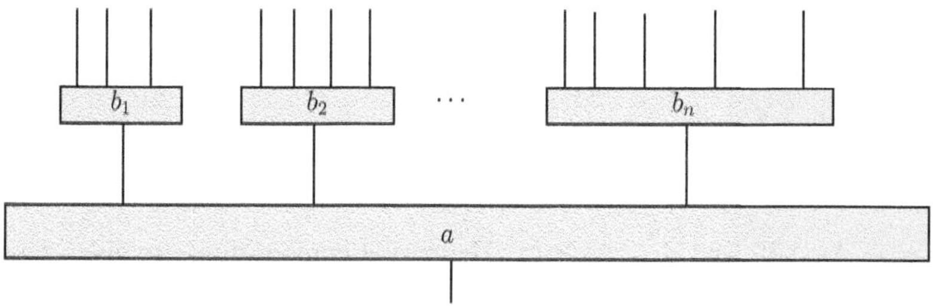

Global composition $\gamma(a; b_1, b_2, \ldots, b_n)$

The operad $\mathcal{P}$ is *augmented* if $\mathcal{P}_0 = \{0\}$ and $\mathcal{P}_1 = k.e$. For any operad $\mathcal{P}$, a $\mathcal{P}$-algebra structure on the vector space V is a morphism of operads from $\mathcal{P}$ to $\text{Endop}(V)$. For any two $\mathcal{P}$-algebras V and W, a morphism of $\mathcal{P}$-algebras is a linear map $f : V \to W$, such that for any $n \geq 0$ and for any $\gamma \in \mathcal{P}_n$, the following diagram commutes,

$$\begin{array}{ccc} V^{\otimes n} & \xrightarrow{f^{\otimes n}} & W^{\otimes n} \\ \downarrow{\scriptstyle \gamma} & & \downarrow{\scriptstyle \gamma} \\ V & \xrightarrow[f]{} & W \end{array}$$

where we have denoted by the same letter γ the element of $\mathcal{P}_n$ and its images in $\text{Endop}(V)_n$ and $\text{Endop}(W)_n$.

Now let V be any k-vector space. The *free $\mathcal{P}$-algebra* is a $\mathcal{P}$-algebra $\mathcal{F}_{\mathcal{P}}(V)$ endowed with a linear map $\iota : V \hookrightarrow \mathcal{F}_{\mathcal{P}}(V)$, such that for any $\mathcal{P}$-algebra A and for any linear map $f : V \to A$, there is a unique $\mathcal{P}$-algebra morphism $\overline{f} : \mathcal{F}_{\mathcal{P}}(V) \to A$, such that $f = \iota \circ \overline{f}$. The free $\mathcal{P}$-algebra $\mathcal{F}_{\mathcal{P}}(V)$ is unique up to isomorphism, and we can prove that a concrete presentation of it is given by:

$$\mathcal{F}_{\mathcal{P}}(V) = \bigoplus_{n\geq 0} \mathcal{P}_n \otimes_{S_n} V^{\otimes n}, \qquad [1.69]$$

with the map ι being obviously defined. When V is of finite dimension d, the corresponding free $\mathcal{P}$-algebra is often called the *free $\mathcal{P}$-algebra with d generators.*

There are several other equivalent definitions for an operad. For more details about operads, see, for example, Loday (1996) and Loday and Vallette (2012).

1.5.4. *A few examples of operads*

1.5.4.1. *The operad* Assoc

This operad governs associative algebras. Assoc_n is given by $k[S_n]$ (the algebra of the symmetric group S_n) for any $n \geq 0$, whereas $\text{Assoc}_0 := \{0\}$. The right action of S_n on Assoc_n is given by linear extension of right multiplication:

$$\left(\sum_i \lambda_i \sigma_i\right) \sigma := \sum_i \lambda_i (\sigma_i \sigma). \qquad [1.70]$$

Let $\sigma \in \text{Assoc}_k$ and $\tau \in \text{Assoc}_l$. The partial compositions are given for any $i = 1, \ldots, k$ by:

$$\sigma \circ_i \tau := \iota_i(\sigma, \tau), \qquad [1.71]$$

with the notations of equation [1.68]. The reader is invited to check the two associativity axioms, as well as the equivariance axiom which reads:

$$(\sigma\sigma') \circ_i (\tau\tau') = (\sigma \circ_{\sigma'(i)} \tau)(\sigma' \circ_i \tau') \qquad [1.72]$$

for any $\sigma, \sigma' \in \text{ASSOC}_k$ and $\tau, \tau' \in \text{ASSOC}_l$. Let us denote by e_k the unit element in the symmetric group S_k. We obviously have $e_k \circ_i e_l = e_{k+l-1}$ for any $i = 1, \ldots, k$. In particular,

$$e_2 \circ_1 e_2 = e_2 \circ_2 e_2 = e_3. \qquad [1.73]$$

Now let V be an algebra over the operad ASSOC, and let $\Phi : \text{ASSOC} \to \text{Endop}(V)$ be the corresponding morphism of operads. Let $\mu : V \otimes V \to V$ be the binary operation $\Phi(e_2)$. In view of equation [1.73] we have:

$$\mu \circ_1 \mu = \mu \circ_2 \mu. \qquad [1.74]$$

In other words, μ is associative. As e_k can be obtained, for any $k \geq 3$, by iteratively composing $k-2$ times the element e_2, we see that any element of ASSOC_k can be obtained from e_2, partial compositions, symmetric group actions and linear combinations. As a result, any k-ary operation on V, which is in the image of Φ, can be obtained in terms of the associative product μ, partial compositions, symmetric group actions and linear combinations. Summing up, an algebra over the operad ASSOC is nothing but an associative algebra. In view of equation [1.69], the free ASSOC-algebra over a vector space W is the (non-unital) tensor algebra $T^+(W) = \bigoplus_{k\geq 1} W^{\otimes k}$.

In the same line of thoughts, the operad governing unital associative algebras is defined similarly, except that the space of 0-ary operations is $k.e_0$, with $e_k \circ_i e_0 = e_{k-1}$ for any $i = 1, \ldots, k$. The unit element $u : k \to V$ of the algebra V is given by $u = \Phi(e_0)$. The free unital algebra over a vector space W is the full tensor algebra $T(W) = \bigoplus_{k\geq 0} W^{\otimes k}$.

1.5.4.2. *The operad* COM

This operad governs commutative associative algebras. COM_n is one-dimensional for any $n \geq 1$, given by $k.\overline{e}_n$ for any $n \geq 0$, whereas $\text{COM}_0 := \{0\}$. The right action of S_n on COM_n is trivial. The partial compositions are defined by:

$$\overline{e}_k \circ_i \overline{e}_l = \overline{e}_{k+l-1} \text{for any} i = 1, \ldots, k. \qquad [1.75]$$

The three axioms of an operad are obviously verified. Let V be an algebra over the operad COM, and let $\Phi : \text{COM} \to \text{Endop}(V)$ be the corresponding morphism of operads. Let $\mu : V \otimes V \to V$ be the binary operation $\Phi(\overline{e}_2)$. We obviously have:

$$\mu \circ_1 \mu = \mu \circ_2 \mu, \qquad \mu = \mu.\tau, \qquad [1.76]$$

where $\tau \in S_2$ is the flip. Hence, μ is associative and commutative. Here, any k-ary operation in the image of Φ can be obtained, up to a scalar, by iteratively composing $\overline{e}_2$ with itself. Hence, an algebra over the operad COM is nothing but a commutative associative algebra. In view of [1.69], the free COM-algebra over a vector space W is the (non-unital) symmetric algebra $S^+(W) = \bigoplus_{k\geq 1} S^k(W)$.

The operad governing unital commutative associative algebras is defined similarly, except that the space of 0-ary operations is $k.\overline{e}_0$, with $\overline{e}_k \circ_i \overline{e}_0 = \overline{e}_{k-1}$ for any $i = 1, \ldots, k$. The unit element $u : k \to V$ of the algebra V is given by $u = \Phi(e_0)$. The free unital algebra over a vector space W is the full symmetric algebra $S(W) = \bigoplus_{k\geq 0} S^k(W)$.

The map $e_k \to \overline{e}_k$ is easily seen to define a morphism of operads Ψ : ASSOC $\to$ COM. Hence, any COM-algebra is also an ASSOC-algebra. This expressed the fact that, forgetting commutativity, a commutative associative algebra is also an associative algebra.

1.5.4.3. *Associative algebras*

Any associative algebra A is some degenerate form of operad: indeed, defining $\mathcal{P}(A)$ by $\mathcal{P}(A)_1 := A$ and $\mathcal{P}(A)_n := \{0\}$ for $n \neq 1$, the collection $\mathcal{P}(A)$ is obviously an operad. An algebra over $\mathcal{P}(A)$ is the same as an A-module.

This point of view leads to a more conceptual definition of operads: an operad is nothing but an associative unital algebra in the category of "S-objects", that is, collections of vector spaces $(\mathcal{P}_n)_{n\geq 0}$ with a right action of S_n on $\mathcal{P}_n$. There is a suitable "tensor product" $\boxtimes$ on S-objects, however not symmetric, such that the global composition γ and the unit $u : k \to \mathcal{P}_1$ (defined by $u(1) = e$) make the following diagrams commute:

$$\begin{array}{ccc} \mathcal{P} \boxtimes \mathcal{P} \boxtimes \mathcal{P} & \xrightarrow{\gamma \boxtimes I} & \mathcal{P} \boxtimes \mathcal{P} \\ \Big\downarrow{\scriptstyle I \boxtimes \gamma} & & \Big\downarrow{\scriptstyle \gamma} \\ \mathcal{P} \boxtimes \mathcal{P} & \xrightarrow{\gamma} & \mathcal{P}, \end{array}$$

$$\begin{array}{ccccc} k \boxtimes \mathcal{P} & \xrightarrow{u \boxtimes I} & \mathcal{P} \boxtimes \mathcal{P} & \xleftarrow{I \boxtimes u} & \mathcal{P} \boxtimes k \\ & \searrow^{\sim} & \Big\downarrow{\scriptstyle \gamma} & \swarrow^{\sim} & \\ & & \mathcal{P}. & & \end{array}$$

These two diagrams commute if and only if e verifies the unit axiom and the partial compositions verify the two associativity axioms and the equivariance axiom (Loday and Vallette 2012).

1.6. Pre-Lie algebras (continued)

1.6.1. *Pre-Lie algebras and augmented operads*

1.6.1.1. *General construction*

We adopt the notations of section 1.5. The sum of the partial compositions yields a right pre-Lie algebra structure on the free $\mathcal{P}$-algebra with one generator, more precisely on $F_{\mathcal{P}}^{+} := \bigoplus_{n\geq 2} \mathcal{P}_n/S_n$, namely:

$$\bar{a} \lhd \bar{b} := \sum_{i=1}^{k} \overline{a \circ_i b}. \qquad [1.77]$$

Following Chapoton (2002), we can consider the pro-unipotent group $G_{\mathcal{P}}^{e}$ associated with the completion of the pre-Lie algebra $F_{\mathcal{P}}^{+}$ for the filtration induced by the grading. More precisely, Chapoton's group $G_{\mathcal{P}}$ is given by the elements $g \in \widehat{F_{\mathcal{P}}}$, such that $g_1 \neq 0$, whereas $G_{\mathcal{P}}^{e}$ is the subgroup of $G_{\mathcal{P}}$ formed by elements g, such that $g_1 = e$.

Any element $a \in \mathcal{P}_n$ gives rise to an n-ary operation $\omega_a : F_{\mathcal{P}}^{\otimes n} \to F_{\mathcal{P}}$, and for any $x, y_1, \ldots, y_n \in F_{\mathcal{P}}^{+}$, we have[1](Manchon and Saidi 2011):

$$\omega_a(y_1, \ldots, y_n) \lhd x = \sum_{j=1}^{n} \omega_a(y_1, \ldots, y_j \lhd x, \ldots, y_n). \qquad [1.78]$$

1.6.1.2. *The pre-Lie operad*

Pre-Lie algebras are algebras over the *pre-Lie operad*, which has been described in detail by Chapoton and Livernet (2001) as follows: $\mathcal{PL}_n$ is the vector space of labeled rooted trees, and the partial composition $s \circ_i t$ is given by summing all of the possible ways of inserting the tree t inside the tree s at the vertex labeled by i. To be precise, the sum runs over the possible ways of branching on t the edges of s, which arrive on the vertex i.

The free left pre-Lie algebra with one generator is then given by the space $\mathcal{T} = \bigoplus_{n\geq 1} T_n$ of rooted trees, as quotienting with the symmetric group actions

---

1. We thank Muriel Livernet for having brought this point to our attention.

amounts to neglect the labels. The pre-Lie operation $(s,t) \mapsto (s \to t)$ is given by the sum of the graftings of s on t at all vertices of t. As a result of [1.78], we have two pre-Lie operations on $\mathcal{T}' = \bigoplus_{n\geq 2} T_n$, which interact as follows (Manchon and Saidi 2011):

$$(s \to t) \lhd u = (s \lhd u) \to t + s \to (t \lhd u). \qquad [1.79]$$

The first pre-Lie operation $\lhd$ comes from the fact that $\mathcal{PL}$ is an augmented operad, whereas the second pre-Lie operation $\to$ comes from the fact that $\mathcal{PL}$ is the pre-Lie operad itself! Similarly:

THEOREM 1.4.– *The free pre-Lie algebra with d generators is the vector space of rooted trees with d colors, endowed with grafting.*

1.6.2. *A pedestrian approach to free pre-Lie algebra*

In this section we give a direct proof of Theorem 1.4 without using operads. It is similar to the proof of the main theorem in Chapoton and Livernet (2001) about the structure of the pre-Lie operad, except that we consider unlabeled trees. We stick to $d = 1$ (i.e. one generator), the proof for several generators being completely analogous. Let $\mathcal{T}$ be the vector space spanned by rooted trees. First, the grafting operation is pre-Lie, because for any trees s, t and u in $\mathcal{T}$, the expression:

$$s \to (t \to u) - (s \to t) \to u \qquad [1.80]$$

is obtained by summing up all of the possibilities of grafting s and t at some vertex of u. As such, it is obviously symmetric in s and t. Now let $(A, \rhd)$ be any left pre-lie algebra, and choose any $a \in A$. In order to prove Theorem 1.4 for one generator, we have to show that there is a unique pre-Lie algebra morphism $F_a : \mathcal{T} \longrightarrow A$, such that $F_a(\bullet) = a$. For the first trees, we easily obtain:

$$F_a(\bullet) = a$$

$$F_a\left(\begin{smallmatrix}\bullet\\ \vert\\ \bullet\end{smallmatrix}\right) = a \rhd a$$

$$F_a\left(\begin{smallmatrix}\bullet\\ \vert\\ \bullet\\ \vert\\ \bullet\end{smallmatrix}\right) = (a \rhd a) \rhd a$$

$$F_a\left(\begin{smallmatrix}\bullet\ \ \bullet\\ \vee\\ \bullet\end{smallmatrix}\right) = a \rhd (a \rhd a) - (a \rhd a) \rhd a.$$

Can we continue like this? We proceed by double induction, first, on the number of vertices, second, on the number of branches, that is, the valence of the root. Write any tree t with n vertices as $t = B_+(t_1, \dots, t_k)$, where the t_js are the branches and B_+ is the operator that grafts the branches on a common extra root. By the induction hypothesis on n, the images $F_a(t_j)$ are well-defined.

Suppose first that $k = 1$, that is $t = B_+(t_1) = t_1 \to \bullet$. Then, we obviously have $F_a(t) = F_a(t_1) \rhd a$. Suppose now that $F_a(s)$ is unambiguously defined for any tree s with n vertices and k' branches with $k' \leq k-1$. The equation:

$$\begin{aligned} t &= B_+(t_1, \ldots, t_k) \\ &= t_1 \to B_+(t_2, \ldots, t_k) - \sum_{j=2}^{k} B_+(t_2, \ldots, t_{j-1}, t_1 \to t_j, t_{j+1}, \ldots, t_n) \end{aligned}$$

shows that, if $F_a(t)$ exists, it is uniquely defined by:

$$\begin{aligned} F_a(t) &= F_a(t_1) \rhd F_a\big(B_+(t_2, \ldots, t_k)\big) \\ &\quad - \sum_{j=2}^{k} F_a\big(B_+(t_2, \ldots, t_{j-1}, t_1 \to t_j, t_{j+1}, \ldots, t_n)\big). \end{aligned} \qquad [1.81]$$

What remains to be shown is that this expression does not depend on the choice of the distinguished branch t_1. In order to see this, choose a second branch (say t_2), and consider the expression:

$$T := t_1 \to \big(t_2 \to B_+(t_3, \ldots, t_k)\big) - (t_1 \to t_2) \to B_+(t_3, \ldots, t_k), \qquad [1.82]$$

which is obtained by grafting t_1 and t_2 on $B_+(t_3, \ldots, t_k)$. This expression is the sum of five terms:

1) T_1, obtained by grafting t_1 and t_2 on the root. It is nothing but the tree t itself.

2) T_2, obtained by grafting t_1 on the root and t_2 elsewhere.

3) T_3, obtained by grafting t_2 on the root and t_1 elsewhere.

4) T_4, obtained by grafting t_1 on some branch and t_2 on some other branch.

5) T_5, obtained by grafting t_1 and t_2 on the same branch.

The terms $F_a(T_2) + F_a(T_3)$, $F_a(T_4)$ and $F_a(T_5)$ are well-defined by the induction hypothesis on the number of branches, and are obviously symmetric in t_1 and t_2. We thus arrive at:

$$\begin{aligned} F_a(t) &= F_a(t_1) \rhd \Big(F_a(t_2) \rhd F_a\big(B_+(t_3, \ldots, t_k)\big)\Big) \\ &\quad - \big(F_a(t_1) \rhd F_a(t_2)\big) \rhd F_a\big(B_+(t_3, \ldots, t_k)\big) \\ &\quad - F_a(T_2) - F_a(T_3) - F_a(T_4) - F_a(T_5), \end{aligned}$$

which is symmetric in t_1 and t_2 thanks to the left pre-Lie relation in A. The expression [1.81] is then the same if we exchange t_1 with the branch t_2 or any other branch; hence, it is invariant by any permutation of the branches $t_1, \ldots, t_n$. This proves Theorem 1.4 for one generator. The general case is proven similarly, except we have to replace $a \in A$ by a collection $\{a_1, \ldots, a_d\}$.

1.6.3. *Right-sided commutative Hopf algebras and the Loday–Ronco theorem*

Loday and Ronco (2010) have found a deep link between pre-Lie algebras and commutative Hopf algebras of a certain type: let $\mathcal{H}$ be a commutative Hopf algebra. Following this, we say that $\mathcal{H}$ is *right-sided* if it is free as a commutative algebra, that is, $\mathcal{H} = S(V)$ for some k-vector space V, and if the reduced coproduct verifies:

$$\widetilde{\Delta}(V) \subset \mathcal{H} \otimes V. \qquad [1.83]$$

Suppose that $V = \bigoplus_{n\geq 0}$ is graded with finite-dimensional homogeneous components. Then, the graded dual $A = V^0$ is a left pre-Lie algebra, and by the Milnor–Moore theorem, the graded dual $\mathcal{H}^0$ is isomorphic to the enveloping algebra $\mathcal{U}(A_{\text{Lie}})$ as graded Hopf algebra. Conversely, for any graded pre-Lie algebra A, the graded dual $\mathcal{U}(A_{\text{Lie}})^0$ is free commutative right-sided (Loday and Ronco 2010, Theorem 5.3).

The Hopf algebra $\mathcal{H}_{CK}$ of rooted forests enters into this framework, and, as it was first explicited in Chapoton (2001), the associated pre-Lie algebra is the free pre-Lie algebra of rooted trees with grafting: to see this, denote by (δ_s) the dual basis in the graded dual $\mathcal{H}^\circ_{CK}$ of the forest basis of $\mathcal{H}_{CK}$. The correspondence $\delta : s \mapsto \delta_s$ extends linearly to a unique vector space isomorphism from $\mathcal{H}_{CK}$ onto $\mathcal{H}^\circ_{CK}$. For any tree t, the corresponding δ_t is an infinitesimal character of $\mathcal{H}_{CK}$, that is, it is a primitive element of $\mathcal{H}^\circ$. We denote by $*$ the (convolution) product of $\mathcal{H}^\circ$. We have:

$$\delta_t * \delta_u - \delta_u * \delta_t = \delta_{t \curvearrowright u - u \curvearrowright t}. \qquad [1.84]$$

Here, $t \curvearrowright u$ is obtained by grafting t on u, namely:

$$t \curvearrowright u = \sum_v N'(t, u, v) v, \qquad [1.85]$$

where $N'(t, u, v)$ is the number of partitions $V(t) = V \amalg W, W < V$, such that $v_{|V} = t$ and $v_{|W} = u$. Another normalization is often employed: considering the normalized dual basis $\widetilde{\delta}_t = \sigma(t)\delta_t$, where $\sigma(t) = |\text{Aut}\, t|$ stands for the symmetry factor of t, we obviously have:

$$\widetilde{\delta}_t * \widetilde{\delta}_u - \widetilde{\delta}_u * \widetilde{\delta}_t = \widetilde{\delta}_{t \to u - u \to t}, \qquad [1.86]$$

where:

$$t \to u = \sum_{v} M'(t,u,v)v, \qquad [1.87]$$

where $M'(t,u,v) = \dfrac{\sigma(t)\sigma(u)}{\sigma(v)} N'(t,u,v)$ can be interpreted as the number of ways to graft the tree t on the tree u in order to get the tree v. The operation $\to$ then coincides with the grafting free pre-Lie operation introduced in section 1.6.1.2[2].

The other pre-Lie operation $\lhd$ of section 1.6.1.2, more precisely its opposite $\rhd$, is associated with another right-sided Hopf algebra of forests $\mathcal{H}$ which has been investigated in Calaque *et al.* (2011) and Manchon and Saidi (2011), and which can be defined by considering trees as Feynman diagrams (without loops): let $\mathcal{T}'$ be the vector space spanned by rooted trees with at least one edge. Consider the symmetric algebra $\mathcal{H} = \mathcal{S}(\mathcal{T}')$, which can be seen as the k-vector space generated by rooted forests with all connected components containing at least one edge. We identify the unit of $\mathcal{S}(\mathcal{T}')$ with the rooted tree $\bullet$. A *subforest* of a tree t is either the trivial forest $\bullet$, or a collection $(t_1, \ldots, t_n)$ of pairwise disjoint subtrees of t, each of them containing at least one edge. In particular, two subtrees of a subforest cannot have any common vertex.

Let s be a subforest of a rooted tree t. Denote by t/s the tree obtained by contracting each connected component of s onto a vertex. We turn $\mathcal{H}$ into a bialgebra by defining a coproduct $\Delta : \mathcal{H} \to \mathcal{H} \otimes \mathcal{H}$ on each tree $t \in \mathcal{T}'$ by :

$$\Delta(t) = \sum_{s \subseteq t} s \otimes t/s, \qquad [1.88]$$

where the sum runs over all possible subforests (including the unit $\bullet$ and the full subforest t). As usual we extend the coproduct Δ multiplicatively onto $\mathcal{S}(\mathcal{T}')$. In fact, coassociativity is easily verified. This makes $\mathcal{H} := \bigoplus_{n \geq 0} \mathcal{H}_n$ a connected graded bialgebra, hence a Hopf algebra, where the grading is defined in terms of the number of edges. The antipode $S : \mathcal{H} \to \mathcal{H}$ is given (recursively with respect to the number of edges) by one of the two following formulae:

$$S(t) = -t - \sum_{s, \bullet \neq s \subsetneq t} S(s)\, t/s, \qquad [1.89]$$

$$S(t) = -t - \sum_{s, \bullet \neq s \subsetneq t} s\, S(t/s). \qquad [1.90]$$

It turns out that $\mathcal{H}_{CK}$ is left comodule-bialgebra over $\mathcal{H}$ (Calaque *et al.* 2011; Manchon and Saidi 2011), in the sense that the following diagram commutes:

2. The two notations $\to$ and $\curvearrowright$ come from Calaque *et al.* (2011), but have been exchanged.

$$\begin{array}{ccc}
\mathcal{H}_{CK} & \xrightarrow{\Phi} & \mathcal{H} \otimes \mathcal{H}_{CK} \\
\Delta_{CK} \downarrow & & \\
\mathcal{H}_{CK} \otimes \mathcal{H}_{CK} & & \downarrow I \otimes \Delta_{CK} \\
\Phi \otimes \Phi \downarrow & & \\
\mathcal{H} \otimes \mathcal{H}_{CK} \otimes \mathcal{H} \otimes \mathcal{H}_{CK} & \xrightarrow[m^{1,3}]{} & \mathcal{H} \otimes \mathcal{H}_{CK} \otimes \mathcal{H}_{CK}
\end{array}$$

Here, the coaction $\Phi : \mathcal{H}_{CK} \to \mathcal{H} \otimes \mathcal{H}_{CK}$ is the algebra morphism given by $\Phi(\mathbf{1}) = \bullet \otimes \mathbf{1}$ and $\Phi(t) = \Delta_{\mathcal{H}}(t)$ for any nonempty tree t. As a result, the group of characters of $\mathcal{H}$ acts on the group of characters of $\mathcal{H}_{CK}$ by automorphisms.

1.6.4. *Pre-Lie algebras of vector fields*

1.6.4.1. *Flat torsion-free connections*

Let M be a differentiable manifold, and let ∇ be the covariant derivation operator associated with a connection on the tangent bundle TM. The covariant derivation is a bilinear operator on vector fields (i.e. two sections of the tangent bundle): $(X, Y) \mapsto \nabla_X Y$, such that the following axioms are fulfilled:

$$\nabla_{fX} Y = f \nabla_X Y,$$
$$\nabla_X (fY) = f \nabla_X Y + (X.f)Y \text{(Leibniz rule)}.$$

The torsion of the connection τ is defined by:

$$\tau(X, Y) = \nabla_X Y - \nabla_Y X - [X, Y], \qquad [1.91]$$

and the curvature tensor is defined by:

$$R(X, Y) = [\nabla_X, \nabla_Y] - \nabla_{[X,Y]}. \qquad [1.92]$$

The connection is *flat* if the curvature R vanishes identically, and *torsion-free* if $\tau = 0$. The following crucial observation by Matsushima (1968, Lemma 1) is an immediate consequence of equation [1.43]:

PROPOSITION 1.14.– *For any smooth manifold M endowed with a flat torsion-free connection ∇, the space $\chi(M)$ of vector fields is a left pre-Lie algebra, with pre-Lie product given by:*

$$X \rhd Y := \nabla_X Y. \qquad [1.93]$$

Note that on $M = \mathbb{R}^n$, endowed with its canonical flat torsion-free connection, the pre-Lie product is given by:

$$(f_i\partial_i) \rhd (g_j\partial_j) = f_i(\partial_i g_j)\partial_j. \qquad [1.94]$$

1.6.4.2. *Relating two pre-Lie structures*

Cayley (1857) discovered a link between rooted trees and vector fields on the manifold $\mathbb{R}^n$, endowed with its natural flat torsion free connection, which can be described in modern terms as follows: let $\mathbb{T}$ be the free pre-Lie algebra on the space of vector fields on $\mathbb{R}^n$. A basis of $\mathbb{T}$ is given by rooted trees with vertices decorated by some basis of $\chi(\mathbb{R}^n)$. There is a unique pre-Lie algebra morphism $\mathcal{Y}$, the *Cayley map*, such that $\mathcal{Y}(\bullet_X) = X$ for any vector field X.

PROPOSITION 1.15.– *For any rooted tree t, with each vertex v being decorated by a vector field X_v, the vector field $\mathcal{Y}(t)$ is given at $x \in \mathbb{R}^n$ by the following recursive procedure (Hairer* et al. *2002): if the decorated tree t is obtained by grafting all of its branches t_k on the root r decorated by the vector field $X_r = \sum_{i=1}^n f_i\partial_i$, that is, if it writes $t = B_+^{X_r}(t_1, \ldots, t_k)$, then:*

$$\mathcal{Y}(\bullet_{X_r}) = X_r, \qquad [1.95]$$

$$\mathcal{Y}(t) = \sum_{i=1}^{n} \mathcal{Y}_i(t)\partial_i \text{ with:}$$

$$\mathcal{Y}_i(t)(x) = f_i^{(k)}(x)\big(\mathcal{Y}(t_1)(x), \ldots, \mathcal{Y}(t_k)(x)\big), \qquad [1.96]$$

where $f_i^{(k)}(x)$ stands for the kth differential of f_i.

PROOF.– From equation [1.94], for any vector field X and any other vector field $Y = \sum_{j=1}^n g_j\partial_j$:

$$X \rhd \sum_{j=1}^{n} g_j\partial_j = \sum_j (X.g_j)\partial_j. \qquad [1.97]$$

In other words, $X \rhd Y$ is the derivative of Y along the vector field X, where Y is viewed as a C^∞ map from $\mathbb{R}^n$ to $\mathbb{R}^n$. We prove the result by induction on the number k of branches: for $k = 1$, we check:

$$\begin{aligned}\mathcal{Y}(s \to \bullet_Y)(x) &= \big(\mathcal{Y}(s) \rhd \mathcal{Y}(\bullet_Y)\big)(x)\\ &= \big(\mathcal{Y}(s) \rhd Y\big)(x)\\ &= \sum_{j=1}^{n} \big(\mathcal{Y}(s).g_j\big)(x)\partial_j = \sum_{j=1}^{n} g_j'(x)\big(\mathcal{Y}(s)(x)\big)\partial_j.\end{aligned}$$

Now, we can compute, using the Leibniz rule and the induction hypothesis (we drop the point $x \in \mathbb{R}^n$ where the vector fields are evaluated):

$$\mathcal{Y}(B_+^Y(t_1, \ldots, t_k)) =$$

$$\mathcal{Y}(t_1 \to B_+^Y(t_2, \ldots, t_k)) - \sum_{r=2}^{k} \mathcal{Y}(B_+^Y(t_2, \ldots, t_{r-1}, t_1 \to t_r, t_{r+1}, \ldots, t_k))$$

$$- \sum_{r=2}^{k} \sum_{j=1}^{n} f_i^{(k-1)}(\mathcal{Y}(t_2), \ldots, \mathcal{Y}(t_{r-1}), \mathcal{Y}(t_1) \rhd \mathcal{Y}(t_r), \mathcal{Y}(t_{r-1}), \ldots, \mathcal{Y}(t_k))\partial_j$$

□

COROLLARY 1.2 (closed formula).– *For any rooted tree t with set of vertices $\mathcal{V}(t)$ and root r, each vertex v being decorated by a vector field $X_v = \sum_{i=1}^n X_v^i \partial_i$, the vector field $\mathcal{Y}(t)$ is given at $x \in \mathbb{R}^n$ by the following formula:*

$$\mathcal{Y}(t)(x) = \sum_{F:\mathcal{V}(t)\to\{1,\ldots,n\}} \left(\prod_{v\in\mathcal{V}(t)-\{r\}} \partial_{I(v)}(X_v^{F(v)})(x) \right) \partial_{I(r)}(X_r^{F(r)})(x)\partial_{F(r)}, \qquad [1.98]$$

with the shorthand notation:

$$\partial_{I(v)} := \prod_{w\to v} \partial_{F(w)}, \qquad [1.99]$$

where the product runs over the incoming vertices of v.

Now fix a vector field X on $\mathbb{R}^n$ and consider the map d_X from undecorated rooted trees to vector field-decorated rooted trees, which decorates each vertex by X. It is obviously a pre-Lie algebra morphism, and $F_X := \mathcal{Y} \circ d_X$ is the unique pre-Lie algebra morphism that sends the one-vertex tree • to the vector field X.

1.6.5. *B-series, composition and substitution*

B-series have been defined by Hairer and Wanner, following the pioneering work of Butcher (1963) on Runge–Kutta methods for the numerical resolution of differential equations. Remarkably enough, rooted trees revealed to be an adequate tool not only for vector fields, but also for the numerical approximation of their integral curves. Butcher discovered that the Runge–Kutta methods formed a group (since then called the Butcher group), which was nothing but the character group of the Connes–Kreimer Hopf algebra $\mathcal{H}_{CK}$ (Brouder 2000).

Consider any left pre-Lie algebra $(A, \rhd)$, and introduce a fictitious unit $\mathbf{1}$, such that $\mathbf{1} \rhd a = a \rhd \mathbf{1} = a$ for any $a \in A$, and consider for any $a \in A$, the unique left pre-Lie algebra morphism $F_a : (\mathcal{T}, \to) \longrightarrow (A, \rhd)$, such that $F_a(\bullet) = a$. A *B-series* is an element of $hA[[h]] \oplus k.\mathbf{1}$ defined by:

$$B(\alpha; a) := \alpha(\emptyset)\mathbf{1} + \sum_{s \in T} h^{v(s)} \frac{\alpha(s)}{\sigma(s)} F_a(s), \qquad [1.100]$$

where α is any linear form on $\mathcal{T} \oplus k\emptyset$ (here, $\sigma(s)$ is the symmetry factor of the tree, that is, the order of its group of automorphisms). It matches the usual notion of B-series (Hairer *et al.* 2002) when A is the pre-Lie algebra of vector fields on $\mathbb{R}^n$ (it is also convenient to set $F_a(\emptyset) = \mathbf{1}$). In this case, the vector fields $F_a(t)$ for a tree t are differentiable maps from $\mathbb{R}^n$ to $\mathbb{R}^n$ called *elementary differentials*. B-series can be composed coefficient wise, as series in the indeterminate h, whose coefficients are maps from $\mathbb{R}^n$ to $\mathbb{R}^n$. The same definition with trees decorated by a set of colors $\mathcal{D}$ leads to straightforward generalizations. For example, the P-series used in partitioned Runge–Kutta methods (Hairer *et al.* 2002) correspond to bi-coloured trees.

A slightly different way of defining B-series is the following: consider the unique pre-Lie algebra morphism $\mathcal{F}_a : \mathcal{T} \to hA[h]$, such that $\mathcal{F}_a(\bullet) = ha$. It respects the gradings given by the number of vertices and the powers of h, respectively; hence, it extends to $\mathcal{F}_a : \widehat{\mathcal{T}} \to hA[[h]]$, where $\widehat{\mathcal{T}}$ is the completion of $\mathcal{T}$ with respect to the grading. We further extend it to the empty tree by setting $\mathcal{F}_a(\emptyset) = \mathbf{1}$. We then have:

$$B(\alpha; a) = \mathcal{F}_a \circ \widetilde{\delta}^{-1}(\alpha), \qquad [1.101]$$

where $\widetilde{\delta}$ is the isomorphism from $\widehat{\mathcal{T}} \oplus k\emptyset$ to $(\mathcal{T} \oplus k\emptyset)^*$ given by the normalized dual basis (see section 1.6.3).

We restrict ourselves to B-series $B(\alpha; a)$ with $\alpha(\emptyset) = 1$. Such αs are in one-to-one correspondence with characters of the algebra of forests (which is the underlying algebra of $\mathcal{H}_{CK}$) by setting:

$$\alpha(t_1 \cdots t_k) := \alpha(t_1) \cdots \alpha(t_k). \qquad [1.102]$$

The Hairer–Wanner theorem (Hairer *et al.* 2002, Theorem III.1.10) says that the composition of B-series corresponds to the convolution product of characters of $\mathcal{H}_{CK}$, namely:

$$B(\beta; a) \circ B(\alpha; a) = B(\alpha * \beta, a), \qquad [1.103]$$

where linear forms α, β on $\mathcal{T} \oplus k\emptyset$ and their character counterparts are identified modulo the above correspondence.

Let us now turn to substitution (Chartier *et al.* 2010). The idea is to replace the vector field a in a B-series $B(\beta; a)$ by another vector field $\widetilde{a}$, which expresses itself as a B-series, that is, $\widetilde{a} = h^{-1}B(\alpha; a)$, where α is now a linear form on $\mathcal{T} \oplus k\emptyset$, such that $\alpha(\emptyset) = 0$. Here, we suppose that $\alpha(\bullet) = 1$. The following proposition is proven in Calaque *et al.* (2011):

PROPOSITION 1.16.– *For any linear forms α, β on $\mathcal{T}$ with $\alpha(\bullet = 1)$, we have:*

$$B\big(\beta; \frac{1}{h}B(\alpha; a)\big) = B(\alpha \star \beta; a), \qquad [1.104]$$

where α is the multiplicatively extended to forests, β is seen as an infinitesimal character of $\mathcal{H}_{CK}$, and $\star$ is the dualization of the left coaction Φ of $\mathcal{H}$ on $\mathcal{H}_{CK}$ defined in section 1.6.3.

The condition $\alpha(\bullet) = 1$ is in fact dropped in Calaque *et al.* (2011, Proposition 15): the price to pay is that one has to replace the Hopf algebra $\mathcal{H}$ by a non-connected bialgebra $\widetilde{\mathcal{H}} = S(\mathcal{T})$ with a suitable coproduct, such that $\mathcal{H}$ is obtained as the quotient $\widetilde{\mathcal{H}}/\mathcal{J}$, where $\mathcal{J}$ is the ideal generated by $\bullet - \mathbf{1}$. The substitution product $\star$ then coincides with the one considered in Chartier *et al.* (2010) via natural identifications.

1.7. Other related algebraic structures

1.7.1. *NAP algebras*

NAP algebras (NAP for Non-Associative Permutative) appear under this name in Livernet (2006), and under the name "left- (right-)commutative algebras" in Dzhumadil'daev and Löfwall (2002). They can be seen in some sense as a "simplified version" of pre-Lie algebras. Saidi showed that the pre-Lie operad is a deformation of the NAP operad in a precise sense, involving the notion of *current-preserving operad* (Saidi 2014).

1.7.1.1. *Definition and general properties*

A *left NAP algebra* over a field k is a k-vector space A with a bilinear binary composition $\blacktriangleright$ that satisfies the left NAP identity:

$$a \blacktriangleright (b \blacktriangleright c) = b \blacktriangleright (a \blacktriangleright c) \qquad [1.105]$$

for any $a, b, c \in A$. Analogously, a *right NAP algebra* is a k-vector space A with a binary composition $\blacktriangleleft$ satisfying the right NAP identity:

$$(a \blacktriangleleft b) \blacktriangleleft c = (a \blacktriangleleft c) \blacktriangleleft b. \qquad [1.106]$$

As any right NAP algebra is also a left NAP algebra with product $a \blacktriangleright b := b \blacktriangleleft a$, we can stick to left NAP algebras, which is what we will do unless specifically indicated.

1.7.1.2. *Free NAP algebras*

The *left Butcher product* $s \circ t$ of two rooted trees s and t is defined by grafting s on the root of t. For example:

$$\bullet \circ \mathbf{I} = \mathbf{V}, \qquad \mathbf{I} \circ \mathbf{I} = \mathbf{\dot{V}}. \qquad [1.107]$$

The following theorem is due to Dzhumadil'daev and Löfwall (2002) (see Livernet (2006) for an operadic approach):

THEOREM 1.5.– *The free NAP algebra with d generators is the vector space spanned by rooted trees with d colors, endowed with the left Butcher product.*

PROOF.– We give the proof for one generator, the case of d generators being entirely similar. The left NAP property for the Butcher product is obvious. Let $(A, \blacktriangleright)$ be any left NAP algebra, and let $a \in A$. We have to prove that there exists a unique left NAP algebra morphism G_a from $(\mathcal{T}, \circ)$ to $(A, \blacktriangleright)$, such that $G_a(\bullet) = a$. As in the pre-Lie case, we proceed by double induction, first on the number n of vertices, and second on the number k of branches. In the case $k = 1$, the tree t writes $B_+(t_1) = t_1 \circ \bullet$; hence, $G_a(t) = G_a(s) \blacktriangleright a$ is the only possible choice. Now a tree with k branches writes:

$$t = B_+(t_1, \ldots, t_k) = t_1 \circ \big(t_2 \circ \cdots \circ (t_k \circ \bullet) \ldots\big). \qquad [1.108]$$

The only possible choice is then:

$$G_a(t) = G_a(t_1) \blacktriangleright \Big(G_a(t_2) \blacktriangleright \cdots \blacktriangleright \big(G_a(t_k) \blacktriangleright a\big) \ldots\Big), \qquad [1.109]$$

and the result is clearly symmetric in t_1 and t_2 due to the left NAP identity in A. Using the induction hypothesis, the result is also invariant under permutation of the branches $2, 3, \ldots, k$. Hence, it is invariant under the permutation of all branches, which proves the theorem. □

Despite the similarity with the pre-Lie situation described in section 1.6.2, the NAP framework is much simpler due to the set-theoretic nature of the Butcher product: for any trees s and t, the Butcher product $s \circ t$ is a tree, whereas the grafting $s \to t$ is a sum of trees. We obtain for the first trees:

$$\begin{aligned} G_a(\bullet) &= a \\ G_a(\mathbf{I}) &= a \blacktriangleright a \\ G_a(\mathbf{\dot{I}}) &= (a \blacktriangleright a) \blacktriangleright a \\ G_a(\mathbf{V}) &= a \blacktriangleright (a \blacktriangleright a). \end{aligned}$$

1.7.1.3. *NAP algebras of vector fields*

We consider the flat affine n-dimensional space E_n although it is possible, through parallel transport, to consider any smooth manifold endowed with a flat torsion-free connection. Fix an origin in E_n, which will be denoted by O. For vector fields $X = \sum_{i=1}^n f_i \partial_i$ and $Y = \sum_{j=1}^n g_j \partial_j$, we set:

$$X \blacktriangleright_O Y = \sum_{j=1}^n (X_O . g_j)\partial_j, \qquad [1.110]$$

where $X_O := \sum_{i=0}^n f_i(O)\partial_i$ is the constant vector field obtained by freezing the coefficients of X at $x = O$.

PROPOSITION 1.17.– *The space $\chi(\mathbb{R}^n)$ of vector fields endowed with product $\blacktriangleright_O$ is a left NAP algebra. Moreover, for any other choice of origin $O' \in E_n$, the conjugation with the translation of vector $\overrightarrow{OO'}$ is an isomorphism from $\big(\chi(\mathbb{R}^n), \blacktriangleright_O\big)$ onto $\big(\chi(\mathbb{R}^n), \blacktriangleright_{O'}\big)$.*

PROOF.– Let $X = \sum_{i=1}^n f_i \partial_i$, $Y = \sum_{j=1}^n g_j \partial_j$ and $Z = \sum_{k=1}^n h_k \partial_k$ be three vector fields. Then:

$$X \blacktriangleright_O (Y \blacktriangleright_O Z) = \sum_{k=1}^n X_O.(Y_O.h_k)\partial_k \qquad [1.111]$$

is symmetric in X and Y, due to the fact that the two constant vector fields X_O and Y_O commute. The second assertion is left as an exercise for the reader. □

With the notations of section 1.6.4, there is a unique NAP algebra morphism

$$\mathcal{Y}_O : (\mathbb{T}, \circ) \longrightarrow \big(\chi(\mathbb{R}^n), \blacktriangleright_O\big), \qquad [1.112]$$

the *frozen Cayley map*, such that $\mathcal{Y}_O(\bullet_X) = X$. By also considering the unique NAP algebra morphism $G_{X,O} = \mathcal{Y}_O \circ d_X : (\mathcal{T}, \circ) \to \big(\chi(\mathbb{R}^n), \blacktriangleright_O\big)$, the maps $G_{X,O}(t) : \mathbb{R}^n \to \mathbb{R}^n$ are called the *frozen elementary differentials*.

PROPOSITION 1.18.– *For any rooted tree t, each vertex v being decorated by a vector field X_v, the vector field $\mathcal{Y}_O(t)$ is given at $x \in \mathbb{R}^n$ by the following recursive procedure: if the decorated tree t is obtained by grafing all of its branches t_k on the root r decorated by the vector field $X_r = \sum_{i=1}^n f_i \partial_i$, that is, if it writes $t = B_+^{X_r}(t_1, \ldots, t_k)$, then:*

$$\mathcal{Y}_O(\bullet_{X_r}) = X_r, \qquad [1.113]$$

$$\mathcal{Y}_O(t) = \sum_{i=1}^{n} \mathcal{Y}_{O,i}(t)\partial_i \text{ with:}$$

$$\mathcal{Y}_{O,i}(t)(x) = f_i^{(k)}(x)\big(\mathcal{Y}_O(t_1)(O), \ldots, \mathcal{Y}_O(t_k)(O)\big), \qquad [1.114]$$

where $f_i^{(k)}(x)$ stands for the kth differential of f_i, evaluated at x.

PROOF.– We prove the result by induction on the number k of branches: for $k = 1$, we check:

$$\begin{aligned}
\mathcal{Y}_O(s \circ \bullet_Y)(x) &= \big(\mathcal{Y}_O(s) \blacktriangleright_O \mathcal{Y}_O(\bullet_Y)\big)(x) \\
&= \big(\mathcal{Y}_O(s) \blacktriangleright_O Y\big)(x) \\
&= \sum_{j=1}^{n} \big(\mathcal{Y}_O(s)_O.g_j\big)(x)\partial_j \\
&= \sum_{j=1}^{n} g_j'(x)\big(\mathcal{Y}_O(s)(O)\big)\partial_j.
\end{aligned}$$

Now we can compute using the induction hypothesis and the fact that the vector fields $\mathcal{Y}_O(t_j)(O)$ are constant:

$$\begin{aligned}
\mathcal{Y}_O\big(B_+^Y(t_1, \ldots, t_k)\big)(x) &= \mathcal{Y}_O\big(t_1 \circ B_+^Y(t_2, \ldots, t_k)\big)(x) \\
&= \Big(\mathcal{Y}_O(t_1) \blacktriangleright_O \sum_{j=1}^{n} f_i^{(k-1)}\big(\mathcal{Y}_O(t_2)(O), \ldots, \mathcal{Y}_O(t_k)(O)\big)\partial_j\Big)(x) \\
&= \sum_{j=1}^{n} f_i^{(k)}(x)\big(\mathcal{Y}_O(t_1)(O), \ldots, \mathcal{Y}_O(t_k)(O)\big)\partial_j.
\end{aligned}$$

□

COROLLARY 1.3 (closed formula).– *With the notations of Corollary 1.2, for any rooted tree t with set of vertices $\mathcal{V}(t)$ and root r, each vertex v being decorated by a vector field $X_v = \sum_{i=1}^{n} X_v^i \partial_i$, the vector field $\mathcal{Y}_O(t)$ is given at $x \in \mathbb{R}^n$ by the following formula:*

$$\mathcal{Y}_O(t)(x) = \sum_{F:\mathcal{V}(t)\to\{1,\ldots,n\}} \left(\prod_{v\in\mathcal{V}(t)-\{r\}} \partial_{I(v)}(X_v^{F(v)})(O) \right) \partial_{I(r)}(X_r^{F(r)})(x)\partial_{F(r)}. \qquad [1.115]$$

1.7.2. *Novikov algebras*

A Novikov algebra is a right pre-Lie algebra which is also left NAP, namely, a vector space A together with a bilinear product $*$, such that, for any $a, b, c \in A$, we have:

$$a * (b * c) - (a * b) * c = a * (c * b) - (a * c) * b, \qquad [1.116]$$

$$a * (b * c) = b * (a * c). \qquad [1.117]$$

Novikov algebras first appeared in hydrodynamical equations (Balinskii and Novikov 1985; Osborn 1992). The prototype is a commutative associative algebra together with a derivation D, with the Novikov product being given by:

$$a * b := (Da)b. \qquad [1.118]$$

The free Novikov algebra on a set of generators has been given in Dzhumadil'daev and Löfwall (2002, section 7) in terms of some classes of rooted trees.

1.7.3. *Assosymmetric algebras*

An assosymmetric algebra is a vector space endowed with a bilinear operation that is both left and right pre-Lie, which means that the associator $a * (b * c) - (a * b) * c$ is symmetric under the permutation group S_3. This notion was introduced by Kleinfeld (1957) (see also Hentzel *et al.* (1996)). All associative algebras are obviously assosymmetric; however, the converse is not true.

1.7.4. *Dendriform algebras*

A *dendriform algebra* (Loday 2001) over the field k is a k-vector space A endowed with two bilinear operations, denoted $\prec$ and $\succ$ and called right and left products, respectively, subject to the three axioms below:

$$(a \prec b) \prec c = a \prec (b \prec c + b \succ c) \qquad [1.119]$$

$$(a \succ b) \prec c = a \succ (b \prec c) \qquad [1.120]$$

$$a \succ (b \succ c) = (a \prec b + a \succ b) \succ c. \qquad [1.121]$$

We readily verify that these relations yield associativity for the product

$$a * b := a \prec b + a \succ b. \qquad [1.122]$$

However, at the same time-ordering, the dendriform relations imply that the bilinear product $\rhd$ defined by:

$$a \rhd b := a \succ b - b \prec a, \qquad [1.123]$$

is left pre-Lie. The associative operation $*$ and the pre-Lie operation $\rhd$ define the same Lie bracket, and this is, of course, still true for the opposite (right) pre-Lie product $\lhd$:

$$[\![a, b]\!] := a * b - b * a = a \rhd b - b \rhd a = a \lhd b - b \lhd a.$$

In the commutative case (commutative dendriform algebras are also named *Zinbiel algebras* (Loday 1995, 2001), the left and right operations are further required to identify, so that $a \succ b = b \prec a$. In this case, both pre-Lie products vanish. A natural example of a commutative dendriform algebra is given by the shuffle algebra in terms of half-shuffles (Schützenberger 1958/1959). Any associative algebra A equipped with a linear integral-like map $I : A \to A$ satisfying the integration by parts rule also gives a dendriform algebra, when $a \prec b := aI(b)$ and $a \succ b := I(a)b$. The left pre-Lie product is then given by $a \rhd b = [I(a), b]$. It is worth mentioning that Zinbiel algebras are also NAP algebras, as shown by the computation below (dating back to Schützenberger (1958/1959)):

$$\begin{aligned} a \succ (b \succ c) &= (a \succ b + a \prec b) \succ c \\ &= (a \succ b + b \succ a) \succ c \\ &= b \succ (a \succ c). \end{aligned}$$

There also exists a twisted version of dendriform algebras, encompassing operators like the Jackson integral I_q (Ebrahimi-Fard and Manchon 2011). Returning to ordinary dendriform algebras, we observe that:

$$a * b + b \rhd a = a \succ b + b \succ a. \qquad [1.124]$$

This identity generalizes to any number of elements, expressing the symmetrization of

$$\Big(...((a_1 \succ a_2) \succ a_3) \cdots \Big) \succ a_n$$

in terms of the associative product and the left pre-Lie product (Ebrahimi-Fard *et al.* 2008). For more on dendriform algebras and the associated pre-Lie structures, see Ebrahimi-Fard *et al.* (2008), Ebrahimi-Fard and Manchon (2009a, 2009b, 2011) and Ebrahimi-Fard's note in the present volume.

1.7.5. *Post-Lie algebras*

Post-Lie algebras have been introduced by Vallette (2007), independent of the introduction of the closely related notion of D-algebra in Munthe-Kaas and Wright

(2008). A left post-Lie algebra on a field k is a k-vector space A together with a bilinear binary product $\triangleright$ and a Lie bracket $[-,-]$, such that

$$a \triangleright [b,c] = [a \triangleright b, c] + [b, a \triangleright c],$$

$$[a,b] \triangleright c = a \triangleright (b \triangleright c) - (a \triangleright b) \triangleright c - b \triangleright (a \triangleright c) + (b \triangleright a) \triangleright c$$

for any $a, b, c \in A$. In particular, a post-Lie algebra A is a pre-Lie algebra if and only if the Lie bracket vanishes. The space of vector fields on a Lie group is a post-Lie algebra, and the free post-Lie algebra with one generator is the free Lie algebra on the linear span of *planar* rooted trees. The binary product $\triangleright$ is given by *left grafting*, which is not pre-Lie anymore because of planarity. This is the starting point to the Lie–Butcher theory (see, for example, Lundervold and Munthe-Kaas (2013); Ebrahimi-Fard *et al.* (2015); Curry *et al.* (2019, 2020)).

1.8. References

Agrachev, A. and Gamkrelidze, R. (1981). Chronological algebras and nonstationary vector fields. *J. Sov. Math.*, 17(1), 1650–1675.

Andruskiewitsch, N. and Cuadra, J. (2013). On the structure of (co-Frobenius) Hopf algebras. *J. Noncommutative Geom.*, 13(1), 83–104.

Andruskiewitsch, N. and Schneider, H.-J. (2002). Pointed Hopf algebras. In *New Directions in Hopf Algebras*, Montgomery, S. and Schneider, H.-J. (eds). MSRI and Cambridge University Press, Cambridge.

Balinskii, A.A. and Novikov, S.P. (1985). Poisson bracket of hydrodynamic type, Frobenius algebras and Lie algebras. *Doklady An. SSSR*, 283(5), 1036–1039.

Brouder, C. (2000). Runge–Kutta methods and renormalization. *Eur. Phys. J.*, C12, 512–534.

Burde, D. (2006). Left-symmetric algebras or pre-Lie algebras in Geometry and Physics. *Central Eur. J. Math.* DOI: 10.2478/s11533-006-0014-9.

Butcher, J.C. (1963). Coefficients for the study of Runge–Kutta integration processes. *J. Austral. Math. Soc.*, 3, 185–201.

Calaque, D., Ebrahimi-Fard, K., Manchon, D. (2011). Two interacting Hopf algebras of trees: A Hopf-algebraic approach to composition and substitution of B-series. *Adv. Appl. Math.*, 47(2), 282–308

Cayley, A. (1857). On the theory of analytical forms called trees. *Phil. Mag.*, 13, 172–176.

Chapoton, F. (2001). Algèbres pré-Lie et algèbres de Hopf liées à la renormalisation. *C. R. Acad. Sci.*, 332(1), 681–684.

Chapoton, F. (2002). Rooted trees and an exponential-like series. arXiv:math/0209104.

Chapoton, F. and Livernet, M. (2001). Pre-Lie algebras and the rooted trees operad. *Int. Math. Res. Not.*, 2001, 395–408.

Chartier, P., Hairer, E., Vilmart, G. (2005). A substitution law for B-series vector fields. Research report, INRIA, Rocquencourt.

Chartier, P., Hairer, E., Vilmart, G. (2010). Algebraic structures of B-series. *Found. Comput. Math.*, 10, 407–427.

Connes, A. and Kreimer, D. (1998). Hopf algebras, renormalization and noncommutative geometry. *Comm. Math. Phys.*, 199, 203–242.

Curry, C., Ebrahimi-Fard, K., Munthe-Kaas, H.Z. (2019). What is a post-Lie algebra and why is it useful in geometric integration. In *Numerical Mathematics and Advanced Applications ENUMATH 2017*, Radu, F., Kumar, K., Berre, I., Nordbotten, J.M., Pop, I.S. (eds). Springer, Cham [Online]. Available at: https://doi.org/10.1007/978-3-319-96415-738.

Curry, C., Ebrahimi-Fard, K., Manchon, D., Munthe-Kaas, H.Z. (2020). Planary branched rough paths and rough differential equations on homogeneous spaces. *J. Diff. Eq.*, 269(11), 9740–9782.

Dăscălescu, S., Năstăsescu, C., Raianu, S. (2001). *Hopf Algebras: An Introduction*. Marcel Dekker, New York.

Dür, A. (1986). *Möbius Functions, Incidence Algebras and Power Series Representations*. Springer, Berlin, Heidelberg.

Dzhumadil'daev, A. and Löfwall, C. (2002). Trees, free right-symmetric algebras, free Novikov algebras and identities. *Homol. Homotopy Appl.*, 4(2), 165–190.

Ebrahimi-Fard, K. and Manchon, D. (2009a). Dendriform equations. *J. Algebra*, 322, 4053–4079.

Ebrahimi-Fard, K. and Manchon, D. (2009b). A Magnus- and Fer-type formula in dendriform algebras. *Found. Comput. Math.*, 9(3), 295–316.

Ebrahimi-Fard, K. and Manchon, D. (2011). Twisted dendriform algebras and the pre-Lie Magnus expansion. *J. Pure Appl. Alg.*, 215(11), 2615–2627.

Ebrahimi-Fard, K., Manchon, D., Patras, F. (2008). New identities in dendriform algebras. *J. Algebra*, 320, 708–727.

Ebrahimi-Fard, K., Lundervold, A., Munthe-Kaas, H.Z. (2015). On the Lie enveloping algebra of a Post-Lie algebra. *J. Lie Theory*, 25(4), 1139–1165.

Gerstenhaber, M. (1963). The cohomology structure of an associative ring. *Ann. Math.*, 78, 267–288.

Guin, D. and Oudom, J.-M. (2005). Sur l'algèbre enveloppante d'une algèbre pré-Lie. *C. R. Acad. Sci.*, 340(1).

Hairer, E., Lubich, C., Wanner, G. (2002). *Geometric Numerical Tntegration, Structure-Preserving Algorithms for Ordinary Differential Equations*. Springer, Berlin.

Hentzel, I.R., Jacobs, D.P., Peresi, A. (1996). A basis for free assosymmetric algebras. *J. Algebra*, 183(2), 306–318.

Kassel, C. (1995). *Quantum Groups*. Springer-Verlag, New York.

Kleinfeld, E. (1957). Assosymmetric rings. *Proc. Amer. Math. Soc.*, 8(5), 983–986.

Kreimer, D. (2002). The combinatorics of (perturbative) quantum field theories. *Phys. Rep.*, 363, 387–424.

Livernet, M. (2006). A rigidity theorem for pre-Lie algebras. *J. Pure Appl. Alg.*, 207(1), 1–18.

Loday, J.-L. (1995). Cup-product for Leibniz cohomology and dual Leibniz algebras. *Math. Scand.*, 77(2), 189–196.

Loday, J.-L. (1996). La renaissance des opérades. In *Séminaire Bourbaki : Volume 1994/95, Exposés 790–804*, Société mathématique de France (ed.). Astérisque, France.

Loday, J.-L. (2001). Dialgebras. In *Dialgebras and Related Operads*, Loday, J.-L., Frabetti, A., Chapoton, F., Goichot, F. (eds). Springer, Berlin.

Loday, J.-L. and Ronco, M. (2010). Combinatorial Hopf algebras. *Clay Math. Proc.*, 11.

Loday, J.-L. and Vallette, B. (2012). *Algebraic Operads*. Springer, Berlin, Heidelberg.

Lundervold, A. and Munthe-Kaas, H.Z. (2013). On post-Lie algebras, Lie-Butcher series and moving frames. *Found. Comput. Math.*, 13(4), 583–613.

Manchon, D. (2008). Hopf algebras and renormalisation. *Handbook of Algebra*, 5, 365–427.

Manchon, D. and Saidi, A. (2011). Lois pré-Lie en interaction. *Comm. Alg.*, 39(10), 3662–3680.

Matsushima, Y. (1968). Affine structures on complex manifolds. *Osaka J. Math.*, 5, 215–222.

May, J.-P. (1972). *The Geometry of Iterated Loop Spaces*. Springer-Verlag, Berlin, Heidelberg.

Montgomery, S. (1993). *Hopf Algebras and their Actions on Rings*. CBMS Regional Conference Series in Mathematics, Los Angeles, CA.

Munthe-Kaas, H.Z. and Wright, W. (2008). On the Hopf algebraic structure of Lie group integrators. *Found. Comput. Math.*, 8(2), 227–257.

Osborn, J.-M. (1992). Novikov algebras. *Nova J. Alg.*, 1, 1–14.

Saidi, A. (2014). The pre-Lie operad as a deformation of NAP. *J. Algebra Appl.*, 13(1), 1350076.

Schützenberger, M.-P. (1958/1959). *Sur une propriété combinatoire des algèbres de Lie libres pouvant être utilisée dans un problème de mathématiques appliquées.* Séminaire Dubreil-Jacotin Pisot, Paris.

Sweedler, M.-E. (1969). *Hopf Algebras.* Benjamin, New York.

Vallette, B. (2007). Homology of generalized partition posets. *J. Pure Appl. Algebra*, 208(2), 699–725.

Vinberg, E.-B. (1963). The theory of homogeneous convex cones. *Transl. Moscow Math. Soc.*, 12, 340–403.

2

From Iterated Integrals and Chronological Calculus to Hopf and Rota–Baxter Algebras

Kurusch EBRAHIMI-FARD[1] and Frédéric PATRAS[2]

[1] *Norwegian University of Science and Technology, Trondheim, Norway*

[2] *CNRS, Côte d'Azur University, Nice, France*

2.1. Introduction

Gian-Carlo Rota mentioned in one of his last articles the problem of developing a theory around the notion of integration algebra, which should be dual to the one of differential algebra. This idea has been developed historically along three lines: using properties of iterated integrals and generalizations thereof, as in the theory of rough paths; using the algebraic structures underlying chronological calculus such as shuffle and pre-Lie products, as they appear in theoretical physics and control theory; and, more generally, using a particular operator identity, which came to be known as the Rota–Baxter relation. Recent developments along each of these lines of research and their various application domains are not always known to other communities in mathematics and related fields. The general aim of this survey is therefore to present a modern and unified perspective on these approaches, featuring among others on noncommutative Rota–Baxter algebras and related Hopf and Lie algebraic structures such as descent algebras, algebras of simplices as well as shuffle, pre- and post-Lie algebras and their enveloping algebras. Accordingly, two viewpoints are proposed: a geometric one that can be derived using combinatorial properties of

Algebra and Applications 2,
coordinated by Abdenacer MAKHLOUF.

Euclidean simplices and the duality between integrands and integration domains in iterated integrals, and an algebraic one, provided by the axiomatization of integral calculus in terms of Rota–Baxter algebra.

Chen's theory of iterated integrals (Chen 1971) and the so-called chronological calculus (the latter being based on the use of time-ordered products of operators $M(t)$ and $N(t)$ such as $\int_0^t \dot{M}(u)N(u)du$, such as, for example, in the seminal works of Agrachev and Gamkrelidze (1981)) are classical topics in analysis, stochastic integration and control theory. They have evolved recently through the introduction of combinatorial Hopf algebra techniques such as the ones related to permutations or (free) pre-Lie algebras and their enveloping algebras.

In the first part of this chapter (sections 2.2 and 2.3), we survey these developments using a simple but not popular enough observation, namely: the fundamental duality between integration domains and integrands governing calculus with iterated integrals. This approach, originating in the combinatorial properties of simplices and its close links with the combinatorics of symmetric groups and Lie algebras (Patras 1991), was developed in Agrachev and Gamkrelidze (1994) and Brouder *et al.* (2010) to deal with operator-valued iterated integrals and in Patras (1997) in the context of Chen's approach to loop space cohomology. It allows us to encode the usual algebraic free Lie and pre-Lie techniques in geometric terms. The philosophy underlying these constructions is in the same vein as Chen's, in that it relates in the end the rules of classical calculus with the ones of simplicial calculus in algebraic topology, allowing us to extend Chen's ideas to various noncommutative and non-associative structures.

The second part of the chapter (section 2.4) extends these ideas and techniques to a broader context, making also explicit the link (following in particular the ideas put forward in Ebrahimi-Fard and Patras (2014)) between the direct approach of the first part (sections 2.2 and 2.3) and the algebra-axiomatic one in the second part. The latter can be traced back to the period 1960–1972, when Pierre Cartier and Gian-Carlo Rota, together with collaborators, set out to develop a general theory of the algebraic and combinatorial structures underlying integral calculus. Nowadays, this approach has been subsumed by what is better known as the theory of Rota–Baxter algebras. However, in spite of the fact that those algebras need not be commutative, a large part of the principal results were first obtained in the context of commutative function algebras. Interest in the general, i.e. noncommutative, case arose around 2000, in areas including, among others, integral and finite difference calculus for operator algebras. Perturbative quantum field theory provided a rather unexpected context in which noncommutative Rota–Baxter algebras came to prominence; more precisely, the Hopf algebraic approach to the so-called BPHZ (Bogoliubov–Parasiuk–Hepp–Zimmermann) renormalization program (Connes and Kreimer 2000) (describing a specific subtraction procedure by which the regularized amplitudes corresponding to Feynman graphs can be rendered finite). Since then, the range of applications of the theory of noncommutative Rota–Baxter algebras has grown steadily. In Rota (1995), Guo (2012) and Kung *et al.* (2009), the reader will find useful introductory surveys.

We should emphasize that the second part (section 2.4) of the work at hand neither strives for an exhaustive presentation of the theory of Rota–Baxter algebras nor seeks to include a complete list of references. The latter restriction is mainly due to the extensive developments that have taken place over the last 10 years, which resulted in numerous ramifications of the theory of Rota–Baxter algebras into other fields. The present survey instead offers some insights into recent progress made in a particular direction – maybe the most significative one from the point of view of integral and differential calculus – namely the extension of the fundamental results of the commutative theory to the Hopf algebraic and noncommutative settings. The results presented in this survey were mainly obtained by the authors, a variety of them together with José Gracia-Bondía and Dominique Manchon, and include, among others, the Bohnenblust–Spitzer identity and the construction of the standard Rota–Baxter algebra (a generalization to the noncommutative setting of Rota's presentation of the free Rota–Baxter algebra in terms of symmetric functions).

This chapter is organized into two parts. The first one (sections 2.2–2.3) begins by examining the Hopf algebraic interpretation of classical iterated integral calculus and its rooting in the geometry of simplices. This part includes the case of tree-shaped iterated integrals, which are relevant in many situations (e.g. time-dependent perturbation theory, non-geometric rough paths, etc.). Chronological calculus and the associated pre-Lie structures are introduced, together with the construction of their enveloping algebras. Fundamental objects are considered, such as the Magnus operator and its inverse, arising from logarithmic and exponential maps.

The second part (section 2.4) begins with a very brief historic account of works by Baxter, Cartier, Chen, Rota and others. Then, the notion of Rota–Baxter algebra is introduced and its physical as well as probabilistic origins are recalled with a presentation of several classical examples of such algebras that stem mostly (but not exclusively) from analysis. Algebraic structures related to the Rota–Baxter identity, such as (quasi-)shuffle, post- and pre-Lie algebras as well as Atkinson's factorization, are discussed. The latter is then linked in the commutative case to Spitzer's classical identity. The approaches of Cartier and Rota to constructions of free commutative Rota–Baxter algebras are then briefly outlined. Links with the theories of symmetric functions, commutative shuffle algebras and fundamental (Spitzer-type) identities are put forward as well. This presentation of the commutative theory is tuned towards its noncommutative extension as well as the inclusion of Hopf algebraic properties; this approach is neither standard nor, in some sense, the most natural one from the point of view of the commutative setting, but it has the advantage of making the ideas underlying the transition from the commutative to the noncommutative realm transparent.

We then enter the core of the subject, i.e. the theory of noncommutative Rota–Baxter algebras, with a view towards Hopf algebraic properties. Our approach is based mainly on three ideas. First, the existence of noncommutative analogues of most of the fundamental commutative constructions. This idea is illustrated by the close relationship between the structure of free Rota–Baxter algebras and noncommutative symmetric functions (Gelfand *et al.* 1995). Then, the existence of noncommutative analogues of most of the fundamental commutative formulas is shown. Eventually, we would like to emphasize that calculus in Rota–Baxter algebras is in some sense "generic" with respect to several theories, in the sense that calculus in free Rota–Baxter algebras models calculus in various domains, very much like tensor calculus, is generic for the theory of iterated integrals and Lie algebras[1]. Theories to which such a remark applies in the noncommutative framework include the one of iterated integrals and sums of operator valued functions (corresponding to noncommutative shuffle and quasi-shuffle products) as well as of derivations and differential operators (corresponding to pre-Lie products). Concretely, all this points to the observation that it is often convenient to work within the setting of Rota–Baxter algebras when it comes to looking for universal formulas solving problems in integral calculus, difference calculus, dynamical systems and related fields.

All algebraic structures are defined over the ground field k of characteristic zero. They are graded if their structure maps are defined in the tensor category of graded vector spaces, i.e. if objects are graded vector spaces and the structure maps respect the grading. Connectedness in the graded sense refers to the degree-zero component being isomorphic to the ground field. We also assume any k-(co)algebra to be (co)associative, if not stated otherwise.

2.2. Generalized iterated integrals

We begin by studying algebraic structures underlying calculus with iterated integrals and show how they lead naturally to the notions of descent (Hopf) algebra as well as permutation Hopf algebra. We feature an elementary but powerful observation: since n-fold iterated integrals are over an n-dimensional simplex; their algebraic properties translate (exactly) into combinatorial and geometric properties of Euclidean

1. In a sense that can be made rigorous, see, for instance, Reutenauer (1993). More precisely, from a technical viewpoint, formulas in tensor algebras can be shown to hold in arbitrary enveloping algebras of graded Lie algebras. Similarly, formulas that hold in free Rota–Baxter algebras hold in any pre-Lie or shuffle algebra.

simplices. With the exception of this geometric point of view, originating in Patras (1991), and the resulting presentation, the material is relatively classical and is to some extent considered "folklore", although seemingly more in algebra and combinatorics than in integral calculus.

The study of descent algebras was initiated to a large extent by the work of Solomon (1976), which gave rise to the theory of noncommutative representations of Coxeter groups. Descents are studied in detail in Reutenauer's classical monograph on free Lie algebras (Reutenauer 1993) and are closely related to the fine structure of graded connected cocommutative Hopf algebras (i.e. enveloping algebras of reduced graded Lie algebras) (Patras 1994). The Hopf algebra of descents was in fact introduced by Malvenuto and Reutenauer in their seminal paper on the duality between descents and quasi-symmetric functions (Malvenuto and Reutenauer 1995), and independently in Gelfand *et al.* (1995) in the context of the theory of noncommutative symmetric functions.

The algebra structure on the linear span of permutations was presented by Patras under the name "ring of simplices" (Patras 1991). Its Hopf algebra structure was unfolded later in Malvenuto and Reutenauer (1995). It is known in the literature as permutation Hopf algebra, Malvenuto–Reutenauer Hopf algebra or Hopf algebra of free quasi-symmetric functions (Duchamp *et al.* 2011).

The tree-shaped expansion of a quadratic differential equation is taken essentially from reference (Brouder *et al.* 2010)[2], where the technique was used in the context of time-dependent perturbation theory. The construction of rooted trees is given mainly in this chapter as a simple illustration of a general process of obtaining combinatorial expansions of solutions of (nonlinear) differential equations. This process was used systematically in recent work by Martin Hairer and collaborators to find (renormalized) solutions of stochastic (partial) differential equations (Hairer 2014). Although simple, the example of rooted trees is interesting in that it allows us to underline various algebraic phenomena. It shows, for example, how integration by parts formula translates in the context of perturbative expansions.

2.2.1. *Permutations and simplices*

Let us write S_n for the group of permutations of the set $[n] := \{1, \dots, n\}$ and use the notation $\sigma = (\sigma(1), \dots, \sigma(n))$ for elements $\sigma \in S_n$. Consider as a starting point the linear $N \times N$ matrix differential equation:

$$\dot{X}(t) = X(t)H(t), \qquad X(0) = Id, \tag{2.1}$$

2. For a concrete application of the technique, although to a problem different from the one we use here as a motivation, we refer to the same article.

where $H(t)$ is, say, C^∞ with bounded coefficients. Here, Id denotes the $N \times N$ identity matrix. The solution of such a linear initial-value problem is given by a series of iterated integrals, usually referred to in the mathematics literature as Picard series and in the physics literature as Dyson series or time-ordered exponential:

$$\begin{aligned} X(t) &= X_0(t) + X_1(t) + X_2(t) + \cdots + X_n(t) + \cdots \\ &= Id + \int\limits_{\Delta_t^1} dt_1 H(t_1) + \int\limits_{\Delta_t^2} dt_1 dt_2 H(t_1) H(t_2) + \\ &\quad \cdots + \int\limits_{\Delta_t^n} dt_1 \cdots dt_n H(t_1) \cdots H(t_n) + \cdots . \end{aligned} \quad [2.2]$$

The integration domains in equation [2.2] are defined to be the "standard" n-simplices

$$\Delta_t^n := \{(t_1, \ldots, t_n) \mid 0 \leq t_1 \leq \cdots \leq t_n \leq t\}.$$

Given an (n-dimensional) convex polytope P in $\mathbb{R}^n$, we will write, in general, $\langle P|H\rangle$ for the integral of the product $H(t_1)\cdots H(t_n)$ over P, $\int\limits_P dt_1 \cdots dt_n H(t_1) \cdots H(t_n)$, so that $X_n(t) = \langle \Delta_t^n | H\rangle$. Note that with this notation

$$\dot{X}_n(t) = X_{n-1}(t) H(t) = \langle \Delta_t^{n-1} | H\rangle \cdot H(t),$$

whereas Fubini's rule reads

$$\langle P \times Q | H\rangle = \langle P|H\rangle\langle Q|H\rangle.$$

In more abstract terms, any time-dependent matrix (or more generally sufficiently regular operator) $H = H(t)$ defines a character, i.e. a multiplicative map from the noncommutative algebra of linear combinations of convex Euclidean polytopes equipped with the cartesian product to the algebra of time-dependent matrices. This rule is compatible with geometric decompositions in the sense that if $P = Q \coprod T$, where P, Q and T are n-dimensional polytopes, and $Q \cap T$ is contained in a hyperplane, then $\langle P|H\rangle = \langle Q|H\rangle + \langle T|H\rangle$. These rules, together with the integration by parts formula, are enough, as we will see, to capture the fundamental algebraic and combinatorial properties of classical calculus.

As a motivational example, consider the classical (continuous Baker–Campbell–Hausdorff) problem of expressing the solution $X(t)$ of equation [2.1] as a true (matrix) exponential

$$X(t) = \exp(\Omega(t)), \quad \Omega(0) = 0.$$

Its answer relies on computing the logarithm of the function $X(t)$, which amounts to calculating the logarithm of the sum of the iterated integrals $X_i(t) = \langle \Delta_t^i | H\rangle$. By

Fubini's rule, the formula for multiplying two such iterated integrals results from the simplicial decomposition of a cartesian product of simplices as a union of simplices

$$\Delta_t^n \times \Delta_t^m = \left\{ (t_1, \dots, t_n, u_1, \dots, u_m) \mid \begin{array}{l} 0 \le t_1 \le \dots \le t_n \le t, \\ 0 \le u_1 \le \dots \le u_m \le t \end{array} \right\}$$
$$= \bigcup \{ (y_{\sigma(1)}, \dots, y_{\sigma(n+m)}) \mid 0 \le y_1 \le \dots \le y_{m+n} \le t \},$$

where the union runs over permutations $\sigma \in S_{m+n}$ such that $\sigma(1) < \dots < \sigma(n)$ and $\sigma(n+1) < \dots < \sigma(m+n)$. This is the classical rule used, among others, in algebraic topology, to define products in simplicial and singular cohomology. Since $\sigma(i) < \sigma(i+1)$ with a possible exception for $i = n$, i.e. it may be that $\sigma(n) > \sigma(n+1)$, such a permutation is said to have at most one descent, in position n. This yields the following:

DEFINITION 2.1.– *The descent set of a permutation $\sigma \in S_n$ is the subset of $[n-1]$ defined by:*

$$Desc(\sigma) := \{ i \mid \sigma(i) > \sigma(i+1),\ 1 \le i < n \}.$$

For example, the descent set of the permutation $(3174526) \in S_7$ is

$$Desc(3174526) = \{1, 3, 5\}.$$

In particular, setting

$$\Delta_{\sigma,t}^n := \{ (t_{\sigma(1)}, \dots, t_{\sigma(n)}) \mid 0 \le t_1 \le \dots \le t_n \le t \}$$

and defining for $\sigma \in S_n$

$$X_\sigma(t) := \langle \Delta_{\sigma,t}^n | H \rangle = \int_{\Delta_{\sigma,t}^n} du_1 \cdots du_n H(u_1) \cdots H(u_n),$$

we obtain for the product of iterated integrals

$$X_i(t) X_j(t) = \sum_{\substack{\sigma \in S_{i+j} \\ Desc(\sigma) \subseteq \{i\}}} X_\sigma(t).$$

Compare this with products of integrals following from the classical integration by parts formula

$$X_1(t) X_1(t) = \int_{0 \le t_1 < t} dt_1 H(t_1) \int_{0 \le t_2 \le t} dt_2 H(t_2)$$
$$= \iint_{0 \le t_1 \le t_2 \le t} dt_1 dt_2 H(t_1) H(t_2) + \iint_{0 \le t_2 \le t_1 \le t} dt_1 dt_2 H(t_1) H(t_2)$$
$$= X_{(12)}(t) + X_{(21)}(t),$$

with the descent in position 1 in the second term. Similarly,

$$\begin{aligned} X_2(t)X_1(t) &= \int_{0\le t_1\le t_2\le t} dt_1dt_2H(t_1)H(t_2) \int_{0\le t_3\le t} dt_3H(t_3) \\ &= \iiint_{0\le t_1\le t_2\le t_3\le t} dt_1dt_2dt_3H(t_1)H(t_2)H(t_3) \\ &\quad + \iiint_{0\le t_1\le t_3\le t_2\le t} dt_1dt_2dt_3H(t_1)H(t_2)H(t_3) \\ &\quad + \iiint_{0\le t_3\le t_1\le t_2\le t} dt_1dt_2dt_3H(t_1)H(t_2)H(t_3) \\ &= X_{(123)}(t) + X_{(132)}(t) + X_{(231)}(t), \end{aligned}$$

with the descent in position 2 in the second and third terms.

Iterating products of iterated integrals naturally leads to computing products of non-standard simplices parametrized by permutations. Recall that if $A := \{a_1, \dots, a_n\}$ is a set of integers, written in the natural order: $a_1 < a_2 < \cdots < a_n$, the standardization map $st_{[n]}$ is the unique strictly increasing map from A to $[n] = \{1, \dots, n\}$: $st_{[n]}(a_i) = i$ (we could write $st^A_{[n]}$ for $st_{[n]}$ to emphasize the dependency on A, but we stick to a light notation since the dependency on A will always be explicited or be clear from the context further on). We simply write st when no confusion can arise. We then get in general

$$\Delta^n_{\sigma,t} \times \Delta^m_{\beta,t} = \bigcup \{(y_{\gamma(1)}, \dots, y_{\gamma(n+m)}) \mid 0 \le y_1 \le \cdots \le y_{m+n} \le t\},$$

where the union runs now over permutations $\gamma \in S_{n+m}$ such that $st_{[n]}(\gamma(i)) = \sigma(i)$ for $i \le n$ with $A := \{\gamma(1), \dots, \gamma(n)\}$ and $st_{[m]}(\gamma(i)) = \beta(i-n)$ for $n+1 \le i \le n+m$ and $A := \{\gamma(n+1), \dots, \gamma(n+m)\}$. Equivalently, writing $\sigma \otimes \beta$ for the permutation in S_{n+m} such that $\sigma \otimes \beta(i) = \sigma(i)$ for $i \le n$ and $\sigma \otimes \beta(i) = n + \beta(i-n)$ else, we get

$$\Delta^n_{\sigma,t} \times \Delta^m_{\beta,t} = \bigcup \Delta^{n+m}_{\nu,t},$$

where ν runs over permutations of the form $\alpha \circ (\sigma \otimes \beta)$ with $Desc(\alpha) \subset \{n\}$.

It is useful to understand these formulas geometrically. Let us denote pr^+_n and pr^-_m the two canonical projections from $\mathbb{R}^{n+m}$ to $\mathbb{R}^n$ and $\mathbb{R}^m$ ($(t_1, \dots, t_{n+m}) \longmapsto (t_1, \dots, t_n)$, respectively, $(t_{n+1}, \dots, t_{n+m})$). Then, from the standard properties of projections of simplices, we have $pr^+_n(\Delta^n_{\sigma,t} \times \Delta^m_{\beta,t}) = \Delta^n_{\sigma,t}$, respectively, $pr^-_m(\Delta^n_{\sigma,t} \times \Delta^m_{\beta,t}) = \Delta^m_{\beta,t}$. In fact, with the same notation as above, we also have (looking, for example, at the boundary of Δ^ν_t)

$$pr^+_n(\Delta^{n+m}_{\nu,t}) = \Delta^n_{\sigma,t}, \quad pr^-_m(\Delta^{n+m}_{\nu,t}) = \Delta^m_{\beta,t};$$

and this identity is another characterization of permutations ν of the form $\alpha \circ (\sigma \otimes \beta)$ with $Desc(\alpha) \subset \{n\}$.

This yields the following definition (Patras 1991, section 1.3):

DEFINITION 2.2.– *The algebra of simplices (or algebra of permutations), denoted $\mathcal{S}$, is the vector space $\bigoplus_{n\in\mathbb{N}} \mathbb{Q}[S_n]$ equipped with the product*

$$\sigma * \beta := \sum_{\substack{\alpha\in S_{n+m}\\ Desc(\alpha)\subseteq\{n\}}} \alpha \circ (\sigma \otimes \beta)$$

$$= \sum_{\substack{\gamma\in S_{n+m}\\ st_{[n]}(\gamma(i))=\sigma(i),i\leq n\\ st_{[m]}(\gamma(i))=\beta(i-n),i>n}} \gamma = \sum_{\substack{\nu\in S_{n+m}\\ pr_n^+(\Delta^{n+m}_{\nu,t})=\Delta^n_{\sigma,t}\\ pr_m^-(\Delta^{n+m}_{\nu,t})=\Delta^m_{\beta,t}}} \nu \qquad [2.3]$$

for $\sigma \in S_n$ and $\beta \in S_m$.

Concretely, $\sigma * \beta$ is the sum of all permutations α in S_{n+m} such that the elements of the sequence $(\sigma(1), \ldots, \sigma(n))$ are ordered as the elements of $(\alpha(1), \ldots, \alpha(n))$ (i.e. $\alpha(i) > \alpha(j)$ if and only if $\sigma(i) > \sigma(j)$) and the elements of the sequence $(\beta(1), \ldots, \beta(m))$ are ordered as the elements of $(\alpha(n+1), \ldots, \alpha(n+m))$.

Note that $\mathcal{S}$ is a graded algebra with unit and that the associativity of the product $*$ in equation [2.3] follows immediately from the associativity of cartesian products in geometry.

The link between the algebraic structure of iterated integrals and the one of simplices and permutations does not only hold when dealing with a single time-dependent operator. Actually, in many applications, say, for example, when expanding the solution of a differential equation such as $\dot{X}(t) = X(t)H_1(t) + H_2(t)X(t)$ (or, more generally, nonlinear differential equations involving several time-dependent operators or that have more complex structures – see Agrachev and Gamkrelidze (1994) as well as Brouder *et al.* (2010), from which the following results are taken, for a concrete example), we have to consider iterated integrals of products of various time-dependent operators. This does not affect the algebraic rules that have just been devised, as we show now.

Let $\mathbf{L} = (L_1, \ldots, L_n)$ be an arbitrary sequence of time-dependent operators, $L_i = L_i(t)$, $1 \leq i \leq n$, sufficiently regular (e.g. as in Brouder *et al.* (2010), assume they are strongly continuous maps from $\mathbb{R}$ into bounded self-adjoint operators on a Hilbert space $\mathcal{H}$). For $\sigma \in S_n$, set:

$$\mathbf{L}_\sigma(t) := \int_{\Delta^n_{\sigma,t}} dt_1 \cdots dt_n L_1(t_1) \cdots L_n(t_n).$$

The notation is extended to linear combinations of permutations, so that for $\mu = \sum_{\sigma \in S_n} \mu_\sigma \sigma$, $\mathbf{L}_\mu(t) := \sum_{\sigma \in S_n} \mu_\sigma \mathbf{L}_\sigma(t)$. For $\mathbf{K} = (K_1, \ldots, K_m)$, another sequence of time-dependent operators, we write $\mathbf{L} \cdot \mathbf{K}$ for the concatenation product $(L_1, \ldots, L_n, K_1, \ldots, K_m)$. Then, for $\alpha \in S_n$ and $\beta \in S_m$, two permutations, it follows from the combinatorial formula for the cartesian product of simplices that:

LEMMA 2.1.– *We have:*

$$\mathbf{L}_\alpha(t)\mathbf{L}_\beta(t) = \mathbf{L}_{\alpha * \beta}(t).$$

We also mention the following lemma, which is useful in nonlinear contexts and which can be generalized to other forms of products of operator sequences:

LEMMA 2.2.– *We have, for* $\mathbf{L}$ *and* $\mathbf{K}$ *as above and* $J = J(t)$ *a time-dependent operator:*

$$\int_0^t ds \mathbf{L}_\alpha(s) J(s) \mathbf{K}_\beta(s) = \sum_\gamma (\mathbf{L} \cdot (J) \cdot \mathbf{K})_\gamma(t),$$

where γ *runs over the permutations in* S_{n+m+1} *with* $\gamma(n+1) = n+1$, $st(\gamma(1), \ldots, \gamma(n)) = \alpha$, $st(\gamma(n+2), \ldots, \gamma(n+m+1)) = \beta$.

The proof follows from the same arguments that lead to the combinatorial expression for a cartesian product of simplices; we refer to Brouder *et al.* (2010) for details.

2.2.2. *Descents, NCSF and the BCH formula*

In many applications in integral calculus, only iterated products of standard simplices Δ_t^n appear. It is therefore not necessary to appeal to the full algebra $\mathcal{S}$, but only to the smaller algebra generated by the identity elements in the various S_n. This algebra is known as the *descent algebra* and plays a key role in the theory of free Lie algebras (Reutenauer 1993) as well as in the structure theory of Hopf algebras (Patras 1994).

DEFINITION 2.3.– *The descent algebra* $\mathcal{D}$ *is the graded unital subalgebra of* $\mathcal{S}$ *generated by the identity permutations* $1_n := (1, \ldots, n) \in S_n$, $n \geq 1$.

Note that, geometrically, elements of $\mathcal{D}$ are therefore linear combinations of cartesian products of standard simplices; the associated integrals are products of iterated integrals over standard simplices.

Generalizing what happens when considering the cartesian product of two standard simplices, to each subset A of the set $[n-1]$ are associated two elements in $\mathbb{Q}[S_n]$.

They are referred to as "Solomon elements" since they appear as a key ingredient in Solomon's theory of noncommutative representations of the symmetric groups (Solomon 1976):

DEFINITION 2.4.– *For $A \subset [n-1]$, the Solomon elements $D_{=A}$ and D_A in $\mathbb{Q}[S_n]$ are defined by:*

$$D_{=A} := \sum_{\substack{\sigma \in S_n \\ Desc(\sigma)=A}} \sigma, \qquad D_A := \sum_{\substack{\sigma \in S_n \\ Desc(\sigma)\subseteq A}} \sigma.$$

To avoid ambiguities, elements will be written $D^{(n)}_{=A}$, respectively, and $D^{(n)}_A$, when their belonging to $\mathbb{Q}[S_n]$ has to be emphasized.

From the definition of the product equation [2.3] in $\mathcal{S} = \bigoplus_{n\in\mathbb{N}} \mathbb{Q}[S_n]$, the permutations that appear in the expansion of $D^{(i_1+\dots+i_n)}_{i_1,i_1+i_2,\dots,i_1+\dots+i_{n-1}}$ parametrize the decomposition into a union of simplices of the product of the simplices $\Delta^{i_1}_t, \dots, \Delta^{i_n}_t$. In particular, their sum belong to $\mathcal{D}_n$, and we have:

LEMMA 2.3.– *In the descent algebra, we have*

$$1_{i_1} * \dots * 1_{i_n} = D^{(i_1+\dots+i_n)}_{i_1,i_1+i_2,\dots,i_1+\dots+i_{n-1}}$$

and

$$D^{(n)}_S * D^{(m)}_T = D^{(n+m)}_{S\cup\{n\}\cup(T+n)}.$$

Since the (non-empty) sets $\{\sigma \in S_n \mid Desc(\sigma) = A\}$, where A runs over subsets of $[n-1]$, define a partition of S_n, the elements $D_{=A}$ are non-zero and linearly independent in $\mathbb{Q}[S_n]$. By Möbius inversion (in the poset of subsets of $[n-1]$), D_A are also linearly independent. Moreover, since the Möbius inversion coefficients of the poset of subsets of $[n-1]$ are $(-1)^{|A|-|B|}$ for $B \subset A$, we have:

PROPOSITION 2.1.– *The two families $D_{=A}$ and D_A are two linear bases of $\mathcal{D}_n$ related by the formulas:*

$$D_A = \sum_{B\subseteq A} D_{=B}, \qquad D_{=A} = \sum_{B\subseteq A} (-1)^{|A|-|B|} D_B.$$

From Proposition 2.1 and Lemma 2.3, we get:

COROLLARY 2.1.– *The descent algebra is a free associative algebra with unit generated by the 1_n, $n \geq 1$.*

REMARK 2.1.– *In the context of the category of graded connected cocommutative Hopf algebras, a notion of descent algebra is defined as an algebra of natural transformations of the forgetful functor to graded vector spaces. It is generated by the projections from such a Hopf algebra to its graded components. The product is induced by the convolution product of linear endomorphisms. The notion of descent algebra defined in that way is isomorphic to* $\mathcal{D}$ *(but the interpretation of its elements as linear combinations of permutations is not available since there is in general no canonical action of symmetric groups on a graded connected cocommutative Hopf algebra) (Patras 1994). As a free associative algebra over a set of graded generators, the descent algebra is canonically isomorphic to the algebra of noncommutative symmetric functions* (NCSF) *(Gelfand* et al. *1995). These two points of view have proven to be very useful to investigate the properties and applications of* $\mathcal{D}$.

A classical application of the rules governing the descent algebra is the explicit description of the so-called continuous Baker–Campbell–Hausdorff (BCH) formula:

THEOREM 2.1.– *For the series equation* [2.2] *we have, using the notation* $X_\sigma(t) =: X_n(t)\cdot\sigma$*:*

$$\Omega(t) := \log(X(t)) = \sum_{n=1}^{\infty} X_n(t)\cdot\Big(\sum_{S\subset[n-1]} \frac{(-1)^{|S|}}{|S|+1} D_S^{(n)}\Big)$$

$$= \sum_{n=1}^{\infty} X_n(t)\cdot\Big(\sum_{S\subset[n-1]} \frac{(-1)^{|S|}}{n}\binom{n-1}{|S|}^{-1} D_{=S}^{(n)}\Big). \quad [2.4]$$

PROOF.– Recall that $X_n(t)\cdot\sigma := X_\sigma(t)$. The first identities follow immediately from the product formulas for standard simplices and iterated integrals. Let us expand D_S in terms of $D_{=S}$. The coefficient of $D_{=T}, T\subset[n-1], |T| = l$, in this expansion, is given by:

$$\sum_{T\subset S\subset[n-1]} \frac{(-1)^{|S|}}{|S|+1} = \sum_{j=0}^{n-l-1} \binom{n-l-1}{j}\frac{(-1)^{l+j}}{l+j+1}$$

$$= (-1)^l \sum_{j=0}^{n-l-1} \int_0^1 (-1)^j \binom{n-l-1}{j} x^{l+j}\,dx$$

$$= (-1)^{l-1}\int_0^1 (1-x)^{n-l-1}x^l\,dx = (-1)^l \frac{1}{n}\binom{n-1}{l}^{-1}.$$

From this, the formula follows. □

REMARK 2.2.– *The series* $\Omega(t)$ *is a Lie series (Reutenauer 1993, Chapter 3). Concretely, this property means that* $\Omega(t)$ *can be expressed as a weighted sum of iterated integrals of nested commutators*

$$[X]_\sigma(t) := X_n(t)\cdot[\sigma(1),[\sigma(2),[\cdots[\sigma(n-1),\sigma(n)]\cdots]]].$$

Here, we identify a permutation $\beta \in S_n$ *with the word* $\beta(1)\cdots\beta(n)$ *and the bracket of two words is defined by* $[w,w'] := ww' - w'w$*, so that, for example, for* $\sigma = (213)$, $[2,[1,3]] = [2,13-31] = 213 - 231 - 132 + 321$. *Note that in the latter expansion, there is a unique word ending with* $3 = \sigma(3)$. *This property holds in general and is straightforward to see. Indeed, when* $[\sigma(1),[\sigma(2),[\cdots[\sigma(n-1),\sigma(n)]\cdots]]]$ *is expanded as a sum of words, the word* $\sigma(1)\cdots\sigma(n)$ *is the only one ending with* $\sigma(n)$ *that appears. This observation has two classical consequences (Reutenauer 1993, section 5.6.2):*

– the brackets $[\sigma(1),[\sigma(2),[\cdots[\sigma(n-1),n]\cdots]]]$*, where* $\sigma \in S_{n-1}$*, are linearly independent in* $\mathbb{Q}[S_n]$ *;*

– since the space P_n *of multilinear Lie polynomials (i.e. the vector space generated by the iterated brackets* $[\sigma(1),[\sigma(2),[\cdots[\sigma(n-1),\sigma(n)]\cdots]]]$*) is of dimension* $(n-1)!$*, the brackets* $[\sigma(1),[\sigma(2),[\cdots[\sigma(n-1),n]\cdots]]]$*,* $\sigma \in S_{n-1}$ *form a basis of* P_n*. For any element* x *of* P_n*,* $x = \sum_{\sigma\in S_n} x_\sigma \cdot \sigma$*, the expansion of* x *in this basis is then*

$$x = \sum_{\sigma\in S_{n-1}} x_{\sigma^+}[\sigma(1),[\sigma(2),[\cdots[\sigma(n-1),n]\cdots]]],$$

where $\sigma^+ = (\sigma(1),\ldots,\sigma(n-1),n)$.

Similar results hold by using left-to-right iterated brackets or by fixing another letter as the last letter (e.g. 1 instead of n*). This result has an interesting consequence as it implies (Arnal* et al. *2018):*

COROLLARY 2.2.– *The series* $\Omega(t)$ *rewrites*

$$\Omega(t) = \sum_{n=1}^{\infty}\sum_{\sigma\in S_{n-1}} \frac{(-1)^{|Desc(\sigma)|}}{n}\binom{n-1}{|Desc(\sigma)|}^{-1}[X]_{\sigma^+}.$$

2.2.3. Rooted trees and nonlinear differential equations

We have seen in the previous sections how to handle algebraically products of iterated integrals parametrized by permutations. In various situations, such as when dealing with nonlinear differential equations, it is useful to consider more general iterated integrals parametrized by rooted trees. Here, we will treat the example of planar binary trees, following reference Brouder *et al.* (2010), whose developments were motivated by time-dependent perturbation theory. As motivation, the reader may consider the perturbative expansion of the solution of the quadratic equation $\dot{X}(t) = X(t)H(t)X(t)$. The equations governing the Hamiltonians and effective Hamiltonians of time-dependent perturbation theory are more complex, but obey the same combinatorics. The following constructions can be generalized to other nonlinear equations; our presentation should be enough to aim at the general principle.

Consider again the time-dependent operator $H(t)$ and, for example, the iterated integral

$$\int\limits_{0\leq t_1,t_3\leq t_2\leq t} dt_1dt_2dt_3H(t_1)H(t_2)H(t_3).$$

The integration domain $\{(t_1,t_2,t_3)\mid 0\leq t_1,t_3\leq t_2\leq t\}$ can be parametrized by the poset structure on the set $\{1,2,3\}$, defined by $2<<1,\ 2<<3$ (in general, the rule to define the poset structure from the iterated integral is $t_i\leq t_j \implies j<<i$). The associated Hasse diagram is a rooted tree (a connected and simply connected graph with a distinguished vertex called the root, the graph is directed according to the ordering) with 2 as a root and two branches relating 2 and 1 as well as 2 and 3. The domain splits, moreover, as the union of two simplices: $\{(t_1,t_2,t_3)\mid 0\leq t_1\leq t_3\leq t_2\leq t\}=\{(t_1,t_3,t_2)\mid 0\leq t_1\leq t_2\leq t_3\leq t\}$ and $\{(t_1,t_2,t_3)\mid 0\leq t_3\leq t_1\leq t_2\leq t\}=\{(t_2,t_3,t_1)\mid 0\leq t_1\leq t_2\leq t_3\leq t\}$.

Let us generalize this construction. Consider a planar binary tree T with n vertices. We recall that planar means that the tree is drawn in the plane with the root at the bottom and that at each vertex the outgoing branches are ordered – say, from left to right. Planar binary means that each vertex has two outgoing branches. At internal vertices v, one of these branches may be empty, by which we mean that it does not link v to another vertex. At leaf vertices (top vertices), the two branches are empty. A planar binary tree can be represented uniquely as the grafting of two planar binary trees, $T=T_1\vee T_2$, where T_1 and T_2 are the branches attached to the root. We recall that empty branches are allowed, but we distinguish the case where the left or the right branch is empty (so that an outgoing branch is always pointing to the left or the right). Notationally, we distinguish between $T_1\vee\emptyset$ and $\emptyset\vee T_1$. The tree is then labeled from left to right and top to bottom by elements from the set $\{1,\ldots,n\}$. When performed inductively, the labeling of a tree $T_1\vee T_2$ with $n=k+1+l$ vertices (T_1 and T_2 having k, respectively, l vertices) by a subset S of the integers of cardinality n is performed as follows. The tree T_1 is labeled by the first k elements in S, the root is labeled by the $k+1$-th element of S and the tree T_2 is labeled by the remaining l elements.

Using such a labeling of the vertices, we can view the tree T as the Hasse diagram of a (well-defined) order on $[n]$, denoted $<_T$. We write Δ_T for the associated subset of the cube $[0,t]^n\subset\mathbb{R}^n$ defined by the inequalities $t_i\leq t_j \iff j\leq_T i$. We then set

$$X_T(t)=\langle\Delta_T|H\rangle=\int_{\Delta_T} dt_1\cdots dt_nH(t_1)\cdots H(t_n).$$

The notation $X_T(t)$ is extended to linear combinations of trees, so that, for example, if $Z=T+2T'$, where T and T' are two arbitrary trees, then $X_Z(t):=X_T(t)+2X_{T'}(t)$.

As the root of the tree is associated with the highest coordinate of the points in Δ_T and since the labels of T_1 and T_2 are incomparable for $<_T$, we get:

LEMMA 2.4.– *For* $T = T_1 \vee T_2$ *we have that*

$$X_T(t) = \int_0^t ds X_{T_1}(s)H(s)X_{T_2}(s),$$

or

$$\dot{X}_T(t) = X_{T_1}(t)H(t)X_{T_2}(t).$$

PROPOSITION 2.2.– *Let* $T = T_1 \vee T_2$, $U = U_1 \vee U_2$ *be two planar binary trees. We have:*

$$X_T(t)X_U(t) = \int_0^t ds X_T(s)X_{U_1}(s)H(s)X_{U_2}(s) + \int_0^t ds X_{T_1}(s)H(s)X_{T_2}(s)X_U(s).$$

In the formula, one or several of the trees T_1, T_2, U_1, U_2 *could be the empty tree.*

PROOF.– The proof follows from the integration by parts formula. □

Proposition 2.2 can be rephrased geometrically: define recursively a product $*$ on planar binary trees by

$$T * U = (T * U_1) \vee U_2 + T_1 \vee (T_2 * U), \qquad [2.5]$$

then

$$X_T(t)X_U(t) = X_{T*U}(t).$$

Using a self-explaining notation for Δ_{T*U}, which is actually a union of polytopes associated to trees, we have

$$\Delta_T(t) \times \Delta_U(t) = \Delta_{T*U}(t).$$

COROLLARY 2.3.– *The product equation* [2.5] *gives a combinatorial description of the cartesian product of the domains* Δ_T *encoded by trees. It provides the linear span of the set of planar binary trees,* PBT, *with the structure of an associative algebra.*

This follows by direct inspection of the formulas; the associativity follows from the associativity of the cartesian product of polytopes.

REMARK 2.3.– *In other terms, the integration by parts formula induces an algebra structure on the set of planar binary trees (about the latter, see Remark 2.4 below). Integration by parts can be axiomatized through the notion of (weight* 0*) Rota–Baxter* *algebra, which will be studied in the second part of this chapter. We will encounter various phenomena reminiscent of what happens here with planar binary trees.*

Iterated integrals parametrized by trees and iterated integrals parametrized by permutations are related through the canonical simplicial decompositions of the domains Δ_T. Let us write S_T for the set of total orders refining $\leq_T$ (i.e. if $<<$ is a total order in S_T, $i \leq_T j \Rightarrow i << j$) and identify a total order on $[n]$ with a permutation according to the rule: $2 >> 4 >> 1 >> 3$ is identified with $\sigma \in S_4$ with $\sigma^{-1}(1) = 2, \sigma^{-1}(2) = 4, \sigma^{-1}(3) = 1, \sigma^{-1}(4) = 3$. In general, the permutation $\beta \in S_n$ corresponds to the total order on $[n]$:

$$\beta^{-1}(1) >> \beta^{-1}(2) >> \cdots >> \beta^{-1}(n).$$

LEMMA 2.5.– *The domain Δ_T is the union of the simplices Δ_σ, where σ runs over S_T, and we get*

$$X_T(t) := \int_{\Delta_T} dt_1 \cdots dt_n H(t_1) \cdots H(t_n) = \sum_{\sigma \in S_T} X_\sigma(t).$$

We invite the reader to check the statement in the lemma. Here is an example for the order $2 <_T 1$, $2 <_T 3$. We find that

$$\begin{aligned}\Delta_T &= \{(t_1, t_2, t_3) \mid 0 \leq t_1, t_3 \leq t_2 \leq t\} \\ &= \{(t_1, t_2, t_3) \mid 0 \leq t_1 \leq t_3 \leq t_2 \leq t\} \cup \{(t_1, t_2, t_3) \mid 0 \leq t_3 \leq t_1 \leq t_2 \leq t\} \\ &= \{(t_1, t_3, t_2) \mid 0 \leq t_1 \leq t_2 \leq t_3 \leq t\} \cup \{(t_2, t_3, t_1) \mid 0 \leq t_1 \leq t_2 \leq t_3 \leq t\}\end{aligned}$$

and $S_T = \{1 >> 3 >> 2, 3 >> 1 >> 2\} = \{(1,3,2), (2,3,1)\}$. That is,

$$X_T(t) = \int_{\Delta_T} dt_1 \cdots dt_n H(t_1)H(t_2)H(t_3) = X_{(1,3,2)}(t) + X_{(2,3,1)}(t).$$

We also get directly, as a byproduct of the geometrical analysis, the following:

COROLLARY 2.4.– *The map κ from PBT to $\mathcal{S}$*

$$T \longmapsto \sum_{\sigma \in S_T} \sigma$$

induced by the simplicial decomposition

$$\Delta_T(t) = \bigcup_{\sigma \in S_T} \Delta_\sigma(t)$$

is an (injective) algebra map.

REMARK 2.4.– *The map from trees to permutations in the last corollary generalizes to arbitrary trees, not necessarily binary ones, and is sometimes referred to as an "arborification process", a term introduced by Jean Ecalle in the theory of dynamical systems (Fauvet and Menous 2017) (although the idea of associating to a tree, viewed as a preorder, the set of all its strict refinements – linearizations – is likely much older). The arborification process allows us to view a planar binary tree as the sum of its linearizations and to identify PBT with a subspace of $\mathbb{S}$. This point of view, which was followed in Loday and Ronco (1998), motivates further the proposition saying that the space PBT is stable by the product in $\mathbb{S}$, and is therefore a subalgebra of the algebra of permutations. The algebra PBT can also be obtained and studied from the point of view of binary search trees (Hivert* et al. *2005).*

REMARK 2.5.– *In analysis, Bloch sequences, Dyck paths, bracketings and non-crossing partitions, which are in bijection with planar binary trees, have also been used to describe the perturbative expansions in time-dependent perturbation theory (see Brouder* et al. *(2012), where the equivalence between the various points of view is detailed). The advantage of using trees in iterated integral calculus together with perturbative expansions with respect to other families of combinatorial objects is that it is relatively straightforward to extend their use to more complex situations, as illustrated, for example, by their recent use in the study of stochastic (partial) differential equations, where perturbative solutions are indexed by complex decorated trees (Hairer 2014).*

2.2.4. *Flows and Hopf algebraic structures*

Recall that a bialgebra H is both an associative algebra and a coassociative coalgebra, such that the coproduct of the latter is a morphism of algebras from H to $H \otimes H$ or, equivalently, such that the product of the former is a morphism of coalgebras from $H \otimes H$ to H. A Hopf algebra is a bialgebra equipped with an antipode, the convolution inverse to the identity map. Commutative Hopf algebras appear naturally as algebras of functions on groups and, conversely, an affine algebraic group is the spectrum of a commutative Hopf algebra. We omit details and refer the reader to the various textbooks and surveys on the subject (Cartier 2007; Radford 2012). In the study of differential equations, Hopf algebra structures appear naturally in relation to the group structures underlying flow maps. This observation is one of the many reasons motivating the definition of a Hopf algebra structure on the descent algebra or on the algebra of simplices/permutations.

Note first that the initial value problem, $\dot{X}(t) = X(t)H(t)$, $X(0) = Id$, can be solved on the interval $[0, t]$ as a succession of solutions: $X(t_1), \ldots, X(t_n) = X(t)$ with $0 < t_1 < \cdots < t_n = t$. The i-th solution is then obtained by solving the differential equation on the subinterval $[t_{i-1}, t_i]$ with initial condition $X(t_{i-1})$. For example, given $0 < s < t$, we have that

$$X(t) = X(s)\Big(Id + \int\limits_{\Delta^1_{t,s}} dt_1 H(t_1) + \cdots + \int\limits_{\Delta^n_{t,s}} dt_1 \cdots dt_n H(t_1) \cdots H(t_n) + \cdots\Big)$$

$$= \Big(Id + \int\limits_{\Delta^1_{s,0}} dt_1 H(t_1) + \cdots + \int\limits_{\Delta^n_{s,0}} dt_1 \cdots dt_n H(t_1) \cdots H(t_n) + \cdots\Big)$$

$$\Big(Id + \int\limits_{\Delta^1_{t,s}} dt_1 H(t_1) + \cdots + \int\limits_{\Delta^n_{t,s}} dt_1 \cdots dt_n H(t_1) \cdots H(t_n) + \cdots\Big),$$

where $\Delta^n_{t,s} := \{(t_1, \ldots, t_n) | s \leq t_1 \leq \cdots \leq t_n \leq t\}$. This observation, known as Chen's rule, is helpful in practice, especially in numerical analysis as well as in the theory of rough paths, where it allows us to "control" the construction of rough paths (Gubinelli 2004).

This idea leads to a rule of thumb for defining coproducts for Hopf algebra structures according to "time-orderings". Starting from a domain parametrized by $0 \leq t_1 \leq \cdots \leq t_n \leq t$ and given $0 \leq i \leq n$ together with an arbitrary $s \leq t$, we define the two subsets $0 \leq t_1 \leq \cdots \leq t_i \leq s$ and $s \leq t_{i+1} \leq \cdots \leq t_n \leq t$. When applied, for example, to a permutation $\sigma \in S_n$, starting from

$$\Delta^n_{\sigma,t} = \{(t_{\sigma(1)}, \ldots, t_{\sigma(n)}) | 0 \leq t_1 \leq \cdots \leq t_n \leq t\},$$

Chen's rule leads to the definition of a coproduct on $\mathcal{S}$:

$$\Delta(\sigma) := \sum_{i=0}^{n} \sigma_{|i} \otimes st(\sigma_{|>i}). \qquad [2.6]$$

Here, we consider a permutation written as the sequence of its values. $\sigma_{|i}$ stands for the subsequence associated with the first i integers and $\sigma_{|>i}$ stands for the subsequence associated with the integers strictly greater than i. The map st shifts the sequence $\sigma_{|>i}$ of integers by adding $-i$. For example, for the permutation $\sigma = (3, 8, 1, 4, 2, 5, 7, 6)$ and $i = 4$, we get $\sigma_{|i} = (3, 1, 4, 2)$, $\sigma_{|>i} = (8, 5, 7, 6)$ and $st(\sigma_{|>i}) = (8-4, 5-4, 7-4, 6-4) = (4, 1, 3, 2)$.

As the operation of dividing a sequence into blocks is coassociative, these rules define a coassociative coproduct on $\mathcal{S}$ (see Malvenuto and Reutenauer (1995) for a combinatorial proof):

THEOREM 2.2.– *The coproduct* [2.6] *equips the algebra* $\mathcal{S}$ *with the structure of a Hopf algebra.*

The Hopf algebra $\mathcal{S}$ is neither commutative nor cocommutative. It is known under several names, i.e. the Hopf algebra of permutations, the Malvenuto–Reutenauer Hopf algebra, or the Hopf algebra of free quasi-symmetric functions.

Since $\mathcal{D}$ is a subalgebra of $\mathcal{S}$ freely generated by the identity permutations 1_n and since

$$\Delta(1_n) = \sum_{i=0}^{n} 1_i \otimes 1_{n-i},$$

it forms a Hopf subalgebra of $\mathcal{S}$.

COROLLARY 2.5.– *The coproduct* [2.6] *restricts to* $\mathcal{D}$ *and equips the descent algebra with the structure of a Hopf algebra.*

Note that, as a Hopf algebra, the descent algebra is cocommutative since it is generated as an algebra by 1_n on which the coproduct is cocommutative.

Similar arguments, using Chen's rule to construct coproducts, apply to planar binary trees. In combinatorial terms, the coproduct obtained that way is simply the restriction to PBT of the one on $\mathcal{S}$, when trees are identified with the sum of their linearizations (Loday and Ronco 1998; Hivert *et al.* 2005). The simplest way to understand the construction of this coproduct on trees is however via the graphical representation. Since the bottom-to-top ordering of the vertices reflects decreasing times, the coproduct is obtained by splitting in all possible ways a tree T as the union of a subtree containing the root and the family of the remaining trees.

More formally, let us say that a family $V = \{v_1, \dots, v_k\}$ of vertices of T distinct from the root is admissible if they cannot be compared with respect to the order $<_T$, i.e. $v_i <_T v_j$ cannot hold for any pair of vertices in V. We assume that the family V is ordered according to the labels put on the vertices (from left to right and top to bottom in T; note that this is a different order from the one obtained by restriction of $<_T$) and write (slightly abusively) T_i for the subtree of T having v_i as its root. As a poset, T_i is the set of all vertices v such that $v_i <_T v$. We write T_V for the subtree containing the root, remaining after the T_i have been pruned, and we denote by T'_V the product $T_1 * \cdots * T_k$. In geometrical terms, this last product corresponds to a cartesian product.

COROLLARY 2.6.– *With the above notations in place, the coproduct equation* [2.6] *restricts to* PBT *and equips the algebra of planar binary trees with the structure of a Hopf algebra. When acting on trees, the coproduct reads*

$$\Delta(T) := 1 \otimes T + \sum_{V} T_V \otimes T'_V,$$

where the sum runs over all admissible subsets of vertices of the tree T.

2.3. Advances in chronological calculus

Chronological algebras and time-ordered products appear in (almost) uncountably many places, especially in theoretical physics and control theory. They can be understood as an abstract way of encoding integration by parts. Chronological calculus augments the tools of classical algebra for analysis (such as associative as well as Lie algebra techniques), by introducing the notions of shuffle and pre-Lie products. The former refines the notion of an associative product, while the latter augments that of the Lie bracket. A powerful framework for chronological calculus is the general theory of – weight zero – Rota–Baxter algebras, which will be developed in the following section.

In a nutshell, Rota–Baxter algebras of weight-zero axiomatize the notion of associative algebra equipped with an integral-type operator R satisfying integration by parts rule:

$$R(x)R(y) = R(xR(y)) + R(R(x)y).$$

The general notion of Rota–Baxter algebra allows for the existence of a remainder term

$$R(x)R(y) = R(xR(y)) + R(R(x)y) - \theta R(xy)$$

weighted by the scalar θ. Examples of maps satisfying the Rota–Baxter relation of weight θ are provided by re-summing series or when dealing with generalized integral operators.

Before turning to the general theory of Rota–Baxter algebras, we will show how the underlying techniques apply in the usual formalism of chronological calculus. Attributing properly the origin of this formalism, which evolved along 70 years of developments in theoretical physics, control theory and combinatorics would be rather difficult and lies outside the scope of the present survey. Among many others, the names of Chen, Feynman, Fliess, Kawski, Rota and Schützenberger should certainly be quoted (we include the latter because it seems that his introduction of half-shuffles in Schützenberger (1958/59) in the context of free Lie algebras was actually an elaboration of the notion of time-ordered products as studied by Russian physicists – hence the notation T for these products in his article[3]).

The development presented in this part of the chapter is mostly obtained by revisiting the seminal work of Agrachev and Gamkrelidze (1981), who introduced and featured the role of pre-Lie algebra structures in integral/differential calculus. This should be seen in the light of later works. Chapoton and Livernet (2001) studied

3. Hoang Ngoc Minh, private communication.

the operad of free pre-Lie algebras, Oudom and Guin constructed their enveloping algebras (Oudom and Guin 2008). Ebrahimi-Fard, together with Dominique Manchon (Ebrahimi-Fard and Manchon 2009), used shuffle and pre-Lie algebras in the theory of differential equations and the Magnus formula. In Ebrahimi-Fard and Patras (2014), we studied the interpretation of the structure of the enveloping algebra of a pre-Lie algebra by means of Feynman's time-ordered products. Patras, together with Frédéric Chapoton (Chapoton and Patras 2013), studied the exponential construction and interpretation of the Magnus operator and its inverse.

In the following, we will extend the results of Patras (1991), which relate the combinatorics of simplices with the theory of free Lie algebras, and show how another fundamental construction, the Eilenberg–MacLane and Schützenberger "half-shuffle" products (Schützenberger 1958/59), provides a way to relate chronological calculus with simplicial geometry.

2.3.1. *Chronological calculus and half-shuffles*

Chronological calculus is based on the idea of time-ordering of operators. Consider, for example, two time-dependent operators, $M(t)$ and $N(t)$, with $M(0) = N(0) = 0$, in a non-unital algebra A of operators – having suitable regularity properties allowing us to compute derivatives, integrals and so on. We say that the product $\dot{M}(s)\dot{N}(u)$ is causal if $s > u$ and anti-causal if $s < u$. Note that we will consider integrals of these products; therefore, the equal-time, or diagonal term, $\dot{M}(s)\dot{N}(s)$, is irrelevant. The classical integration by parts rule

$$\begin{aligned} M(t)N(t) &= \int_0^t ds \int_0^s du \dot{M}(s)\dot{N}(u) + \int_0^t ds \int_0^s du \dot{M}(u)\dot{N}(s) \\ &=: (M \nwarrow N)(t) + (M \nearrow N)(t) \end{aligned}$$

can then be viewed as a way to decompose the usual operator product into the sum of a "causal" product and an "anti-causal" product. In this sense, the $\nwarrow$ and $\nearrow$ symbols indicate in what direction time increases, i.e. from right to left in $(M \nwarrow N)(t) := \int_0^t dM(s)N(s)$ and from left to right in $(M \nearrow N)(t) := \int_0^t M(s)dN(s)$. Let us mention that the idea of time-ordering seems to originate in theoretical physics, for example, when defining Feynman propagators, respectively, Feynman's time-ordered product (further below we will return to the latter).

It is useful to understand this construction of operator products in more abstract terms. Recall that a left (respectively, right) B-module, or representation of an associative algebra B, is a vector space V equipped with a left (respectively, right) action of B (written as $\cdot$-operation) such that $a\cdot(b\cdot v) = (ab)\cdot v$ (respectively, $(v\cdot b)\cdot a = v \cdot (ba)$) for arbitrary elements $a, b \in B$ and $v \in V$. Returning to time-dependent operators, the associativity of the operator product in A, i.e. $M(NP) = (MN)P$, can then be formulated by considering A as a bimodule, i.e. as a left and right module, over itself for the operator product. The integration by parts rule leads to the decomposition of the product into two commuting right and left actions of A on itself, denoted as above in terms of the symbols $\nwarrow$ and $\nearrow$. Besides indicating the direction of increasing time, the symbols also point in the direction of the action, i.e. in $M \nwarrow N$, the operator N acts from the right on M. Indeed, we verify quickly that

$$((M \nwarrow N) \nwarrow P)(t) = \int_0^t ds\dot{M}(s)N(s)P(s) = (M \nwarrow NP)(t)$$

$$(P \nearrow (N \nearrow M))(t) = \int_0^t dsP(s)N(s)\dot{M}(s) = (PN \nearrow M)(t)$$

$$((P \nearrow M) \nwarrow N)(t) = \int_0^t dsP(s)\dot{M}(s)N(s) = (P \nearrow (M \nwarrow N))(t).$$

In particular, using this notation, the associativity, $(MN)P = M(NP)$, can be formulated as

$$\begin{aligned}&(M \nwarrow N) \nwarrow P + (M \nearrow N) \nwarrow P + (MN) \nearrow P \\ &= M \nwarrow (NP) + M \nearrow (N \nwarrow P) + M \nearrow (N \nearrow P),\end{aligned}$$

and splits into the three operator identities

$$\begin{aligned}(M \nwarrow N) \nwarrow P &= M \nwarrow (NP) \\ (M \nearrow N) \nwarrow P &= M \nearrow (N \nwarrow P) \\ (MN) \nearrow P &= M \nearrow (N \nearrow P).\end{aligned} \qquad [2.7]$$

It turns out that the above notions of causal and anti-causal products, $\nwarrow$ and $\nearrow$, together with these three relations happen to coincide with the axiomatization of shuffle products in algebraic topology. The latter was one (among many) of the seminal ideas of Eilenberg and MacLane (1953, section 5), who used the three relations above to give an abstract proof of the associativity of the shuffle product in a simplicial framework.

This connection between chronological calculus and algebraic topology is, in our opinion, illuminating. Indeed, it explains how the causal and anti-causal products are computed when dealing with iterated integrals and reads as follows. We adapt here the definition of half-shuffle products of general simplicial objects to the context of this chapter, i.e. we define them in the (geometrical realization of the) algebra of simplices. We recall that given two permutations $\sigma \in S_n$, $\beta \in S_m$, the shuffle product of the corresponding simplices induced by the decomposition of their cartesian product reads

$$\Delta^n_{\sigma,t} \times \Delta^m_{\beta,t} = \bigcup_{\substack{\gamma \in S_{n+m} \\ st_{[n]}(\gamma(i))=\sigma(i), i \leq n \\ st_{[m]}(\gamma(i))=\beta(i-n), i>n}} \Delta^{n+m}_{\gamma,t} = \sum_{\substack{\nu \in S_{n+m} \\ pr^+_n(\Delta^{n+m}_{\nu,t})=\Delta^n_{\sigma,t} \\ pr_m(\Delta^{n+m}_{\nu,t})=\Delta^m_{\beta,t}}} \Delta^{n+m}_{\nu,t}.$$

The so-called left and right half-shuffle products are then defined, respectively, by (summing up to the full shuffle product)

$$\Delta^n_{\sigma,t} < \Delta^m_{\beta,t} := \bigcup_{\substack{\gamma \in S_{n+m},\, \gamma^{-1}(n+m)=\sigma^{-1}(n) \\ st_{[n]}(\gamma(i))=\sigma(i),\, i\leq n \\ st_{[m]}(\gamma(i))=\beta(i-n),\, i>n}} \Delta^{n+m}_{\gamma,t},$$

and

$$\Delta^n_{\sigma,t} > \Delta^m_{\beta,t} = \bigcup_{\substack{\gamma \in S_{n+m},\, \gamma^{-1}(n+m)=n+\beta^{-1}(m) \\ st_{[n]}(\gamma(i))=\sigma(i),\, i\leq n \\ st_{[m]}(\gamma(i))=\beta(i-n),\, i>n}} \Delta^{n+m}_{\gamma,t},$$

such that, on iterated integrals (with the same notations as in the previous section) the left and right half-shuffle products correspond to the causal and anti-causal products. For example,

$$(X_\sigma \nwarrow X_\beta)(t) = \langle \Delta^n_{\sigma,t} < \Delta^m_{\beta,t} | H\rangle.$$

It is easy to check that they define two commuting left and right actions of the algebra of simplices on itself or, equivalently, that the three identities [2.7] above still hold when considering simplices instead of time-dependent operators. For example, given $\sigma \in S_n$, $\beta \in S_m$ and $\nu \in S_p$,

$$\Delta^n_{\sigma,t} < (\Delta^m_{\beta,t} \times \Delta^p_{\nu,t}) = \bigcup_{\substack{\gamma \in S_{n+m+p},\, \gamma^{-1}(n+m+p)=\sigma^{-1}(n) \\ st_{[n]}(\gamma(i))=\sigma(i),\, i\leq n \\ st_{[m]}(\gamma(i))=\beta(i-n),\, n<i\leq n+m \\ st_{[p]}(\gamma(i))=\nu(i-n-m),\, i>n+m}} \Delta^{n+m+p}_{\gamma,t}$$

$$= (\Delta^n_{\sigma,t} < \Delta^m_{\beta,t}) < \Delta^p_{\nu,t}.$$

In other terms, the algebra of time-dependent operators, with the binary maps $\nwarrow$ and $\nearrow$, and the algebra of simplices, with $<$ and $>$, are both examples of so-called shuffle algebras, where the latter notion is given in the following:

DEFINITION 2.5 (Shuffle algebra).– *A shuffle algebra is a (non-unital) associative algebra* $(A, \cdot_A)$ *whose product splits into the sum of two so-called half-shuffle products,* $\cdot_A = \prec + \succ$, *such that:*

– *the two binary operations* $\prec$ *and* $\succ$ *define, respectively, a right and a left action of* A *on itself, commuting with each other; or, equivalently,*

– *the left and right half-shuffle products hold the following shuffle relations:*

$$(M \prec N) \prec P = M \prec (N \cdot_A P), \quad [2.8]$$

$$(M \succ N) \prec P = M \succ (N \prec P), \quad [2.9]$$

$$(M \cdot_A N) \succ P = M \succ (N \succ P). \quad [2.10]$$

REMARK 2.6.– *When dealing with functions instead of operators (or with scalar-valued iterated integrals), the algebra product* $\cdot_A$ *is commutative, i.e.* $\int_0^t ds\dot{M}(s)N(s) = \int_0^t dsN(s)\dot{M}(s)$. *This amounts to the relation* $(M \nwarrow N)(t) = (N \nearrow M)(t)$. *The axioms simplify then to the sole relation* $(M \nwarrow N) \nwarrow P = M \nwarrow (N \cdot_A P)$ *that defines the notion of commutative shuffle algebras. See Schützenberger (1958/59) for details.*

REMARK 2.7.– *A shuffle algebra* A *can be augmented with a unit: set for* $x \in A$, $1 \nwarrow x = x \nearrow 1 := 0, x \nwarrow 1 = 1 \nearrow x := x$. *However, the half-products* $1 \nwarrow 1$ *and* $1 \nearrow 1$ *cannot be defined consistently. The trick seems to originate in Schützenberger (1958/59) (in a commutative setting).*

REMARK 2.8.– *A theorem of Foissy (2007) shows that the shuffle algebra of simplices is free as a shuffle algebra. The proof of the theorem is rather involved and may be considered among the nicest achievements resulting from the theory of shuffle algebras. It is however already interesting to look at the subalgebra spanned by* Δ_t^1 *(Loday and Ronco 1998).*

We set $a \succ b \prec c$ for $(a \succ b) \prec c = a \succ (b \prec c)$. Then, we have the following:

LEMMA 2.6.– *The free shuffle algebra* $Sh(x)$ *generated by a single letter* x *then has a basis* B *defined inductively by:*

– $x \in B$

– *If* $y, z \in B$, *then* $y \succ x \prec z$, $y \succ x$ *and* $x \prec z$ *are in* B.

Indeed, by induction on the length of the elements (given by the number of x appearing), the identities

$$(y \succ x \prec z) \prec (y' \succ x \prec z') = y \succ x \prec (z \cdot_A (y' \succ x \prec z'))$$

and

$$(y \succ x \prec z) \succ (y' \succ x \prec z') = ((x \succ x \prec z) \cdot_A y') \succ x \prec z')$$

imply that the elements of B span linearly $Sh(x)$.

To prove that they are linearly independent, it is enough to check that, in the shuffle algebra spanned by simplices $\Delta_{\sigma,t}^n$, the elements constructed recursively when Δ_t^1 is substituted for x in the definition of B contain each a simplex indexed by a permutation that does not appear in the expansion of the other terms. This follows recursively from the definition of the products $<$ and $>$ and the identities $p_n^+(\Delta_{\sigma,t}^n > \Delta_t^1 < \Delta_{\beta,t}^m) = \Delta_{\sigma,t}^n$ and $p_m^-(\Delta_{\sigma,t}^n > \Delta_t^1 < \Delta_{\beta,t}^m) = \Delta_{\beta,t}^m$, where $\sigma \in S_n$, $\beta \in S_m$.

In particular, the shuffle subalgebra of the shuffle algebra of simplices generated by Δ_t^1 is a free shuffle algebra.

Foissy's freeness properties imply that computations with iterated integrals are "generic" for shuffle calculus: an algebraic identity involving the symbols $\nwarrow$ and $\nearrow$, or $<$ and $>$, that holds for iterated integrals of an arbitrary time-dependent operator holds for arbitrary shuffle algebras.

REMARK 2.9.– *The terminology* chronological product/algebra *comes from Agrachev and Gamkrelidze (1981) and is synonymous in their work with* pre-Lie product/algebra. *The terminology* dendriform relations *is often used for the three identities* [2.8]–[2.10] *encoding the splitting of the associativity of the shuffle product, and* dendriform algebra *for the corresponding notion of algebra. We prefer, however, the terminology* shuffle relations/algebra *for conceptual and historical reasons. Indeed, the notion of* shuffle *is natural in the light of the idea of splitting a product into two, i.e. left and right,* half-shuffle *products corresponding to a right and left action.*

2.3.2. *Chronological calculus and pre-Lie products*

In this section, we survey the Lie-theoretic perspective of chronological calculus, following Agrachev and Gamkrelidze (1981). Starting from the representation theoretic point of view, we consider a vector space A with a binary product $\rhd\colon A \otimes A \to A$ and the associated bracket product $[a, b]_\rhd := a \rhd b - b \rhd a$. Write L_x for the linear endomorphism of A defined by left multiplication:

$$L_x(y) := x \rhd y,$$

and define the usual commutator bracket of linear endomorphisms of A, $[L_x, L_y] := L_x \circ L_y - L_y \circ L_x$.

DEFINITION 2.6.– *The pair $(A, \rhd)$ is a left pre-Lie algebra if and only if for any $x, y \in A$, the identity $[L_x, L_y] = L_{[x,y]_\rhd}$ holds, which is equivalent to the left pre-Lie relation*

$$x \rhd (y \rhd z) - (x \rhd y) \rhd z = y \rhd (x \rhd z) - (y \rhd x) \rhd z.$$

The canonical example of a pre-Lie algebra, which can be traced back to Caley's work on rooted trees, is given in terms of derivations. We verify, for example, that the space of vector fields generated by the derivations $x^n \partial_x$ is a pre-Lie algebra with product $(x^n \partial_x) \rhd (x^m \partial_x) := m \cdot x^{n+m-1} \partial_x$. We refer the reader to Chapoton and Livernet (2001), Cartier (2011) and Manchon (2011) for other examples and more details on the notion of pre-Lie algebra.

REMARK 2.10.– *The notion of right pre-Lie algebra is defined similarly in terms of the product* $\lhd$: $A \otimes A \to A$, *together with the right pre-Lie relation* $x \lhd (y \lhd z) - (x \lhd y) \lhd z = x \lhd (z \lhd y) - (x \lhd z) \lhd y$. *We will abbreviate left pre-Lie to pre-Lie from now on.*

We can always augment A with a unit. The pre-Lie product on $A \oplus k$ is defined by $(a + l) \rhd (b + l') := a \rhd b + l'a + lb + ll'$ for $a, b \in A,\ l, l' \in k$. Using the map $x \longmapsto L_x$, this observation allows us to embed $(A, [-, -]_\rhd)$ into the vector space of linear endomorphisms of $A \oplus k$ equipped with its canonical Lie bracket. In particular, there is a forgetful functor from pre-Lie to Lie algebras. In other terms:

LEMMA 2.7.– *The pair* $(A, [-, -]_\rhd)$ *is a Lie algebra.*

The link with classical chronological calculus is as follows. As in the previous section, we denote by $(A, \cdot_A)$ an algebra of time-dependent operators and consider the Lie bracket $[M, N](t)$. The integration by parts formula implies that

$$[M, N](t) = \int_0^t dsM(s)\dot{N}(s) + \int_0^t ds\dot{M}(s)N(s) - \int_0^t dsN(s)\dot{M}(s) - \int_0^t ds\dot{N}(s)M(s),$$

which can be written as the difference of:

$$\begin{aligned}(M \rhd N)(t) :&= (M \nearrow N)(t) - (N \nwarrow M)(t) \\ &= \int_0^t dsM(s)\dot{N}(s) - \int_0^t ds\dot{N}(s)M(s)\end{aligned}$$

and $(N \rhd M)(t)$ so that $[M, N](t) = [M, N]_\rhd(t)$. That the algebra A is indeed a pre-Lie algebra follows from the Jacobi identity:

$$\begin{aligned}([M, N]_\rhd \rhd P)(t) &= ([M, N] \rhd P)(t) = \int_0^t ds[[M(s), N(s)], \dot{P}(s)] \\ &= - \int_0^t ds[[N(s), \dot{P}(s)], M(s)] + \int_0^t ds[[M(s), \dot{P}(s)], N(s)] \\ &= (M \rhd (N \rhd P))(t) - (N \rhd (M \rhd P))(t).\end{aligned}$$

More generally, there is a forgetful functor from shuffle to pre-Lie algebras (Aguiar 2000):

PROPOSITION 2.3.– *Any shuffle algebra* $(A, \prec, \succ)$ *has the structure of a pre-Lie algebra with*

$$x \rhd y := x \succ y - y \prec x. \qquad [2.11]$$

Note that $x \triangleleft y := x \prec y - y \succ x = -y \triangleright x$ is a right pre-Lie product. We omit the proof that follows from the calculations above in the particular case of time-dependent operators.

2.3.3. *Time-ordered products and enveloping algebras*

One of the central ideas of chronological calculus is to develop a group theoretic approach adapted to the study of flows of differential equations, based on the finer structure provided by pre-Lie algebras rather than Lie algebras. A key step is the construction of enveloping algebras of pre-Lie algebras.

We recall that the enveloping algebra $U(L)$ of a Lie algebra L is an associative algebra (uniquely defined up to isomorphism) with the following properties:

– the Lie algebra L embeds in $U(L)$ (as a Lie algebra, where the Lie algebra structure on $U(L)$ is induced by the associative product, i.e. in terms of the commutator bracket $[x, y] := xy - yx$);

– for any associative algebra A (we write g_A for its associated Lie algebra, i.e. A equipped with the commutator bracket), there is a natural bijection between the Lie algebra maps from L to g_A and the associative algebra maps from $U(L)$ to A.

Usually, the algebra $U(L)$ is constructed as the quotient of the free associative algebra over L by the ideal generated by relations $[l, l'] - ll' - l'l = 0,\ l, l' \in L$. It is a cocommutative Hopf algebra with a coproduct determined by its action on the elements of L, i.e. for $l \in L$, $\Delta(l) := l \otimes \mathbf{1} + \mathbf{1} \otimes l$.

Since pre-Lie algebras are Lie algebras with extra structure, their enveloping algebras also have additional structures that have been described in the works of Lada and Markl (2005) and Oudom and Guin (2008). In the first part of this section, we briefly recall the results of these works. We recall first that the algebra of polynomials $k[V]$ over a vector space V is equipped with a bi-commutative Hopf algebra structure such that given a monomial $v_1 \cdots v_n$, $v_i \in V$, the coproduct reads

$$\Delta(v_1 \cdots v_n) = \sum_{I \coprod J = [n]} v_I \otimes v_J,$$

where $v_I := v_{i_1} \cdots v_{i_l}$, for $I = \{i_1, \ldots, i_l\} \subset [n] := \{1, \ldots, n\}$.

From now on, $(V, \triangleright)$ denotes a pre-Lie algebra.

DEFINITION 2.7.– *The brace map on* V

$$k[V] \otimes V \longrightarrow V$$

$$P \otimes v \longmapsto \{P\}v$$

is defined inductively by

$$\{w\}v := w \rhd v,\ w \in V$$

$$\{w_1, \dots, w_n\}v := \{w_n\}(\{w_1, \dots, w_{n-1}\}v)$$

$$- \sum_{i=1}^{n-1} \{w_1, \dots, \{w_n\}w_i, \dots, w_{n-1}\}v.$$

We define now a product $*$ on $k[V]$ in terms of the brace map: for $v_1, \dots, v_n$ and $w_1, \dots, w_m$ in V,

$$(w_1 \cdots w_m) * (v_1 \cdots v_n) = \sum_f W_0(\{W_1\}v_1) \cdots (\{W_n\}v_n), \qquad [2.12]$$

where the sum is over all maps f from $\{1, \dots, m\}$ to $\{0, \dots, n\}$ and $W_i := \prod_{j \in f^{-1}(i)} w_j$. For example, $w * v = wv + \{w\}v$. The central result of Oudom–Guin is the following:

THEOREM 2.3.– *The product $*$ is associative and unital. It equips the vector space of polynomials $k[V]$ with the structure of an enveloping algebra of V and makes $(k[V], \Delta)$ a Hopf algebra (the Hopf algebra structure being the one associated with the enveloping algebra structure of $k[V]$).*

Following Ebrahimi-Fard and Patras (2014), we will provide an interpretation of the aforementioned result of Oudom–Guin in terms of chronological calculus. Note that the results in Ebrahimi-Fard and Patras (2014) are stated in the more general framework of Rota–Baxter algebras. Assume from now on that the pre-Lie algebra $(V, \rhd)$ is furthermore a shuffle algebra with the induced pre-Lie product [2.11] – for example, an algebra of time-dependent operators. We recall that the time-ordered product of two elements in V is given by $T[v, w] := v \prec w + w \prec v$. In the context of time-dependent operators, the product is obtained by symmetrizing the causal product. In general, for $v_1, \dots, v_n \in V$,

$$T[v_1, v_2, \dots, v_n] := \sum_{\sigma \in S_n} v_{\sigma(1)} \prec (v_{\sigma(2)} \prec (\cdots \prec (v_{\sigma(n-1)} \prec v_{\sigma(n)}) \cdots)).$$

The starting points for the following developments are:

– the observation that a pre-Lie algebra V has in some sense more structure than a Lie algebra, which is reflected in the fact that its polynomial algebra, $k[V]$, is naturally equipped with two Hopf algebra structures (sharing the same underlying coproduct), i.e. the one associated with the associative product $*$ making $k[V]$ an enveloping algebra and the usual product of polynomials;

– the fact that, since V is a shuffle algebra, it is equipped with an associative product, denoted $\cdot$, whose bracket identifies with the Lie bracket $[-,-]_\rhd$. By the universal properties of enveloping algebras, the identity map from $(V, [-,-]_\triangleright)$ to $(V, [-,-])$ induces a map of associative algebras, $\iota\colon (k[V]^+, *) \to (V, \cdot)$, where $k[V]^+$ stands for the augmentation ideal of $k[V]$ consisting of polynomials without constant terms.

In degree two, we then have:

$$\begin{aligned}\iota(wv) = \iota(w * v) - \iota(\{w\}v) &= w \cdot v - w \triangleright v \\ &= w \prec v + w \succ v - (w \succ v - v \prec w) \\ &= w \prec v + v \prec w = T[w, v].\end{aligned}$$

This is a general phenomenon (see Ebrahimi-Fard and Patras (2014, p. 1291) for proof):

THEOREM 2.4.– *The image in V of a monomial $v_1 \cdots v_n$ in $k[V]$ by the canonical map ι is the time-ordered product of the v_is:*

$$\iota(v_1 \cdots v_n) = T[v_1, \ldots, v_n]. \qquad [2.13]$$

2.3.4. *Formal flows and Hopf algebraic structures*

There are two classical ways to study the formal properties of the flow associated with a differential equation, such as, for instance, the evolution operator $X(t)$ in the notation of Theorem 2.1, when dealing with the linear differential equation $\dot{X}(t) = X(t)H(t)$. By formal, we mean in this section that questions related to convergence of series are not addressed, i.e. infinite sums are treated as formal sums.

The first approach to study flows relies on classical group theory and was described earlier in this chapter in terms of the discrete version of the Baker–Campbell–Hausdorff formula (Theorem 2.1 above). It expresses the logarithm of the product of two exponentials and, as such, the formal link between a continuous group and its Lie algebra in the neighborhood of the identity. The two are locally in bijection through the logarithm and exponential map, and this bijection can be formally understood by studying the structure of the (suitably completed) enveloping algebras of free Lie algebras. For details, see, for instance, Reutenauer's monograph (Reutenauer 1993), where this approach is developed.

The other way to express the link between the solution $X(t)$ (of $\dot{X}(t) = X(t)H(t)$) and $\Omega(t) := \log(X(t))$ is given by the Magnus formula (Magnus 1954), which describes the latter in terms of a highly nonlinear differential equation

$$\dot{\Omega}(t) = \frac{ad_\Omega}{e^{ad_\Omega} - 1} H(t) = H(t) + \sum_{n>0} \frac{B_n}{n!} ad^n_{\Omega(t)}(H(t)).$$

Here, ad denotes the adjoint representation, where $ad^n_M(N) := [M, ad^{n-1}_M(N)]$, $ad^0_M(N) := N$, and B_n are the Bernoulli numbers (the numbers $\frac{B_n}{n!}$ are the coefficients of the expansion of $\frac{x}{\exp(x)-1}$ as a formal power series). Its pre-Lie structure was investigated in Ebrahimi-Fard and Manchon (2009) and follows from the observation that, for time-dependent operators, the pre-Lie product

$$(M \rhd N)(t) = \int_0^t du[M(u), \dot{N}(u)]$$

satisfies $\frac{d}{dt}(M \rhd N)(t) = ad_{M(t)}\dot{N}(t)$. The Magnus formula then reads:

$$\dot{\Omega}(t) = \frac{d}{dt}\left(\{\frac{\Omega}{\exp(\Omega)-1}\}H\right)(t) \qquad [2.14]$$

where $\frac{\Omega}{\exp(\Omega)-1}$ is computed using the associative product of the enveloping algebra of the pre-Lie algebra of time-dependent operators.

We explain in this section how this kind of phenomenon can be accounted for. Our presentation is based on the seminal work of Agrachev and Gamkrelidze (1981) and later developments by Chapoton and Livernet (2001), Oudom and Guin (2008), Ebrahimi-Fard and Manchon (2009) and Chapoton and Patras (2013).

We recall first from Chapoton and Livernet (2001) and Oudom and Guin (2008) the following result:

THEOREM 2.5.– *Let $B = (v_n)_{n\in I}$ be a basis of a vector space L. Then, non-planar rooted trees with vertices indexed by elements of the set I form a basis of $V := \mathcal{PL}(L)$, the free pre-Lie algebra over V. A non-planar rooted tree is interpreted recursively as an element of V as follows: if T is the tree obtained by attaching the subtrees $T_1, \dots, T_n$ (viewed as elements of L) to a root vertex indexed by v_i, then*

$$T = \{T_1 \cdots T_n\}v_i.$$

The pre-Lie product on V is obtained as follows: given two trees T, T', the pre-Lie product $T \rhd T'$ is the sum of the trees obtained by summing over all graftings of the root of T successively onto every vertex of T'. The set of forests over non-planar rooted trees forms a basis of the polynomial algebra $k[V]$.

REMARK 2.11.– *A forest F being a set $\{T_1, \dots, T_k\}$ of non-planar rooted trees is interpreted as the monomial $T_1 \cdots T_k$ in $k[V]$. As a Hopf algebra, $(k[V], *, \Delta)$ is known as the Grossman–Larson Hopf algebra (Grossman and Larson 1989).*

From now on, V denotes a free pre-Lie algebra equipped with a basis of non-planar rooted trees, as in the previous theorem. Its enveloping algebra is the associated Grossman–Larson Hopf algebra $k[V]$. It is graded by the number of vertices of the trees in a forest. We denote by $\widehat{k[V]}$ its completion with respect to this grading, which is compatible with the two products on $k[V]$, i.e. the commutative product of monomials and the associative product of the enveloping algebra.

Recall the following general properties of the completion $\hat{H}$ of a graded connected cocommutative Hopf algebra H (Milnor and Moore 1965). The vector space of its primitive elements, denoted $Prim(\hat{H}) := \{x \in \hat{H},\ \Delta(x) = x \otimes 1 + 1 \otimes x\}$, is a Lie algebra for the Lie bracket associated with the product in $\hat{H}$. The set $G(\hat{H}) := \{x \in \hat{H},\ \Delta(x) = x \otimes x\}$ of group-like elements forms instead a group for the product in $\hat{H}$. The exponential and logarithm maps define two inverse bijections between $Prim(\hat{H})$ and $G(\hat{H})$. Note that $\widehat{k[V]}$ is such a Hopf algebra with respect to its two natural Hopf algebra laws.

The vector space $Prim(\widehat{k[V]}) = \hat{V}$ is the completion of V as a graded vector space. It carries two Lie algebra laws. The one associated with the commutative product, which we denote from now on as $\circ$, is trivial (making it a commutative Lie algebra), and the other one is the Lie bracket associated with the pre-Lie product on V.

As a set, the group $G(\widehat{k[V]}) =: G(V)$ consists of exponentials of elements in $\hat{V}$ (using any of the two products). We denote by $\exp^\circ$, $\log^\circ$ the isomorphism between $\hat{V}$ and $G(V)$ coming from the commutative product ($\circ$) and by $\exp^*$, $\log^*$ the isomorphism defined in terms of the associative product ($*$).

REMARK 2.12.– *Note that $\hat{V}$ is equipped with two inverse set automorphisms: $\log^\circ \circ \exp^*$ and $\log^* \circ \exp^\circ$. We are going to show that the second encodes the Magnus formula. The first one identifies with the map W in Agrachev and Gamkrelidze (1981).*

DEFINITION 2.8.– *Let us call 2-group (respectively, 2-pre-Lie algebra) a set equipped with two group (respectively, pre-Lie) structures. The 2-group (respectively, 2-pre-Lie algebra) of formal flows associated with V is the group $(G(V), \circ, *)$ of group-like elements in $\widehat{k[V]}$ (respectively, the triple $(\hat{V}, 0, \triangleleft)$, where 0 stands for the null pre-Lie product).*

Let us focus here on the application of this formalism to the Magnus formula.

DEFINITION 2.9.– *Let $x \in \hat{V}$. The Magnus element $\Omega(x) \in \hat{V}$ is the (necessarily unique) solution $\Omega(x)$ to the equation:*

$$\{\frac{\Omega(x)}{\exp(\Omega(x)) - 1}\}x = \Omega(x).$$

Note that the unicity follows by induction from grading arguments.

The link with the classical Magnus formula is as follows. Consider $\int_0^t dsH(s)$ in an algebra A of time-dependent operators (e.g. matrices $M(t)$ with $M(0) = 0$ and with bounded C^∞ coefficients, to avoid the discussion of convergence phenomena). Define then V to be the free pre-Lie algebra over a variable x. There is a canonical pre-Lie algebra map from V to A induced by the map $x \longmapsto \int_0^t dsH(s)$ as well as an induced map from the Grossman–Larson algebra $k[V]$ to A. These two maps send $\Omega(x)$ to $\Omega(t)$.

Note that, by Theorem 2.4, the last map also sends $\exp^\circ(x) - 1$ to $X(t) - 1$, where $X(t)$ is the solution of $\dot{X}(t) = X(t)H(t)$, $X(0) = 1$. Since $\Omega(t) = \log(X(t))$, we get that the images in A of the two elements $\Omega(x)$ and $\log^* \circ \exp^\circ(x)$ in $\widehat{k[V]}$ coincide. In fact, more generally:

THEOREM 2.6.– *The Magnus element $\Omega(x)$ identifies with* $\log^* \circ \exp^\circ(x)$.

PROOF.– Indeed, we have:

$$\log^*(\exp^\circ(x)) = \sum_{n>0} (-1)^{n-1} \frac{(\exp^\circ(x) - 1)^{*n}}{n}.$$

Let us write $u = v + o(1)$, meaning that u and v in $\widehat{k[V]}$ are equal up to a formal linear combination of forests in $\widehat{k[V]}$ with at least two trees. Denoting $\widehat{k[V]}_{\geq 2}$ the latter space, we have in particular $\widehat{k[V]} = k \oplus V \oplus \widehat{k[V]}_{\geq 2}$. Note that, by definition of the $*$-product,

$$\widehat{k[V]} * \widehat{k[V]}_{\geq 2} \subset \widehat{k[V]}_{\geq 2}.$$

We then have:

$$\begin{aligned}\log^*(\exp^\circ(x)) &= \left(\sum_{n>0} (-1)^{n-1} \frac{(\exp^\circ(x) - 1)^{*n-1}}{n}\right) * \big(x + o(1)\big) \\ &= \left(\sum_{n>0} (-1)^{n-1} \frac{(\exp^\circ(x) - 1)^{*n-1}}{n}\right) * x + o(1).\end{aligned}$$

We recall now that for any $y \in \widehat{k[V]}$, $y * x = \{y\}x + o(1)$. Since $\log^*(\exp^\circ(x)) \in V$, and for $v, w \in V$, $(v = w + o(1)) \implies (v = w)$, we obtain finally

$$\log^*(\exp^\circ(x)) = \{\sum_n (-1)^{n-1} \frac{(\exp^\circ(x) - 1)^{*n-1}}{n}\}x = \{\frac{\log^*(\exp^\circ(x))}{\exp^\circ(x) - 1}\}x,$$

from which the statement of the theorem follows. □

2.4. Rota–Baxter algebras

The mathematician Gian-Carlo Rota once lamented in a 1998 paper (Rota 1998) that:

> Whereas algebraists have devoted a lot of attention to derivations, the algebraic theory of the indefinite integral has been strangely neglected. The shuffle identities are only the tip of the iceberg for algebra and combinatorics of the indefinite integral operator which remains unexplored. The profound work of K. T. Chen bears witness to the importance of the algebra of the indefinite integral. J. Milnor once told me that he considers K. T. Chen's work to be some of the most important and most neglected mathematics of the latter half of this century. However, not even Milnor's opinion has so far made a dent into the uncontrollable forces of fashion and fad.

Rota refers above to the unital associative shuffle algebra with its commutative shuffle product, which plays a critical role in Chen's seminal works (Chen 1957, 1971) (see also Ree's article (Ree 1958)). As we know from the previous sections, the shuffle product encodes indeed in abstract algebraic terms the classical integration by parts rule and its extension to the product of iterated integrals. As we will see in this section, which forms the second part of this chapter, the notion of Rota–Baxter algebra defined through the Rota–Baxter relation provides a far reaching generalization of Chen's shuffle algebra.

In the following, we will outline the main algebraic aspects of the contemporary theory of Rota–Baxter algebras. We should remark though that we refrain from giving an exhaustive account. Let us emphasize the underlying aim. Rota–Baxter algebra is representative for a whole class of theories, including iterated integrals of scalar and operator valued functions – corresponding to commutative and noncommutative shuffle products – as well as summation and projection operators – corresponding to commutative and noncommutative quasi-shuffle products. The main distinction between the commutative and noncommutative realms is encoded in the presence of pre-Lie and post-Lie products. Furthermore, since the axioms of Rota–Baxter algebra encode more general "integral-type operators" than the Riemann integral of chronological calculus, the identities that are obtained in Rota–Baxter algebra have a much wider range of applications.

2.4.1. *Origin*

The Rota–Baxter identity and together with it the notion of (commutative) Rota–Baxter algebra first appeared in the 1960 work of the American mathematician Glen Baxter[4] (1960). Baxter's paper was motivated by a fundamental result in

4. Note that this Baxter should not be confused with the Australian theoretical physicist Rodney Baxter – the latter is known, among other things, for the (quantum) Yang–Baxter equation.

probability theory due to Frank Spitzer (1956), on which we will comment further below. Various proofs of the so-called Spitzer identity had been obtained before. Baxter's approach, however, which is based on the identity that bears his name, succeeded in unveiling the algebraic and combinatorial structures underlying Spitzer's identity.

About 10 years later, Gian-Carlo Rota realized that Spitzer's identity can be deduced using classical results from the theory of symmetric functions (Rota 1969, 1995; Rota and Smith 1972). Indeed, Rota showed that Spitzer's identity is *equivalent* to the Waring identity. Since the algebra of symmetric functions is naturally equipped with a Hopf algebra structure, Rota's observation could have paved the way to the use of Hopf algebra techniques in the theory of Rota–Baxter algebras. However, this does not seem to have been the case: Hopf algebra structures were not put forward in the classical theory of Rota–Baxter algebras and their applications – as far as we know. Here, we will indicate briefly how some classical properties can be given Hopf-theoretic interpretations. We will also build a connection with the results in the first part of this chapter, where Hopf algebra structures emerged in relation to iterated integrals and Chen's rule.

Beside the links with the theory of symmetric functions, Rota and his school advocated the field of algebraic combinatorics, both mathematically and on epistemological grounds – an attitude that can be recognized in the introduction to his 1969 publication:

> The spectacular results in the fluctuation theory of sums of independent random variables [...] have gradually led to the realisation that the nature of the problem, as well as that of the methods of solution, is algebraic and combinatorial [...]. It is the present purpose to carry this algebraisation to the limit: the result we present amounts to a solution of the word problem for Baxter algebras. (Rota 1969)

Another essential component of the mathematical foundation of the theory of Rota–Baxter algebras can be traced back to Pierre Cartier's 1972 paper (Cartier 1972) on the structure of free commutative Rota–Baxter algebras. Similarly to Rota, Cartier considered the word problem, which amounts to explicitly constructing a basis of the free commutative Rota–Baxter algebra. However, he suggested a rather different approach, which led him to the notion of commutative quasi-shuffle product. The latter generalizes Chen's classical shuffle product (Chen 1957, 1971), which was studied in the previous sections of this chapter. As we will see, this extension from shuffles to quasi-shuffles is one of the (many) reasons to work in the axiomatic framework of Rota–Baxter algebra.

Cartier's algebra product appeared independently a few years later in a 1979 paper by Newman and Radford (1979), where the authors endow the free coalgebra over an associative algebra with a Hopf algebra structure. A quarter of a century later, Gaines (1994) introduced a recursive definition of the quasi-shuffle product – similar to Cartier's – in the context of the multiplication of iterated stochastic integrals in the case of Wiener processes. After, Li and Liu (1997) considered Gaines' quasi-shuffle product while studying the algebraic structure of multiple stochastic integrals with respect to Brownian motion and standard Poisson processes. Around the same time as Gaines' work, Robin Hudson and collaborators described a combinatorial formula for the product of iterated quantum stochastic integrals (Hudson and Parthasarathy 1994; Cohen *et al.* 1995). This product was coined sticky-shuffle (Hudson 2012) and is equivalent to Gaines' quasi-shuffle product. In the light of the Hopf algebraic perspective alluded to earlier, we should also mention that Hudson described the sticky-shuffle (or quasi-shuffle) Hopf algebra in Hudson (2009). Eventually, and independently of the aforementioned developments, the notion of the quasi-shuffle product was formalized by Michael Hoffman in Hoffman (2000) by embedding it in a Hopf algebraic framework and detailing its relation with Chen's shuffle product. See also the following references: (Foissy *et al.* 2016; Hoffman and Ihara 2017; Foissy and Patras 2020). We remark in passing that Hoffman's main motivation came from understanding algebraic and combinatorial structures underlying multiple zeta values (MZVs) (Cartier 2001). This rough (and likely incomplete) synopsis of the history of the commutative quasi-shuffle product suggests that Rota was right about the iceberg of algebra and combinatorics that comes with the notion of the commutative shuffle product. Indeed, we will see that the latter is an abelian example of a special case of a larger structure known as noncommutative quasi-shuffle algebra.

The central aim of this section will indeed be to advocate the noncommutative theory of Rota–Baxter algebras. In spite of multiple good reasons, theoretical as well as applied ones (e.g. integration and finite difference calculus with respect to operator algebra valued functions), interest in noncommutative Rota–Baxter algebra has been for a long time rather sporadic, compared to the amount of work that erupted shortly after Baxter's paper came out. This changed only in the last 15 years, starting largely with three papers by Aguiar (2000), Guo and Keigher (2000) and Ebrahimi-Fard (2002), which (re)initiated interest for the algebraic structures and properties of (non-)commutative Rota–Baxter algebra. Since then, the theory has grown in both systematic developments and applications.

The momentum behind this interest stems initially from the use of noncommutative Rota–Baxter algebra in the context of renormalization in perturbative quantum field theory. We recall that the renormalization program allows us to give sense to integrals appearing in perturbative computations in quantum field theory, which are otherwise divergent. The main idea is to identify and then extract and eliminate those parts that cause the divergencies in a coherent way, which is moreover compatible with the underlying physics (Caswell and Kennedy 1982; Collins 1984). The extraction and elimination part, i.e. the renormalization process, is combinatorial in nature. It was essentially understood and described by Bogoliubov and others (better known as the

Bogoliubov–Parasiuk–Hepp–Zimmermann (BPHZ) renormalization method (Collins 1984)) and has been known in physics for a long time[5]. See Bogoliubov and Shirkov (1959) for one of the earliest textbooks on the subject. The theoretical framework underlying renormalization plays a fundamental role in the so-called standard model and the spectacular results in particle physics that followed from it. Let us remark that from a mathematical point of view[6], Bogoliubov's subtraction scheme still carries a veil of mystery.

In the late 1990s, Connes and Kreimer came up with a rather unexpected approach to the BPHZ renormalization method in terms of a combinatorial Hopf algebra of Feynman graphs (Kreimer 1998; Connes and Kreimer 2000). In this approach, the recursive subtraction algorithm of Bogoliubov *et al.* is considered in the context of dimensional regularization together with minimal subtraction. It was understood that the subtraction scheme map has to satisfy a particular set of identities; it was pointed out by C. Brouder that these identities could be subsumed into a single relation, which was henceforth called *multiplicativity constraint* (Kreimer 1999). A few years later, it was realized by Ebrahimi-Fard that the identity proposed by Brouder coincides with the Rota–Baxter identity (of weight $\theta = 1$). Subsequent works clarified the role of the Rota–Baxter relation in the aforementioned work of Connes and Kreimer and allowed us to relate their results to the structural properties of Rota–Baxter algebras, such as Atkinson's decomposition (Atkinson 1963). We refer the reader interested in details to Manchon's article in *La Gazette des mathématiciens* (Manchon 2009), where a concise and elegant account of the algebraic structures involved in the Connes–Kreimer theory, including the Rota–Baxter algebra aspect, is presented.

This chapter is concerned with two phenomena that are of crucial importance. On the one hand, Bogoliubov's recursive algorithm and the algebraic Birkhoff decomposition derived by Connes and Kreimer are an instance of a general result, known as Atkinson's factorization, which is valid in any Rota–Baxter algebra. On the other hand, the work of Connes and Kreimer triggered considerable research into what is nowadays known as the theory of combinatorial Hopf algebras as well as into the theory of pre-Lie algebras. This also leads to a renewed interest in noncommutative Rota–Baxter algebras.

The rest of this chapter is organized as follows. In the next section, we recall the definition of Rota–Baxter algebra together with related algebraic structures of Lie

5. In fact, we remark that Bogoliubov's and collaborators' original article (Bogoliubov and Parasiuk 1957) appeared around the same time as Magnus' (Magnus 1954) and Spitzer's (Spitzer 1956) work.

6. Despite the fact that other approaches to the renormalization problem, like, for instance, the Epstein–Glaser method (Epstein and Glaser 1973) or Wilson's renormalization group (Wilson 1975), allow for both a comprehensive understanding of the origin of divergencies as well as the very nature of the renormalization program.

theoretic type. Then, Atkinson's factorization for Rota–Baxter algebras is presented, followed by the classical, i.e. commutative Spitzer and Bohnenblust–Spitzer, identities. In section 2.4.8, we briefly describe Cartier's and Rota's constructions of free commutative Rota–Baxter algebras. The generalization of Spitzer-type formulas to noncommutative Rota–Baxter algebras is presented in section 2.4.7. In this context, the notion of pre-Lie algebra is central. We indicate how, for noncommutative Rota–Baxter algebras, the theory of noncommutative symmetric functions plays a role analogous to that of symmetric functions for the commutative Rota–Baxter algebras described by Rota.

2.4.2. *Definition and examples*

DEFINITION 2.10.– *A* Rota–Baxter algebra of weight $\theta \in k$ *consists of an associative k-algebra A equipped with a linear operator $R\colon A \to A$ satisfying the* Rota–Baxter relation of weight θ*:*

$$R(x)R(y) = R\big(R(x)y + xR(y) - \theta xy\big) \qquad \forall x, y \in A. \tag{2.15}$$

A Rota–Baxter algebra is said to be commutative if it is commutative as an associative algebra. Note that if R is a Rota–Baxter map of weight θ, then the map $R' := \beta R$ for $\beta \in k$ different from zero is of weight $\beta\theta$. This allows us to rescale the original weight $\theta \neq 0$ to the new $\theta' = -1$ (or $\theta' = +1$). Rota–Baxter maps come in pairs, i.e. the linear map

$$\tilde{R} := \theta Id - R$$

satisfies as well the Rota–Baxter relation of weight θ [2.15]. Here, Id denotes the identity map on A.

LEMMA 2.8.– *Let (A, R) be a Rota–Baxter algebra of weight θ, the pair $(A, \tilde{R})$ is also a weight θ Rota–Baxter algebra.*

We leave it to the reader to check that

$$\begin{aligned} R(x)\tilde{R}(y) &= R\big(x\tilde{R}(y)\big) + \tilde{R}\big(R(x)y\big) \\ \tilde{R}(x)R(y) &= \tilde{R}\big(xR(y)\big) + R\big(\tilde{R}(x)y\big). \end{aligned} \tag{2.16}$$

REMARK 2.13.– *Rota–Baxter algebras do not necessarily have a unit as associative algebras. When they do not, we augment them with a unit as follows. Let B be a Rota–Baxter algebra over a field k, we write $\overline{B}$ for the augmented unital algebra $k \oplus B$ (with augmentation the projection onto k orthogonally to B) and call it the augmented Rota–Baxter algebra (of B). However, we do not* extend in general the operator R to $\overline{B}$, so that R is then defined only on B.

We consider first some examples by starting with the case when the weight θ equals zero and the algebra A is commutative. The most natural interpretation of a commutative weight-zero Rota–Baxter map is given in terms of the integration by parts relation for the indefinite Riemann integral. Indeed, take, for example, A to be the algebra $C^\infty(\mathbb{R})$ of smooth functions on the real line and set $R(f)(t) := \int_0^t f(x)dx$. For two functions $f, g \in C^\infty(\mathbb{R})$, we obtain the weight-zero Rota–Baxter identity:

$$R(f)R(g) = R\big(R(f)g + fR(g)\big). \qquad [2.17]$$

Even in this somewhat degenerate case, the theory is not without interest since it embraces, among other aspects, Chen's commutative iterated integral calculus, Schützenberger's theory of commutative shuffle algebras (Schützenberger 1958/59) as well as a large part of Fliess' approach to control theory (Fliess *et al.* 1983) (related to chronological algebras in the sense of Kawski (1994)). Indeed, for R being the integral operator, we recognize in $R(f_1R(f_2\cdots R(f_n)\cdots))$ the iterated integral of functions $f_1,\dots,f_n$.

We recall that the shuffle product $(f_1,\dots,f_n) \sqcup\!\sqcup (g_1,\dots,g_m)$ of two *sequences* $(f_1,\dots,f_n)$ and $(g_1,\dots,g_m)$ is the multiset of sequences $(h_1,\dots,h_{m+n})$ made of the letters f_i and g_j, where the partial orders of f_i and g_j are preserved. For example, $(f_1,g_1,f_2,g_2,f_3,g_3,f_4)$ or $(f_1,f_2,g_1,g_2,f_3,f_4,g_3)$ are among the shuffles of (f_1,f_2,f_3,f_4) and (g_1,g_2,g_3). However, $(f_1,f_4,f_2,g_1,g_2,f_3,g_3)$ is not, since f_4 appears on the left of f_2 and f_3. For example,

$$\begin{aligned}(f_1,f_2) \sqcup\!\sqcup (g_1,g_2) &= (f_1,f_2,g_1,g_2) + (f_1,g_1,f_2,g_2) + (f_1,g_1,g_2,f_2)\\ &\quad + (g_1,g_2,f_1,f_2) + (g_1,f_1,g_2,f_2) + (g_1,f_1,f_2,g_2).\end{aligned}$$

When letters are repeated, sequences are counted with multiplicities:

$$(f,g) \sqcup\!\sqcup (f) = \{(f,f,g),(f,f,g),(f,g,f)\}.$$

From equation [2.17] follows

LEMMA 2.9.– *In a weight-zero Rota–Baxter algebra, we have:*

$$\begin{aligned}&R\big(f_1R(f_2\cdots R(f_n)\cdots)\big)R\big(g_1R(g_2\cdots R(g_m)\cdots)\big)\\ &= \sum_{\substack{(h_1,\dots,h_{m+n})\in\\ (f_1,\dots,f_n)\sqcup\!\sqcup(g_1,\dots,g_m)}} R\big(h_1R(h_2\cdots R(h_{n+m})\cdots)\big). \qquad [2.18]\end{aligned}$$

Relation [2.18] yields in particular the well-known shuffle product of two iterated integrals due to Chen.

The case of a noncommutative weight-zero Rota–Baxter algebra includes the algebras of chronological calculus (algebras of time-dependent operators stable under a suitable integral map satisfying the integration by parts formula). We will return to this particular case in a specific section below, making explicit the links between the formalism of Rota–Baxter algebras and chronological calculus *à la* Agrachev–Gamkrelidze.

Next, we should address the obvious question: are there are interesting examples of non-zero-weight Rota–Baxter maps? From the point of view of analysis and for the theory of discrete dynamical systems, the fundamental example of non-zero weight Rota–Baxter maps are summation operators.

EXAMPLE 2.1 (Finite summation operators).– *On functions f defined on $\mathbb{N}$ and with values in an associative algebra A, the summation operator $R(f)(n) := \sum_{k=0}^{n-1} f(k)$ is a Rota–Baxter map of weight $\theta = -1$. It is the right inverse of the finite difference operator $\Delta(f)(n) := f(n+1) - f(n)$.*

Another example traces the Rota–Baxter relation back to its origin in classical fluctuation theory. The latter deals with extrema of sequences of real-valued random variables. Their distribution can be studied using operators on random variables such as $X \to \max(0, X)$.

EXAMPLE 2.2 (Fluctuation theory).– *The operator R is defined on the characteristic function $F(t) := \mathbb{E}[\exp(itX)]$ of a real-valued random variable X:*

$$R(F)(t) := \mathbb{E}[\exp(itX^+)],$$

where $X^+ := \max(0, X)$. We can show that R is indeed a Rota–Baxter map of weight $\theta = 1$. This example motivated Spitzer's and hence Baxter's original works (Spitzer 1956; Baxter 1960; Rota 1969; Kung et al. *2009). We will return to it further below.*

EXAMPLE 2.3 (Split algebras).– *Another, more algebraic, example derives from an associative algebra A (not necessarily commutative), which decomposes into a direct sum of two (non-unital) subalgebras, $A = A^+ \oplus A^-$. The two orthogonal projectors $\pi_\pm : A \to A^\pm$ are Rota–Baxter maps of weight $\theta = 1$. In fact, any projector which satisfies relation [2.15] is of weight $\theta = 1$. A rather well-known example for such an algebra are Laurent series $\mathbb{C}[\epsilon^{-1}, \epsilon]]$, which decompose into a "divergent" part $\epsilon^{-1}\mathbb{C}[\epsilon^{-1}]$ and a "regular" part $\mathbb{C}[[\epsilon]]$. Further below we will comment on its role in the modern approach to renormalization in perturbative quantum field theory due to Alain Connes and Dirk Kreimer (Connes and Kreimer 2000).*

EXAMPLE 2.4 (Matrix algebras).– *An interesting class of examples is provided by $n \times n$ matrices together with certain projectors. We denote by π_+ the projector that maps a matrix α to the upper triangular matrix α^+ defined by replacing all entries in α below the main diagonal by zeros. The projector $\pi_- := Id - \pi_+$ maps α to $\pi_-(\alpha) = \alpha^-$, the corresponding strictly lower triangular matrix. It is not hard to see that $\pi_\pm$ satisfies equation* [2.15] *for weight $\theta = 1$. Note that the projector defined* *by mapping* α *to the difference* $\alpha^- - (\alpha^-)^\dagger$, *where* $(\alpha^-)^\dagger$ *is transpose of* α^-, *which satisfies the forthcoming relation* [2.21], *but not equation* [2.15]. *It therefore defines a Lie Rota–Baxter algebra structure (to be defined below) on the* $n \times n$ *matrices – corresponding to the* QR*-factorization of matrices.*

Some of these examples are commutative Rota–Baxter algebras with a non-zero weight. It is interesting to see, as a warm up for the forthcoming developments, how the classical product formulas for iterated integrals transform to take into account the weight.

The product formula in [2.18] holds indeed only in the commutative weight-zero case: it is already wrong when computing $R(x)R(y)$. Indeed, looking closer at the weight θ Rota–Baxter relation [2.15] reveals that it has a shuffle part [2.18] and an additional θ-weight part. In the following example, we separate the parts coming from "pure" shuffling from the additional components that result from the weight θ term using brackets to identify the latter.

$$R(x)R(y) = R(xR(y)) + R(yR(x)) - [\theta R(xy)],$$

$$\begin{aligned} R(x)R(yR(z)) &= R(xR(yR(z))) + R(yR(xR(z))) + R(yR(zR(x))) \\ &\quad - [\theta R(xyR(z))] + [\theta^2 R(yR(xz))]. \end{aligned}$$

Formulas for larger products follow from a recursive application of equation [2.15]. A closed formula follows from the weight-zero [2.18] one by replacing shuffle permutations by quasi-shuffle surjections. A $(m,n;r)$-quasi-shuffle of type $\max(m,n) \leq r < m+n$ is a surjection

$$\sigma : \{1,\ldots,m+n\} \to \{1,\ldots,r\},$$

such that $\sigma(1) < \cdots < \sigma(m)$ and $\sigma(m+1) < \cdots < \sigma(m+n)$. We denote by $\mathrm{Sh}(m,n;r)$ the set of $(m,n;r)$-quasi-shuffles of type $\max(m,n) \leq r \leq m+n$.

LEMMA 2.10.– *The product of iterated Rota–Baxter maps in a commutative weight* θ *Rota–Baxter algebra* A *is given by*

$$\begin{aligned} &R(x_1R(\cdots R(x_m)\cdots))R(x_{m+1}R(\cdots R(x_{m+n})\cdots)) \qquad [2.19] \\ &\quad = \sum_{r=\max(m,n)}^{m+n} (-\theta)^{m+n-r} \sum_{\sigma\in\mathrm{Sh}(m,n;r)} R(y_1^\sigma R(\cdots R(y_r^\sigma)\cdots)) \end{aligned}$$

with $y_i^\sigma := \prod_{j\in\sigma^{-1}(\{i\})} x_i$.

Note that the set $\sigma^{-1}(\{i\})$ contains one or two elements. As in the pure shuffle case, these identities are valid only in the commutative case.

2.4.3. *Related algebraic structures*

What set Rota–Baxter algebras apart are the rich algebraic structures induced by relation [2.15]. In the weight-zero case, we will show that we recover the structures familiar in chronological calculus. In general, however, new structures show up involving quasi-shuffles.

It is natural and useful to extend the notion of the Rota–Baxter operator to more general situations that the commutative or associative case:

DEFINITION 2.11.– *Let A be a vector space equipped with a bilinear product denoted $m_{\bullet} : A \times A \to A$. A linear operator on A is called a Rota–Baxter operator of weight θ if and only if, for any x, y in A*

$$m_{\bullet}(R(x), R(y)) = R(m_{\bullet}(x, R(y)) + m_{\bullet}(R(x), y) - \theta m_{\bullet}(x, y)). \qquad [2.20]$$

When the product defines a given algebraic structure (say of $\mathcal{P}$-algebra) on A (e.g. a Lie algebra), $(A, m_{\bullet}, R)$ is called a Rota–Baxter $\mathcal{P}$-algebra (a Rota–Baxter Lie algebra, for example).

Note that with these conventions, a Rota–Baxter algebra (respectively, a commutative Rota–Baxter algebra) as we defined it is equivalently a Rota–Baxter associative algebra (respectively, a Rota–Baxter commutative and associative algebra). These shortcuts should cause no harm in practice.

EXAMPLE 2.5.– *Let (A, R) be a Rota–Baxter algebra. We verify quickly that on the Lie algebra L_A, defined on A in terms of the usual commutator bracket, the map R satisfies for all elements $x, y \in A$ the Lie analogue of the Rota–Baxter relation of weight θ:*

$$[R(x), R(y)] = R\big([R(x), y] + [x, R(y)] - \theta[x, y]\big), \qquad [2.21]$$

making $(L_A, [-,-], R)$ a Rota–Baxter Lie algebra of weight θ. In other terms, the forgetful functor from associative to Lie algebras (given by the construction of the bracket as an anti-commutator) goes over to a forgetful functor from associative Rota–Baxter to Rota–Baxter Lie algebras. These phenomena have been studied in general in Bai et al. *(2013).*

For an associative Rota–Baxter algebra A of weight θ, the operator $B := R - \tilde{R}$ satisfies the modified Rota–Baxter identity of weight θ:

$$B(x)B(y) = B(B(x)y + xB(y)) - \theta^2 xy. \qquad [2.22]$$

On the level of Lie algebra, the corresponding Lie analogue of equation [2.22] is

$$[B(x), B(y)] = B([B(x), y] + [x, B(y)]) - \theta^2[x, y]. \qquad [2.23]$$

REMARK 2.14.– *Relations* [2.21] *and* [2.23] *are well known in the context of classical integrable systems, where they are called modified classical Yang–Baxter equations (Semenov-Tian-Shansky 1983). Semenov-Tian-Shansky observed that equation* [2.22] *defines an associative analogue of the modified classical Yang–Baxter equation.*

We now introduce one of the key properties of Rota–Baxter algebras.

DEFINITION 2.12 (Rota–Baxter product).– *The argument of the map R on the right-hand side of equation* [2.15] *consists of the sum of three terms; we can show that it defines a new associative product on A. We denote this new product by m_{*_θ} and call it the Rota–Baxter associative product:*

$$m_{*_\theta}(x, y) = x *_\theta y := R(x)y + xR(y) - \theta xy. \qquad [2.24]$$

It is obvious that equation [2.15] implies that $R(x *_\theta y) = R(x)R(y)$ and $\tilde{R}(x *_\theta y) = -\tilde{R}(x)\tilde{R}(y)$. In particular:

LEMMA 2.11.– *The Rota–Baxter map R is a morphism of algebras from $(A, *_\theta)$ to $(A, \cdot)$.*

We remark that in terms of the modified Rota–Baxter map $B := R - \tilde{R}$, we find that $x *_\theta y = \frac{1}{2}(B(x)y + xB(y))$ and $[\![x, y]\!]_\theta = \frac{1}{2}([B(x), y] + [x, B(y)])$.

EXAMPLE 2.6 (Commutative shuffles).– *Consider a weight-zero commutative Rota–Baxter algebra and its Rota–Baxter product $*_0 = *$ defined in equation* [2.24]. *We can look at it as a splitting of an associative and commutative product into two components (the so-called left and right half-shuffle products):*

$$f * g = f \succ g + f \prec g$$

where

$$f \prec g := fR(g), \quad f \succ g := R(f)g. \qquad [2.25]$$

The two abstract products, $\prec$ and $\succ$, satisfy the axiomatic characterization of commutative shuffle algebras that appeared independently in algebraic topology and in the Lie theory in the 1950s (Eilenberg and MacLane 1953; Schützenberger 1958/59):

$$a \prec b = b \succ a, \qquad (a \prec b) \prec c = a \prec (b \prec c + c \prec b). \qquad [2.26]$$

These two relations are actually enough to insure that the product defined by the sum $\prec + \succ$ is associative as well as commutative (Schützenberger 1958/59). Note that knowing one of the two products $\prec, \succ$ is enough since they determine each other.

Schützenberger computed also the free commutative shuffle algebra over an alphabet X (the algebra generated by X and a product $\prec$ such that setting $x \succ y := y \prec x$, the two products satisfy the above relations). He showed that this algebra is (up to isomorphism) the tensor algebra over X, where, using a word notation for tensor products:

$$x_1 \cdots x_n \prec y_1 \cdots y_m := x_1(x_2 \cdots x_n \prec y_1 \cdots y_m + y_1 \cdots y_m \prec x_2 \cdots x_n),$$

so that the product $*$ identifies with the usual shuffle product of words. In control theory, this algebra would (sometimes) be called the free chronological algebra over X (in the sense of Kawski whose terminology differs from that used by Agrachev–Gamkrelidze): this is one of the (many) relations between Rota–Baxter algebras and chronological calculus. In the theory of operads, this algebra is very often called the free Zinbiel algebra over X (Loday and Vallette 2012), a tribute to Cuvier's dual notion of Leibniz algebras (Cuvier 1991, 1994).

EXAMPLE 2.7 (Commutative quasi-shuffles).– *Let us now assume that the weight $\theta = 1$. The idea is similar to the weight-zero case (i.e. commutative shuffles). In the product [2.24], we then have three terms:*

$$x *_1 y = R(x)y + xR(y) - xy.$$

The first two terms correspond to the aforementioned "half-shuffles". The last part, xy, is just the original commutative and associative algebra product. Using the notations $x \prec y := xR(y)$, $x \succ y := R(x)y$ and $x \cdot y := xy$ yields:

$$x \succ y = y \prec x, \qquad (x \prec y) \prec z = x \prec (y \prec z + z \prec y - y \cdot z). \qquad [2.27]$$

This is the commutative quasi-shuffle product (Hoffman 2000). When compared to the ordinary shuffle product (of weight zero), the last term, $y \cdot z$ may be considered a deformation of the shuffle product. We recall that the quasi-shuffle product is very present in the context of the theory of multiple zeta values (MZVs) and other generalizations of special functions (see Cartier 2001).

EXAMPLE 2.8.– *In a weight-zero noncommutative Rota–Baxter algebra A, we no longer have $a \prec b = b \succ a$, but since*

$$aR(b *_0 c) = a(R(b)R(c)) = (aR(b))R(c),$$
$$(R(a)b)R(c) = R(a)bR(c) = R(a)(bR(c)),$$
$$R(a)(R(b)c) = R(a)R(b)c = (R(a)R(b))c = R(a *_0 b)c,$$

the maps $a \longmapsto a \prec b := aR(b)$ and $a \longmapsto a \succ b := R(a)b$ define two commuting right and left representations of the algebra $(A, *_0)$ on itself. That is, the half-shuffles *satisfy the system of axioms, which characterizes the notion of noncommutative shuffles (Aguiar 2000; Manchon 2009) (see also Definition 2.5).*

PROPOSITION 2.4 (Noncommutative shuffles).– *In a weight-zero Rota–Baxter algebra A, setting $a \prec b := aR(b)$, $a \succ b := R(a)b$, the following three shuffle relations are satisfied:*

$$\begin{aligned} (a \prec b) \prec c &= a \prec (b \prec c + b \succ c) \\ a \succ (b \prec c) &= (a \succ b) \prec c \\ a \succ (b \succ c) &= (a \prec b + a \succ b) \succ c, \end{aligned} \quad [2.28]$$

making A a shuffle algebra. The shuffle product $\sqcup\!\sqcup = \succ + \prec$ is the Rota–Baxter product, i.e. $\sqcup\!\sqcup = *_0$.

REMARK 2.15.– *In a weight $\theta = 1$ noncommutative Rota–Baxter algebra A, we still have*

$$\begin{aligned} aR(b *_1 c) &= a(R(b)R(c)) = (aR(b))R(c), \\ (R(a)b)R(c) &= R(a)bR(c) = R(a)(bR(c)), \\ R(a)(R(b)c) &= R(a)R(b)c = (R(a)R(b))c = R(a *_1 b)c, \end{aligned}$$

and the maps $a \longmapsto aR(b)$, $a \longmapsto R(b)a$ define two commuting right and left representations of the algebra $(A, *_1)$ on itself. However, the two half-shuffles $a \prec b := aR(b)$ and $a \succ b := R(a)b$ do not add up to the $*_1$ product, so that A is not a shuffle algebra.

The previous identities imply that the half-shuffles satisfy the system of axioms that characterizes noncommutative quasi-shuffle algebras (also called tridendriform algebras in the literature). These relations seem to have appeared first in stochastic analysis, i.e. in the work of Karandikar on Itô calculus for semimartingales (Karandikar 1982); see also (Curry *et al.* 2019).

THEOREM 2.7 (Noncommutative quasi-shuffles).– *Setting $* := *_1$ for the Rota–Baxter product, and writing $a \cdot b := ab$ for the usual associative product in A (so that $* := \prec + \succ - \cdot$), the left and right half-shuffles, $a \prec b := aR(b)$ and $a \succ b := R(a)b$, satisfy the following identities defining a (noncommutative) quasi-shuffle algebra*

$$\begin{array}{ll} (a \prec b) \prec c = a \prec (b * c), & (a \succ b) \cdot c = a \succ (b \cdot c) \\ a \succ (b \succ c) = (a * b) \succ c, & (a \prec b) \cdot c = a \cdot (b \succ c) \\ (a \succ b) \prec c = a \succ (b \prec c), & (a \cdot b) \prec c = a \cdot (b \prec c). \end{array} \quad [2.29]$$

PROPOSITION 2.5 (Ebrahimi-Fard 2002).– *Let (A, R) be a Rota–Baxter algebra of weight $\theta = 1$. Define $a \prec b := aR(b) - ab = -a\tilde{R}(b)$ and $a \succ b := R(a)b$. Then, $(A, \prec, \succ)$ satisfies the shuffle algebra axioms [2.28].*

One checks easily that R is still a Rota–Baxter operator for the new product $*_\theta$, i.e. the Rota–Baxter relation still holds when we replace the initial product on A by the product $*_\theta$:

LEMMA 2.12.– *Let $(A, \cdot, R)$ be a Rota–Baxter algebra of weight θ, the algebra $(A, *_\theta, R)$ is again a weight θ Rota–Baxter algebra, denoted A^θ.*

Moreover, in light of equation [2.21], we verify that:

LEMMA 2.13.– *The product*

$$[\![x, y]\!]_\theta := [R(x), y] + [x, R(y)] - \theta[x, y] \qquad [2.30]$$

defines another Lie bracket on the Lie algebra $(L_A, [-,-])$. The corresponding Lie algebra is denoted $(L^\theta_A, [\![-,-]\!]_\theta)$.

Analogously to the associative case, we have that R becomes a Lie morphism from L^θ_A to L_A, i.e. $R([\![x, y]\!]_\theta) = [R(x), R(y)]$.

REMARK 2.16.– *Hence, a Rota–Baxter algebra is a vector space with two associative products (respectively, Lie brackets) related by equation* [2.24] *(respectively,* [2.30]*) and morphisms of algebras. This leads to another, equivalent, characterization of Rota–Baxter algebras since the fact that R is a map of algebras from A^θ to A is precisely the Rota–Baxter relation, from which properties of the two Lie brackets follow.*

Equations [2.24] and [2.30] relate two associative products, respectively, two Lie brackets. Saying this, we may wonder whether the terms on the right-hand side of both identities, i.e. $R(x)y$ and $xR(y)$, respectively, $[R(x), y]$ and $[x, R(y)]$, possess more properties. Regarding the former, we have seen in Theorem 2.4 and Theorem 2.7 that they define shuffle (weight zero), respectively, quasi-shuffle algebras (non-zero weight). Let us now consider the Lie algebra case [2.30], i.e. $\theta \neq 0$. It turns out that for $x, y \in A$, the bilinear product

$$x \rhd y := [R(x), y]$$

does not satisfy any relation, meaning that it is magmatic in nature. However, in Bai *et al.* (2013), it was shown that on the Lie algebra $L_A(\theta)$ with Lie bracket $[x, y]^\theta := -\theta[x, y]$, the following relations are satisfied

$$x \rhd [y, z]^\theta = [x \rhd y, z]^\theta + [y, x \rhd z]^\theta \qquad [2.31]$$

$$[x, y]^\theta \rhd z = \mathrm{a}_\rhd(x, y, z) - \mathrm{a}_\rhd(y, x, z), \qquad [2.32]$$

where the associator

$$\mathrm{a}_\rhd(x, y, z) := x \rhd (y \rhd z) - (x \rhd y) \rhd z.$$

Identity [2.31] follows immediately from the Jacobi identity. A quick computation yields equation [2.32]:

$$\begin{aligned}
[x,y]^\theta \rhd z &= -\theta[R([x,y]),z]\\
&= -[R([R(x),y]),z] - [R([x,R(y)]),z] + [[R(x),R(y)],z]\\
&= [R(x),[R(y),z]] - [R([R(x),y]),z] - [R(y),[R(x),z]] + [R([R(y),x]),z]\\
&= x \rhd (y \rhd z) - (x \rhd y) \rhd z - y \rhd (x \rhd z) + (y \rhd x) \rhd z.
\end{aligned}$$

Relations [2.31] and [2.32] yield what is called a post-Lie algebra structure on $L_A(\theta)$.

DEFINITION 2.13 (Post-Lie algebra).– *Let* $(\mathfrak{g},[-,-])$ *be a Lie algebra. A* post-Lie algebra *structure on* $\mathfrak{g}$ *is given in terms of a triple* $(\mathfrak{g},[-,-],\rhd)$ *with the operator* $\rhd : \mathfrak{g}\otimes\mathfrak{g} \to \mathfrak{g}$ *satisfying the post-Lie relations*

$$x \rhd [y,z] = [x \rhd y, z] + [y, x \rhd z] \qquad [2.33]$$

$$[x,y] \rhd z = \mathrm{a}_\rhd(x,y,z) - \mathrm{a}_\rhd(y,x,z). \qquad [2.34]$$

For any post-Lie algebra, we can show that the bracket

$$[\![x,y]\!] := x \rhd y - y \rhd x - [x,y]$$

defines a new Lie bracket on $\mathfrak{g}$ and the corresponding Lie algebra is denoted $(\mathfrak{g},[\![-,-]\!])$. Hence, in a (Lie) Rota–Baxter algebra, identity [2.30] is an example of such a new Lie bracket. We refer the reader to Ebrahimi-Fard and Mencattini (2018, and references therein) for more details on the notion of post-Lie algebras and its relevance in other fields, for example, geometric numerical integration on manifolds and classical integrable systems.

COROLLARY 2.7 (Pre-Lie algebra).– *In the case of an abelian Lie algebra, the post-Lie algebra* $(\mathfrak{g},[-,-],\rhd)$ *reduces to a left pre-Lie algebra* $(\mathfrak{g},\rhd)$ *with left pre-Lie identity:*

$$0 = (x \rhd y) \rhd z - x \rhd (y \rhd z) - (y \rhd x) \rhd z + y \rhd (x \rhd z). \qquad [2.35]$$

As any pre-Lie algebra is a Lie algebra, the bracket $[\![x,y]\!] := x \rhd y - y \rhd x$ *defines a Lie algebra* $(\mathfrak{g},[\![-,-]\!])$.

Note that in the weight $\theta = 0$ case, we have that the Lie bracket $[-,-]^\theta = 0$.

COROLLARY 2.8.– *The product* $x \rhd y := [R(x),y]$ *defines a left pre-Lie algebra on any weight-zero Rota–Baxter algebra* (A,R). *The pre-Lie relation* [2.35] *implies that the commutator*

$$[x,y]_0 := x \rhd y - y \rhd x$$

defines a Lie bracket on A.

In general, any noncommutative Rota–Baxter algebra comes naturally with a post-Lie algebra structure. It reduces to a pre-Lie algebra one when the weight is zero. However, the following result shows that the picture is more involved. Returning to Proposition 2.5, we observe:

PROPOSITION 2.6.– *Let (A, R) be an associative Rota–Baxter algebra of weight θ. For all $x, y \in A$, the product*

$$x \bullet_\theta y := R(x)y + y\tilde{R}(x) \qquad [2.36]$$

satisfies the left pre-Lie identity [2.35].

We leave the proof as a simple exercise. Note that $x \bullet_\theta y = [R(x), y] + \theta yx$ and that this pre-Lie product gives the Lie bracket in equation [2.30]. In fact, both left and right pre-Lie products can be defined, since $x\,{}_\theta\!\bullet y := -y \bullet_\theta x$ is right pre-Lie. For $\theta = 0$, we have $x \bullet_0 y = [R(x), y] := R(x)y - yR(x)$. Obviously, in the commutative case, the pre-Lie product in equation [2.36] simplifies to the original product of the algebra (up to a sign and scaling by the weight θ). Also, note that in a proper Lie Rota–Baxter algebra $(\mathfrak{g}, R)$ the statement of Proposition 2.6 is not available.

2.4.4. *Atkinson's factorization and Bogoliubov's recursion*

We turn now to typical problems in the theory of Rota–Baxter algebras, closely related to the behavior of time-ordered exponentials – we recall that when dealing with weight-zero Rota–Baxter algebras, the latter are the sums of iterated integrals that we investigated at the beginning of this chapter.

Let A be a unital Rota–Baxter algebra of weight θ. Both Atkinson (1963) and Baxter (1960) studied the following fixpoint equations for $x \in A$:

$$\begin{aligned} \overrightarrow{\exp}_R(\lambda x) &= 1 + \lambda R\big(\overrightarrow{\exp}_R(\lambda x)\, x\big), \\ \overleftarrow{\exp}_{\tilde{R}}(\lambda x) &= 1 + \lambda \tilde{R}\big(x\, \overleftarrow{\exp}_{\tilde{R}}(\lambda x)\big). \end{aligned} \qquad [2.37]$$

Note that setting $\lambda = 1$, taking for R the indefinite Riemann integral (a weight-zero Rota–Baxter operator) and replacing x by $H(t)$, a time-dependent matrix (with, for example, bounded C^∞ coefficients), we get by a perturbative expansion for $\overrightarrow{\exp}_R(\lambda x)$ the time-ordered exponential associated with the differential equation

$$\dot{M}(t) - M(t)H(t).$$

The formal parameter λ is introduced to circumvent any discussions regarding convergence or invertibility issues. Identity [2.16] implies the central result in $A[[\lambda]]$:

$$\overrightarrow{\exp}_R(\lambda x)\overleftarrow{\exp}_{\tilde{R}}(\lambda x) = 1 + \lambda\theta\, \overrightarrow{\exp}_R(\lambda x)\, x\, \overleftarrow{\exp}_{\tilde{R}}(\lambda x),$$

which we check directly

$$\begin{aligned}
\overrightarrow{\exp}_R(\lambda x)\overleftarrow{\exp}_{\tilde{R}}(\lambda x) &= \big(1+\lambda R(\overrightarrow{\exp}_R(\lambda x)\,x)\big)\big(1+\lambda\tilde{R}(x\,\overleftarrow{\exp}_{\tilde{R}}(\lambda x))\big)\\
&= 1+\lambda R(\overrightarrow{\exp}_R(\lambda x)\,x)+\lambda\tilde{R}(x\,\overleftarrow{\exp}_{\tilde{R}}(\lambda x))+\lambda^2 R(\overrightarrow{\exp}_R(\lambda x)\,x)\,\tilde{R}(x\,\overleftarrow{\exp}_{\tilde{R}}(\lambda x))\\
&= 1+\lambda R(\overrightarrow{\exp}_R(\lambda x)\,x)+\lambda\tilde{R}(x\,\overleftarrow{\exp}_{\tilde{R}}(\lambda x))\\
&\quad +\lambda^2 R(\overrightarrow{\exp}_R(\lambda x)\,x\,\tilde{R}(x\,\overleftarrow{\exp}_{\tilde{R}}(\lambda x)))+\lambda^2\tilde{R}(R(\overrightarrow{\exp}_R(\lambda x)\,x)\,x\,\overleftarrow{\exp}_{\tilde{R}}(\lambda x))\\
&= 1+\lambda R(\overrightarrow{\exp}_R(\lambda x)\,x\,\overleftarrow{\exp}_{\tilde{R}}(\lambda x))+\lambda\tilde{R}(\overrightarrow{\exp}_R(\lambda x)\,x\,\overleftarrow{\exp}_{\tilde{R}}(\lambda x))\\
&= 1+\lambda\theta\,\overrightarrow{\exp}_R(\lambda x)\,x\,\overleftarrow{\exp}_{\tilde{R}}(\lambda x).
\end{aligned}$$

This then yields Atkinson's factorization:

$$1-\lambda\theta x = \overrightarrow{\exp}_R(\lambda x)^{-1}\overleftarrow{\exp}_{\tilde{R}}(\lambda x)^{-1}. \tag{2.38}$$

Moreover, note that the inverses are given through the following formulas:

$$\overrightarrow{\exp}_R(\lambda x)^{-1} = 1-\lambda R(x\,\overleftarrow{\exp}_{\tilde{R}}(\lambda x))$$

and

$$\overleftarrow{\exp}_{\tilde{R}}(\lambda x)^{-1} = 1-\lambda\tilde{R}(\overrightarrow{\exp}_R(\lambda x)\,x).$$

Let us check the first statement:

$$\begin{aligned}
\overrightarrow{\exp}_R(\lambda x)\big(1-\lambda R(x\,\overleftarrow{\exp}_{\tilde{R}}(\lambda x))\big) &= 1+\lambda R(\overrightarrow{\exp}_R(\lambda x)\,x)-\lambda R(x\,\overleftarrow{\exp}_{\tilde{R}}(\lambda x))\\
&\qquad -\lambda^2 R(\overrightarrow{\exp}_R(\lambda x)\,x)\,R(x\,\overleftarrow{\exp}_{\tilde{R}}(\lambda x))\\
&= 1+\lambda R(\overrightarrow{\exp}_R(\lambda x)\,x)-\lambda R(x\,\overleftarrow{\exp}_{\tilde{R}}(\lambda x))-\lambda^2 R(\overrightarrow{\exp}_R(\lambda x)\,x\,R(x\,\overleftarrow{\exp}_{\tilde{R}}(\lambda x)))\\
&\quad -\lambda^2 R(R(\overrightarrow{\exp}_R(\lambda x)\,x)\,x\,\overleftarrow{\exp}_{\tilde{R}}(\lambda x))+\lambda^2\theta R(\overrightarrow{\exp}_R(\lambda x)\,x^2\,\overleftarrow{\exp}_{\tilde{R}}(\lambda x))\\
&= 1+\lambda R(\overrightarrow{\exp}_R(\lambda x)\,x)-\lambda R(\overrightarrow{\exp}_R(\lambda x)\,x\,\overleftarrow{\exp}_{\tilde{R}}(\lambda x))\\
&\quad +\lambda^2 R(\overrightarrow{\exp}_R(\lambda x)\,x\,\tilde{R}(x\,\overleftarrow{\exp}_{\tilde{R}}(\lambda x)))\\
&= 1-\lambda R(\overrightarrow{\exp}_R(\lambda x)\,x\,\overleftarrow{\exp}_{\tilde{R}}(\lambda x))+\lambda R(\overrightarrow{\exp}_R(\lambda x)\,x\,\overleftarrow{\exp}_{\tilde{R}}(\lambda x))\\
&= 1.
\end{aligned}$$

The other identities follow by similar computations

Bogoliubov's recursion was devised completely independently of the theory of Rota–Baxter algebras as a way to compute the so-called counter-term and renormalized amplitudes in perturbative quantum field theory (pQFT). From the work of Connes and Kreimer, it follows that it can be rephrased in terms of Rota–Baxter algebras – see, for example, Ebrahimi-Fard *et al.* (2007a) for an account of the relations between the theory of renormalization and Rota–Baxter algebra, Hopf algebras as well as group and Lie-theoretical structures.

Let us assume, to fit with pQFT, that $\theta = 1$ and that R is an idempotent operator ($R^2 = R$) such that $R(1) = 0$. We also assume that the Rota–Baxter algebra is graded as an algebra, with R of degree 0 as a linear endomorphism and $x = x_1 + \cdots + x_n + \cdots$ and x_i of degree i. The standard example of such a situation, relevant for renormalization, is when A is an algebra of linear forms on a graded connected coassociative coalgebra with values in Laurent series and R is the projection onto the regular part (the product of linear forms being the convolution product induced by the coalgebra coproduct).

We also omit the parameter λ and set $f := \overrightarrow{\exp}_R(x)$, $h := \overleftarrow{\exp}_{\tilde{R}}(x)$. Bogoliubov's formulas compute the inverse h^{-1} as well as f (the latter is called "counter-term" in renormalization). We sketch it briefly. Note first that from $R^2 = R$, we get $-R\tilde{R} = R(R - Id) = 0 = (R - Id)R = -\tilde{R}R$. So that from $f = 1 + R(fx)$, we obtain $\tilde{R}(f) = 1$. From $f(1 - x) = h^{-1}$, we get:

$$h^{-1} = -fx + 1 + R(fx) = 1 - \tilde{R}(fx).$$

LEMMA 2.14 (Bogoliubov recursion).– *The two resulting coupled equations*

$$h^{-1} = 1 - \tilde{R}(fx), \qquad f = 1 + R(fx)$$

can then be solved recursively since, for $n > 0$,

$$h_n^{-1} = -\tilde{R}(x_n) - \sum_{0<i<n} \tilde{R}(f_i x_{n-i}), \qquad f_n = R(x_n) + \sum_{0<i<n} R(f_i x_{n-i}),$$

where we denote with an index i the i-th term of each series.

In low degrees, we get $h_1^{-1} = -\tilde{R}(x_1)$, $f_1 = R(x_1)$, $h_2^{-1} = -\tilde{R}(f_1 x_1 + x_2)$, $f_2 = R(f_1 x_1 + x_2)$ and so on.

2.4.5. *Spitzer's identity: commutative case*

The interest in Rota–Baxter algebras results from the possibility to describe interesting and useful universal combinatorial identities that hold in a wide variety of contexts. The most famous example of such a universal combinatorial result is known as Spitzer's classical identity (Spitzer 1956).

We have seen that the indefinite Riemann integral operator

$$R(f)(t) := \int_0^t f(s)ds,$$

is a natural weight-zero Rota–Baxter map, say, on scalar-valued functions. We recall that the solution to the integral fixpoint equation

$$y(t) = 1 + R(fy)(t)$$

is given in terms of the exponential function

$$y(t) = \exp\big(R(f)(t)\big).$$

Going back to Atkinson's factorization in terms of the fixpoint equations in equation [2.37], we may wonder how the simple exponential solution in the particular weight-zero case generalizes to Rota–Baxter maps of arbitrary weight? Baxter (1960) understood that Spitzer's classical identity (Spitzer 1956) provides the answer in the case of commutative Rota–Baxter algebra of arbitrary weight.

Spitzer's original result provides a way to compute the characteristic function of the extrema of a discrete process $S_i := X_1 + \cdots + X_i$, defined as a sequence of partial sums of a sequence of independent and identically distributed real-valued random variables X_i. More precisely, considering the new sequence of random variables $Y_i := \sup(0, S_1, S_2, \dots, S_i)$, Spitzer's formula allows us to calculate the characteristic function of Y_i in terms of the positive part of the partial sums, denoted S_j^+, $j \leq i$. Let $F := \mathbb{E}[\exp(itX)]$ be the characteristic function of $X := X_1$ and $R(F) := \mathbb{E}[\exp(itX^+)]$, where $X^+ := Sup(0, X)$ (and similarly for other random variables as well as on the algebra generated by their characteristic functions). In the setting of Rota–Baxter algebra, we then find:

$$\mathbb{E}[\exp(itY_k)] = R(F \cdot R(F \cdot \cdots R(F \cdot R(F)) \cdots)) =: R^{(k)}(F),$$

and

$$\mathbb{E}[\exp(itS_k^+)] = R(F^k).$$

From this, we deduce a formal identity, which is true in every commutative Rota–Baxter algebra of any weight θ, namely:

PROPOSITION 2.7 (Spitzer identity).– *In a weight θ commutative Rota–Baxter algebra, we have (formally)*

$$\log\Big(1 + \sum_{n>0} R^{(m)}(F)\Big) = R\Big(\sum_{n>0} \frac{\theta^{n-1}F^n}{n}\Big). \qquad [2.39]$$

We note that the argument of the Rota–Baxter map R on the right-hand side of equation [2.39] is a logarithm

$$\Omega_0'(F) := \sum_{n>0} \frac{\theta^{n-1}F^n}{n} = -\theta^{-1}\log(1 - \theta F).$$

Since $F = \theta^{-1}(\exp(\theta\Omega'_\theta(F)) - 1)$, we can express $\Omega'_\theta(F)$ in terms of the generating function of the Bernoulli numbers $\frac{x}{e^x-1}$:

$$\Omega'_\theta(F) = \frac{\ell_{\theta\Omega'_\theta(F)}}{e^{\ell_{\theta\Omega'_\theta(F)}} - 1}(F) = F + \sum_{n>0} \frac{B_n}{n!}\ell^n_{-\theta\Omega'_\theta(F)}(F). \qquad [2.40]$$

Here, ℓ is the left multiplication operator, $\ell_x(y) := xy$ and B_n are the Bernoulli numbers.

Going back to equation [2.37] and using that $1 + \sum_{n>0} R^{(m)}(\lambda x)$ is the perturbative expansion of the solution of the fixpoint equation $\overrightarrow{\exp}_R(\lambda x) = 1 + \lambda R(\overrightarrow{\exp}_R(\lambda x)x)$, this result implies for a general commutative Rota–Baxter algebra (A, R) that

$$\overrightarrow{\exp}_R(\lambda x) = \exp\big(R(\Omega'_\theta(\lambda x))\big) \qquad \overrightarrow{\exp}_R(\lambda x)^{-1} = \exp\big(-R(\Omega'_\theta(\lambda x))\big),$$

and similarly for $\overleftarrow{\exp}_{\tilde{R}}(\lambda x)$. In the commutative case, Atkinson's factorization then becomes an obvious statement

$$\begin{aligned} 1 - \lambda\theta x &= \exp\big(-R(\Omega'_\theta(\lambda x))\big)\exp\big(-\tilde{R}(\Omega'_\theta(\lambda x))\big) && [2.41]\\ &= \exp\big(-R(\Omega'_\theta(\lambda x)) - \tilde{R}(\Omega'_\theta(\lambda x))\big)\\ &= \exp\big(-\theta\Omega'_\theta(\lambda x)\big)\\ &= \exp\big(\log(1-\lambda\theta x)\big). && [2.42] \end{aligned}$$

Expression [2.40] will remind the reader of Magnus' classical formula for the logarithm of the fundamental solution of a first-order matrix or operator differential equation [2.14]. See Blanes *et al.* (2009) for a comprehensive review on the Magnus expansion and applications. Although the expression appears to be somewhat pointless in the commutative case, it will allow later for a straightforward generalization of the above identity to noncommutative Rota–Baxter algebras.

Deeper into the combinatorial nature of commutative Rota–Baxter algebras goes the following proposition.

PROPOSITION 2.8 (Bohnenblust–Spitzer formula).– *In a commutative weight θ Rota–Baxter algebra, we have the following identity*

$$\sum_{\sigma\in S_n} R(\cdots R(F_{\sigma(1)})F_{\sigma(2)}\cdots)F_{\sigma(n)} = \sum_{\pi\in\mathcal{P}_n} \theta^{n-|\pi|}\prod_{\pi_i\in\pi}^{*_\theta}(b_i-1)!\Big(\prod_{j\in\pi_i}F_j\Big),$$

where S_n is the symmetric group of order n. On the right-hand side, $\mathcal{P}_n$ denotes set partitions of $[n] := \{1,\ldots,n\}$ and $|\pi|$ is the number of blocks of the partition $\pi\in\mathcal{P}_n$. The size of the ith block $\pi_i \in \pi$ is denoted b_i.

Note that the product $\prod^{*\theta}$ is defined in terms of the Rota–Baxter product in [2.24]. For instance

$$\sum_{\sigma\in S_2} R(F_{\sigma(1)})F_{\sigma(2)} = R(F_1)F_2 + R(F_2)F_1 = \theta F_1F_2 + F_1 *_\theta F_2,$$

which is the weight θ Rota–Baxter identity up to applying R on both sides. In the weight-zero case, i.e. for $\theta = 0$, the Bohnenblust–Spitzer formula simplifies to the obvious identity:

$$\sum_{\sigma\in S_n} R(\cdots R(R(F_{\sigma(1)})F_{\sigma(2)})\cdots)F_{\sigma(n)} = \prod_{i=1}^{n}{}^{*_0} F_i.$$

With the goal to generalize the Bohnenblust–Spitzer identity in Proposition 2.8 to noncommutative Rota–Baxter algebras, we again propose a slightly more involved rewriting. We recall the canonical cycle decomposition $c_1\cdots c_k$ of a permutation $\sigma\in S_n$. Each cycle is written starting with its maximal element, and the cycles are written in increasing order of their first entries. For example, $(32)(541)(6)(87)$ is a canonical cycle decomposition. The jth cycle is denoted $c_j = (a_{j_0}a_{j_1}\cdots a_{j_{m_j-1}})$, where m_j is the size of this cycle and $a_{j_0} > a_l$, $j_1 \le l \le j_{m_j-1}$.

The identity in Proposition 2.8 can then be restated as follows:

$$\sum_{\sigma\in S_n} R(\cdots R(R(F_{\sigma(1)})F_{\sigma(2)})\ldots)F_{\sigma(n)} = \sum_{\sigma\in S_n} \mathfrak{D}^\theta_\sigma(F_1,\ldots,F_n), \qquad [2.43]$$

where for each permutation σ written in its canonical cycle decomposition $c_1\cdots c_k$, we define:

$$\mathfrak{D}^\theta_\sigma(F_1,\ldots,F_n) := \prod_{j=1}^{k}{}^{*\theta}\Big(\big(\prod_{i=1}^{m_j-1}{}^{\circ} r_{\theta F_{a_{j_i}}}\big)(F_{a_{j_0}})\Big). \qquad [2.44]$$

Now the operator r denotes right multiplication, i.e. $r_x(y) := yx$. Note that the second product on the right-hand side is with respect to the composition of these multiplication operators. For instance, the permutation $\sigma := (43)(512)\in S_5$ yields the term:

$$\mathfrak{D}^\theta_\sigma(F_1,\ldots,F_5) = r_{\theta F_3}(F_4) *_\theta (r_{\theta F_2}r_{\theta F_1})(F_5) = \theta^3(F_4F_3) *_\theta (F_5F_1F_2).$$

In the commutative setting, the product $(\prod_{i=1}^{m_j-1} r_{\theta F_{a_{j_i}}})(F_{a_{j_0}})$ is independent of the order of a_{j_i}, $i\ge 1$. Therefore, several of the terms $\mathfrak{D}^\theta_\sigma(F_1,\ldots,F_n)$ on the right-hand side of equation [2.43] may coincide. The resulting coefficients $\prod_{j=1}^{k}(m_j-1)!$ lead to the compact form of the identity in Proposition 2.8 written in terms of set partitions. Indeed, choose a block π_j, say, of size m_j, of a partition π. Its first entry is fixed to be its maximal element. This block then corresponds to $(m_j-1)!$ different cycles of size m_j.

2.4.6. *Free commutative Rota–Baxter algebras*

The construction of free Rota–Baxter algebras was obtained independently by Cartier and Rota. One of the main motivations underlying their works, and especially the one by Rota, was to deduce universal identities such as Spitzer's. Their two approaches are rather complementary. Cartier's construction (Cartier 1972) of a free commutative Rota–Baxter algebra is a forerunner of the modern notion of the quasi-shuffle algebra (Example 2.15) and provides an abstract inductive way to obtain equation [2.19]. Cartier's construction was made more explicit in Guo and Keigher (2000).

We feature here Rota's approach (Rota 1969; Rota and Smith 1972), which naturally has a Hopf algebraic flavor. It is based on the observation that for any commutative algebra A, the algebra $A^{\mathbf{N}}$ of functions from the non-negative integers into A, i.e. sequences of elements of A with pointwise product, is naturally equipped with a weight $\theta = -1$ Rota–Baxter algebra structure defined by the operator:

$$R(a_1, \ldots, a_n, \ldots) = (0, a_1, a_1 + a_2, \ldots, a_1 + \cdots + a_n, \ldots).$$

Taking A to be the algebra $k[x_1, \ldots, x_n, \ldots]$ of polynomials in an infinite number of indeterminates, we can show that the Rota–Baxter subalgebra generated by the sequence $x := (x_1, \ldots, x_n, \ldots)$ is free. This establishes an immediate link to the well-understood classical theory of symmetric functions. In fact, terms such as $R(x^n)$ and $R^{(n)}(x) := R(R^{(n-1)}(x)x)$, $R^{(1)}(x) := R(x)$, $R^{(0)}(x) := 1$, correspond in the theory of symmetric functions to power sums ($h_n := \sum_i x_i^n$) and elementary symmetric functions ($e_n := \sum_{i_1 < \cdots < i_n} x_{i_1} \cdots x_{i_n}$). From this, Rota was able to conclude that Spitzer's identity is equivalent to the Waring formula, which expresses power sums as polynomials in elementary symmetric functions.

The algebra of symmetric functions $\mathcal{S}$ is a free commutative algebra over the e_n (or the h_n). Rota's construction shows that there is a canonical embedding of $\mathcal{S}$ into $RB(x)$, the free Rota–Baxter on one generator. His embedding maps e_n to $R^{(n)}(x)$ and therefore $RB(x)$ contains as a free commutative subalgebra the algebra freely generated by the $R^{(n)}(x)$. Moreover, $\mathcal{S}$ carries naturally a Hopf algebra structure for which the power sums are primitive ($\Delta(h_n) = h_n \otimes 1 + 1 \otimes h_n$) or, equivalently, the sum of all elementary symmetric functions is group-like (as $\Delta(e_n) = \sum_{i=0}^n e_i \otimes e_{n-i}$). This Hopf algebra structure has many nice and important properties surveyed in the work by Macdonald (1998) – for example, it is inherited from induction and restriction among Young subgroups of symmetric groups.

DEFINITION 2.14.– *The commutative Spitzer algebra is the bicommutative Hopf algebra that, as a commutative algebra, is the free commutative algebra generated by $R^{(n)}(x)$ in $RB(x)$. The coproduct is given by*

$$\Delta(R^{(n)}(x)) := \sum_{i=0}^{n} R^{(i)}(x) \otimes R^{(n-i)}(x).$$

Equivalently, the commutative Spitzer algebra is freely generated by $R^{(n)}(x)$ *and the coproduct defined by*

$$\Delta(R(x^n)) := R(x^n) \otimes 1 + 1 \otimes R(x^n).$$

The terminology "Spitzer algebra" was introduced in Ebrahimi-Fard *et al.* (2007b, 2009) in the noncommutative setting; we extend it here to the commutative one.

It is interesting to analyze this Hopf algebra structure from the point of view of the Spitzer formula, which says that:

$$\log\big(1 + \sum_{n>0} R^{(n)}(x)\big) = R\big(- \sum_{n>0} \frac{(-1)^n x^n}{n}\big).$$

It is a general phenomenon in Hopf algebra that the logarithm of a group-like element is primitive (and conversely, the exponential of a primitive element is group-like). It follows therefore from the Spitzer formula (without appealing to the theory of symmetric functions) that the two Hopf algebra structures agree. Moreover, Spitzer's formula gains now a group-theoretical flavor, i.e. $1 + \sum_{n>0} R^{(n)}(x)$ belongs to the group of group-like elements of the commutative Spitzer algebra, whereas $R(\frac{x^n}{n})$ belongs to the associated Lie algebra.

Here, the group is commutative and the Lie algebra structure is trivial (the bracket is null). The interest of these observations is therefore limited. However, these phenomena will be more interesting in the noncommutative case, where they relate, for example, to the Baker–Campbell–Hausdorff and the Magnus formulas.

Let us return briefly to the first part of this article: the elements $R^{(n)}(x)$ in the free Rota–Baxter algebra give rise to iterated integrals when R is interpreted as an integral operator and x as a smooth function. From Chen's rule, we know that it would be natural to define a Hopf algebra structure in this context by requiring

$$\Delta(R^{(n)}(x)) := \sum_{i=0}^{n} R^{(i)}(x) \otimes R^{(n-i)}(x).$$

This is indeed the structure we defined using Rota's embedding of symmetric functions into the free Rota–Baxter algebra. We will deepen this point of view and use it systematically later, when studying noncommutative Rota–Baxter algebras.

2.4.7. *Spitzer's identity: noncommutative case*

Thanks to the particular way we presented the commutative case of Spitzer's identity, we do not have to dwell too much on the details in the noncommutative setting. Instead, we will try to sketch the techniques as well as theoretical ingredients that allow a simple transition from the commutative to the noncommutative Spitzer identity.

The natural question is to understand the expression on the left-hand side of Spitzer's identity, i.e. the logarithm of $\overrightarrow{\exp}_R(\lambda x)$ in equation [2.37]. Note that, depending on the algebra A and Rota–Baxter operator R under consideration, we may either view this solution as the counter-term in renormalization theory or the solution of a first-order linear differential equation in matrix algebra, or the linear fixed-point equation of a discrete dynamical system. In particular, in view of the penultimate example, we would expect a "noncommutative Spitzer formula" to solve, among other things, the Baker–Campbell–Hausdorff problem (to compute the logarithm of the solution of a first-order linear differential equation). We will see that this is indeed the case.

Key in the theory of noncommutative Rota–Baxter algebras is the existence of another algebraic structure beside the Rota–Baxter product $*_\theta$. Indeed, in equations [2.31]–[2.32] and Proposition 2.6, we saw that any associative Rota–Baxter algebra of weight θ inherits both post-Lie and (left) pre-Lie algebra structures.

To understand the role played by the pre-Lie product, we return to the Bohnenblust–Spitzer formula, using the description presented in equation [2.43], which appeared somewhat artificial in the commutative case. For $n = 2$, we see quickly that:

$$F_1 *_\theta F_2 + \theta F_2 F_1 = R(F_1)F_2 + R(F_2)F_1$$

provided that the Rota–Baxter algebra is commutative. In the noncommutative case, the slightly less obvious identity holds:

$$F_1 *_\theta F_2 + F_2 \bullet_\theta F_1 = R(F_1)F_2 + R(F_2)F_1.$$

This simple procedure of replacing the algebra product with the pre-Lie product, $\bullet_\theta$ defined in Proposition 2.6, generalizes to all orders. Indeed, it is "sufficient" to redefine in equation [2.43] the operator $\mathcal{D}^\theta_\sigma$ by substituting in definition [2.44] the operator $r_{\theta x}$ (right product by θx) with the operator $r_{\bullet_\theta x}$, again a right multiplication operator, but defined now in terms of the pre-Lie product, $r_{\bullet_\theta x}(y) := y \bullet_\theta x$.

Using the example following definition [2.44], we substitute in the Bohnenblust–Spitzer formula the expression:

$$\mathcal{D}^\theta_\sigma(F_1, \dots, F_5) - \theta^3(F_4F_3) *_\theta (F_5F_1F_2),$$

with the one defined in terms of the pre-Lie product:

$$\begin{aligned}\mathcal{D}^{\bullet_\theta}_\sigma(F_1, \dots, F_5) &= r_{\bullet_\theta F_3}(F_4) *_\theta r_{\bullet_\theta F_2}(r_{\bullet_\theta F_1}(F_5)) \\ &= (F_4 \bullet_\theta F_3) *_\theta ((F_5 \bullet_\theta F_1) \bullet_\theta F_2).\end{aligned}$$

PROPOSITION 2.9.– *With this redefinition, the Bohnenblust–Spitzer identity [2.43] holds in any noncommutative Rota–Baxter algebra of weight θ (Ebrahimi-Fard* et al. *2009):*

$$\sum_{\sigma\in S_n} R(\cdots R(R(F_{\sigma(1)})F_{\sigma(2)})\cdots)F_{\sigma(n)} = \sum_{\sigma\in S_n} \mathcal{D}_\sigma^{\bullet\theta}(F_1,\ldots,F_n), \qquad [2.45]$$

where for each permutation σ written in its canonical cycle decomposition $c_1\cdots c_k$, we define:

$$\mathcal{D}_\sigma^{\bullet\theta}(F_1,\ldots,F_n) := \prod_{j=1}^{k}{}^{*\theta}\Big(\Big(\prod_{i=1}^{m_j-1}{}^{\circ} r_{\bullet\theta F_{a_{ji}}}\Big)(F_{a_{j0}})\Big). \qquad [2.46]$$

Note that the identity has even a q-analogue (Novelli and Thibon 2009).

To approach the Baker–Campbell–Hausdorff problem generalized to Rota–Baxter algebra, it is useful to recall Spitzer's classical formula and its rewriting in terms of the generating series of the Bernoulli numbers given in equation [2.40], and to compare it to the Magnus classical solution of the problem. We already recalled that in his seminal 1954 paper (Magnus 1954), Magnus considered the (say, matrix-valued) solution $Y(t)$ of the linear differential equation $\dot{Y}(t) = A(t)Y(t)$, $Y(0) = 1$ and showed that the logarithm $\Omega(A)(t) := \log(Y(t))$ is the solution of the differential equation:

$$\dot{\Omega}(A) = \frac{ad_{\Omega(A)}}{e^{ad_{\Omega(A)}} - 1}(A),$$

where ad is the ordinary Lie adjoint action ($ad_x(y) = [x,y] = xy - yx$). We remark that, since $\Omega(A)(0) = 0$, the fact that the indefinite Riemann integral is a weight-zero Rota–Baxter map, implies that:

$$ad_{\Omega(A)}(A) = [\Omega(A), A] = \dot{\Omega}(A) \bullet_0 A = \ell_{\dot{\Omega}(A)\bullet_0}(A),$$

where $\ell_{x\bullet_0}(y) := x \bullet_0 y$.

The generalization to arbitrary Rota–Baxter algebras unifies Spitzer's classical formula with Magnus' expansion thanks to the pre-Lie product:

PROPOSITION 2.10.– *Let A be a Rota–Baxter algebra of weight θ. If $\overrightarrow{\exp}_R(\lambda x)$ is a solution in $A[[\lambda]]$ of the left fixpoint equation in* [2.37]*, then the element $\Omega'(\lambda x)$ defined such that $R(\Omega'(\lambda x)) = \log(\overrightarrow{\exp}_R(\lambda x))$ satisfies (Ebrahimi-Fard and Manchon 2009):*

$$\Omega'(\lambda x) = \frac{\ell_{-\Omega'(\lambda x)\bullet_\theta}}{e^{\ell_{-\Omega'(\lambda x)\bullet_\theta}} - 1}(\lambda x) = \lambda x + \sum_{n>0} \frac{B_n}{n!}\ell^n_{-\Omega'(\lambda x)\bullet_\theta}(\lambda x), \qquad [2.47]$$

where $\ell_{x\bullet_\theta}(y) := x \bullet_\theta y$. This series is called the pre-Lie Magnus expansion.

Here, the prime notation will remind the reader of Magnus' original differential equation. As a remark, we mention that a similar approach applies to Fer's expansion (Ebrahimi-Fard and Manchon 2009). Frédéric Chapoton showed in his work (Chapoton 2009) that [2.47] is of significant interest in the context of the theory of Lie idempotents as well as the theory of operads. Let us make the first few terms of $\Omega'(\lambda x)$ explicit:

$$\Omega'(\lambda x) = \lambda x + \frac{1}{2}\lambda^2 x \bullet_\theta x + \frac{1}{4}\lambda^3 (x \bullet_\theta x) \bullet_\theta x + \frac{1}{12}\lambda^3 x \bullet_\theta (x \bullet_\theta x) +$$
$$- \lambda^4 \frac{1}{8}((x \bullet_\theta x) \bullet_\theta x) \bullet_\theta x - \lambda^4 \frac{1}{24}\Big((x \bullet_\theta (x \bullet_\theta x)) \bullet_\theta x$$
$$+ x \bullet_\theta ((x \bullet_\theta x) \bullet_\theta x) + (x \bullet_\theta x) \bullet_\theta (x \bullet_\theta x)\Big) + \cdots .$$

Note that the pre-Lie identity implies a reduction of the number of terms at order four:

$$-\frac{1}{6}\big((x \bullet_\theta x) \bullet_\theta x\big) \bullet_\theta x - \frac{1}{12} x \bullet_\theta \big((x \bullet_\theta x) \bullet_\theta x\big).$$

We conclude this section with an observation. The attentive reader may wonder about the role of the post-Lie algebra structure shown in equations [2.31]–[2.32]. The post-Lie product $x \rhd y := [R(x), y]$ induced on the Lie Rota–Baxter algebra $\mathfrak{g}$ allows us to solve the Lie bracket flow equation

$$\dot{x}(t) = -x \rhd x, \quad x(0) = x_0.$$

Indeed, by general results of post-Lie algebra, we can show that

$$x(t) = \exp\big(-R(\chi(x_0 t))\big) x_0 \exp\big(R(\chi(x_0 t))\big).$$

Here, $\chi(x_0 t) \in \mathfrak{g}[[t]]$ is the post-Lie Magnus expansion. We refrain from giving more details and refer to Ebrahimi-Fard and Mencattini (2018) and Mencattini *et al.* (2019) and references therein, where the far more intricate notion of post-Lie Magnus expansion and related topics are reviewed in detail.

2.4.8. *Free Rota–Baxter algebras*

We conclude this chapter with the study of free Rota–Baxter algebras, the word problem (find a basis for these algebras) and insights into Hopf algebra structures in relation to the developments in the first part of this chapter. The noncommutative analogue of Cartier's theory of free commutative Rota–Baxter algebras is intimately related to noncommutative analogues of shuffle and quasi-shuffle algebras. A natural way to represent these structures is in terms of graphical, i.e. combinatorial objects, such as planar rooted trees (Aguiar and Moreira 2006; Ebrahimi-Fard and Guo 2008).

However, the description of the noncommutative analogue of Rota's classical solution to the word problem is surprisingly straightforward (Ebrahimi-Fard *et al.* 2007b, 2009). We simply replace in Rota's original work as accounted for earlier in this chapter the algebra of polynomials in an infinite number variables, $k[x_1, \ldots, x_n, \ldots]$, by its associative analogue, i.e. the free associative algebra (or tensor algebra) over the alphabet $X := \{x_1, \ldots, x_n, \ldots\}$. Rota's results continue to hold in this noncommutative setting, i.e. the operator $R(x_1, \ldots, x_n, \ldots) := (0, x_1, x_1+x_2, \ldots, x_1+\cdots+x_n, \ldots)$ still is a Rota–Baxter map, and the Rota–Baxter subalgebra generated by the sequence $x = (x_1, \ldots, x_n, \ldots)$ gives a presentation of the free Rota–Baxter algebra over one generator x.

Besides the fact that this gives a simple answer to the word problem, it generalizes Rota's seminal insight, unveiling the link between free commutative Rota–Baxter algebras and symmetric functions, to one of the fundamental notions in modern algebraic combinatorics, i.e. the theory of noncommutative symmetric functions – the latter has been developed in the last 25 years by J.-Y. Thibon and his collaborators (G. Duchamp, F. Hivert, J.-C. Novelli *inter alia*) in a series of important articles from Gelfand *et al.* (1995) to Duchamp *et al.* (2011).

Indeed, we recall that, if we calculate the terms on the left-hand side of Spitzer's classical identity in the commutative case, i.e. $R^{(n)}(x) = R(R^{(n-1)}(x)x)$, using Rota's presentation of the free Rota–Baxter algebra on one generator, then we find a sequence of elementary symmetric functions $\sum_{0<i_1<\cdots<i_n<k} x_{i_1}\cdots x_{i_n}$ – this is the basis for Rota's proof of the Spitzer formula, as a corollary of Waring's identity. In the noncommutative case, the same observation holds, but this time, noncommutative variables enter the picture. As a result, Florent Hivert's theory of quasi-symmetric functions in noncommutative variables applies. This approach shares the same advantages with Rota's original work. Indeed, it allows us to apply a whole range of results and techniques from the theory of noncommutative symmetric functions.

Let us conclude by briefly explaining how these ideas connect to the notion of descent algebra introduced at the beginning of the chapter. We recall that the construction of the descent algebra was motivated by the combinatorial formulas for products of iterated integrals. It resulted that the descent algebra is a free associative algebra over generators corresponding to iterated integrals over standard simplices. In the Rota–Baxter context, these ideas generalize as follows. First, as in the commutative case, we can show that the $R^{(n)}(x)$ are free generators of a subalgebra of $RB(x)$, the free Rota–Baxter algebra over x (Ebrahimi-Fard *et al.* 2007b, 2009). We can equip this algebra with a Hopf algebra structure (from the point of view of iterated integrals, the coproduct is induced by the Chen rule).

DEFINITION 2.15.– *The Spitzer algebra is the cocommutative Hopf algebra that, as an associative algebra, is the free associative algebra generated by the* $R^{(n)}(x)$ *in* $RB(x)$*. The coproduct is given by*

$$\Delta(R^{(n)}(x)) := \sum_{i=0}^{n} R^{(i)}(x) \otimes R^{(n-i)}(x).$$

Note that the map $1_n \longmapsto R^{(n)}(x)$ *induces an isomorphism between the descent algebra and the Spitzer algebra (as Hopf algebras).*

Equivalently, the Spitzer algebra is freely generated by $R^{(n)}(x)$ *and the coproduct defined by*

$$\Delta(R(x^n)) := R(x^n) \otimes 1 + 1 \otimes R(x^n).$$

This construction allows us to, for example, interpret the pre-Lie Magnus expansion for Rota–Baxter algebras as surveyed in the previous section in group and Lie theoretical terms (since $\overrightarrow{\exp}_R(\lambda x)$ is a group-like element in the Spitzer algebra, and its logarithm a primitive element).

Examples of applications of descent and free Lie algebra techniques to the study of Rota–Baxter algebras can be found in Ebrahimi-Fard *et al.* (2007a, 2007b, 2009).

2.5. References

Agrachev, A. and Gamkrelidze, R. (1981). Chronological algebras and nonstationary vector fields. *J. Sov. Math.*, 17(1), 1650.

Agrachev, A. and Gamkrelidze, R. (1994). The shuffle product and symmetric groups. In *Differential Equations, Dynamical Systems and Control Science*, Elworthy, K.D., Everitt, W.N., Lee, E.B. (eds). Marcel Dekker Inc., New York.

Aguiar, M. (2000). Pre-Poisson algebras. *Lett. Math. Phys.*, 54(4), 263.

Aguiar, M. and Moreira, W. (2006). Combinatorics of the free Baxter algebra. *Electron. J. Comb.*, 13(1), R17.

Arnal, A., Casas, F., Chiralt, C. (2018). A general formula for the Magnus expansion in terms of iterated integrals of right-nested commutators. *J. Phys. Commun.*, 2, 035024.

Atkinson, F.V. (1963). Some aspects of Baxter's functional equation. *J. Math. Anal. Appl.*, 7, 1.

Bai, C., Bellier, O., Guo, L., Ni, X. (2013). Splitting of operations, Manin products, and Rota–Baxter operators. *Int. Math. Res. Notices*, 2013, 485.

Baxter, G. (1960). An analytic problem whose solution follows from a simple algebraic identity. *Pacific J. Math.*, 10, 731.

Blanes, S., Casas, F., Oteo, J., Ros, J. (2009). Magnus expansion: Mathematical study and physical applications. *Phys. Rep.*, 470, 151.

Bogoliubov, N.N. and Parasiuk, O.S. (1957). On the multiplication of causal functions in the quantum theory of fields. *Acta Math.*, 97, 227.

Bogoliubov, N.N. and Shirkov, D. (1959). *Introduction to the Theory of Quantized Fields*. Wiley-Interscience, Hoboken, NJ.

Brouder, C., Mestre, A., Patras, F. (2010). Tree expansion in time-dependent perturbation theory. *J. Math. Phys.*, 51, 072104.

Brouder, C., Duchamp, G., Patras, F., Tóth, G. (2012). The Rayleigh–Schrödinger perturbation series of quasi-degenerate systems. *Int. J. Quantum Chem.*, 112(10), 2256.

Cartier, P. (1972). On the structure of free Baxter algebras. *Adv. Math.*, 9, 253.

Cartier, P. (2001). Fonctions polylogarithmes, nombres polyzêtas et groupes prounipotents. *Séminaire Bourbaki*, 885.

Cartier, P. (2007). A primer of Hopf algebras. *Frontiers in Number Theory, Physics, and Geometry*, 2, 107.

Cartier, P. (2011). Vinberg algebras, Lie groups and combinatorics. *Clay Math. Proc.*, 11, 107.

Caswell, W.E. and Kennedy, A.D. (1982). A simple approach to renormalisation theory. *Phys. Rev. D*, 25, 392.

Chapoton, F. (2009). A rooted-trees q-series lifting a one-parameter family of Lie idempotents. *Algebra Number Theory*, 3, 611.

Chapoton, F. and Livernet, M. (2001). Pre-Lie algebras and the rooted trees operad. *Int. Math. Res. Notices*, 2001, 395.

Chapoton, F. and Patras, F. (2013). Enveloping algebras of pre-Lie algebras, Solomon idempotents and the Magnus formula. *Int. J. Algebra Comput.*, 23(4), 853.

Chen, K. (1957). Integration of paths, geometric invariants and a generalized Baker–Hausdorff formula. *Ann. Math.*, 65, 163.

Chen, K. (1971). Algebras of iterated path integrals and fundamental groups. *Trans. Am. Math. Soc.*, 156, 359.

Cohen, P.B., Eyre, T.W.M., Hudson, R.L. (1995). Higher order Itô product formula and generators of evolutions and flows. *Int. J. Theor. Phys.*, 34, 1.

Collins, J.C. (1984). *Renormalization. Cambridge Monographs in Mathematical Physics*. Cambridge University Press, Cambridge.

Connes, A. and Kreimer, D. (2000). Renormalization in quantum field theory and the Riemann–Hilbert problem I: The Hopf algebra structure of graphs and the main theorem. *Commun. Math. Phys.*, 210, 249.

Curry, C., Ebrahimi-Fard, K., Patras, F. (2019). On non-commutative stochastic exponentials. In *Numerical Mathematics and Advanced Applications*, Radu, F., Kumar, K., Berre, I., Nordbotten, J., Pop, I. (eds). Springer, Basel.

Cuvier, C. (1991). Homologie des algèbres de Leibniz. PhD Thesis, Université de Strasbourg, Strasbourg.

Cuvier, C. (1994). Algèbres de Leibniz : définitions, propriétés. *Ann. Sci. Ecole Norm. S., Sér. 4*, 27(1), 1.

Duchamp, G., Hivert, F., Novelli, J.-C., Thibon, J.-Y. (2011). Non-commutative symmetric functions VII: Free quasi-symmetric functions revisited. *Ann. Comb.*, 15, 655.

Ebrahimi-Fard, K. (2002). Loday-type algebras and the Rota–Baxter relation. *Lett. Math. Phys.*, 61, 139.

Ebrahimi-Fard, K. and Guo, L. (2008). Free Rota–Baxter algebras and rooted trees. *J. Algebra and Its Appl.*, 7, 167.

Ebrahimi-Fard, K. and Manchon, D. (2009). A Magnus- and Fer-type formula in dendriform algebras. *Found. Comput. Math.*, 9, 295.

Ebrahimi-Fard, K. and Mencattini, I. (2018). Post-Lie algebras, factorization theorems and isospectral flows. In *Discrete Mechanics, Geometric Integration and Lie–Butcher Series*, Ebrahimi-Fard, K. and Linán, M.B. (eds). Springer, Basel.

Ebrahimi-Fard, K. and Patras, F. (2014). The pre-Lie structure of the time-ordered exponential. *Lett. Math. Phys.*, 104, 1281.

Ebrahimi-Fard, K., Gracia-Bondía, J.M., Patras, F. (2007a). A Lie theoretic approach to renormalisation. *Commun. Math. Phys.*, 276, 519.

Ebrahimi-Fard, K., Gracia-Bondía, J.M., Patras, F. (2007b). Rota–Baxter algebras and new combinatorial identities. *Lett. Math. Phys.*, 81, 61.

Ebrahimi-Fard, K., Manchon, D., Patras, F. (2009). A non-commutative Bohnenblust–Spitzer identity for Rota–Baxter algebras solves Bogoliubov's counterterm recursion. *J. Non-Commut. Geom.*, 3, 181.

Eilenberg, S. and MacLane, S.M. (1953). On the groups $h(\pi, n)$. *Ann. Math., Second Series*, 58(1), 55.

Epstein, H. and Glaser, V. (1973). The role of locality in perturbation theory. *Ann. Inst. Henri Poincaré, Section A*, XIX(3), 211.

Fauvet, F. and Menous, F. (2017). Ecalle's arborification-coarborification transforms and Connes–Kreimer Hopf algebra. *Ann. Sci. Éc. Norm. Supér.*, 50, 39.

Fliess, M., Lamnabhi, M., Lamnabhi-Lagarrigue, F. (1983). An algebraic approach to nonlinear functional expansions. *IEEE Trans. Circuits Systems CAS*, 30, 554.

Foissy, L. (2007). Bidendriform bialgebras, trees, and free quasi-symmetric functions. *J. Pure Appl. Algebra*, 209(7), 439.

Foissy, L. and Patras, F. (2020). Lie theory for quasi-shuffle bialgebras. In *Periods in Quantum Field Theory and Arithmetic*, Gil, J.B., Ebrahimi-Fard, K., Gangl, H. (eds). Springer, Basel.

Foissy, L., Patras, F., Thibon, J.-Y. (2016). Deformations of shuffles and quasi-shuffles. *Ann. I. Fourier*, 66(1), 209.

Gaines, J. (1994). The algebra of iterated stochastic integrals. *Stochast. Stochast. Rep.*, 49, 169.

Gelfand, I.M., Krob, D., Lascoux, A., Leclerc, B., Retakh, V., Thibon, J.-Y. (1995). Non-commutative symmetric functions. *Adv. Math.*, 112, 218.

Grossman, R. and Larson, R. (1989). Hopf-algebraic structures of families of trees. *J. Algebra*, 126(1), 184.

Gubinelli, M. (2004). Controlling rough paths. *J. Funct. Anal.*, 216(1), 86.

Guo, L. (2012). *An Introduction to Rota–Baxter Algebra*. International Press of Boston, Boston, MA.

Guo, L. and Keigher, W. (2000). Baxter algebras and shuffle products. *Adv. Math.*, 150, 117.

Hairer, M. (2014). A theory of regularity structures. *Invent. Math.*, 198(2), 269.

Hivert, F., Novelli, J.-C., Thibon, J.-Y. (2005). The algebra of binary search trees. *Theor. Comput. Sci.*, 339, 129.

Hoffman, M.E. (2000). Quasi-shuffle products. *J. Algebr. Combinator.*, 11(1), 49.

Hoffman, M.E. and Ihara, K. (2017). Quasi-shuffle products revisited. *J. Algebra*, 481, 293.

Hudson, R.L. (2009). Hopf-algebraic aspects of iterated stochastic integrals. *Infin. Dimension. Anal. Quantum Probab. Relat. Top.*, 12, 479.

Hudson, R.L. (2012). Sticky shuffle product Hopf algebras and their stochastic representations. In *New Trends in Stochastic Analysis and Related Topics*, Zhao, H. and Truman, A. (ed.). World Scientific Publishing, Singapore.

Hudson, R.L. and Parthasarathy, K.R. (1994). Casimir chaos in a Boson Fock space. *J. Funct. Anal.*, 119, 319.

Karandikar, R.L. (1982). Multiplicative decomposition of non-singular matrix valued continuous semimartingales. *Ann. Probab.*, 10, 1088.

Kawski M. (1994). Chronological algebras and nonlinear control. *Proc. Asian Conf. Control*, Tokyo.

Kreimer, D. (1998). On the Hopf algebra structure of perturbative quantum field theories. *Adv. Theor. Math. Phys.*, 2, 303.

Kreimer, D. (1999). Chen's iterated integral represents the operator product expansion. *Adv. Theor. Math. Phys.*, 3, 627.

Kung, J.P.S., Rota, G.-C., Yan, C.H. (2009). *Combinatorics: The Rota Way*. Cambridge University Press, Cambridge.

Lada, T. and Markl, M. (2005). Symmetric brace algebras. *Applied Categorical Structures*, 13(4), 351.

Li, C. and Liu, X. (1997). Algebraic structure of multiple stochastic integrals with respect to Brownian motions and Poisson processes. *Stochast. Stoch. Rep.*, 61, 107.

Loday, J. and Ronco, M. (1998). Hopf algebra of the planar binary trees. *Adv. Math.*, 139, 293.

Loday, J. and Vallette, B. (2012). *Algebraic Operads*. Springer-Verlag, Berlin, Heidelberg.

Macdonald, I.G. (1998). *Symmetric Functions and Hall Polynomials*. Oxford University Press, Oxford.

Magnus, W. (1954). On the exponential solution of differential equations for a linear operator. *Commun. Pure Appl. Math.*, 7, 649.

Malvenuto, C. and Reutenauer, C. (1995). Duality between quasi-symmetric functions and the Solomon descent algebra. *J. Algebra*, 177, 967.

Manchon, D. (2009). Quelques aspects combinatoires de la renormalisation. *Gazette des Mathématiciens*, 119.

Manchon, D. (2011). A short survey on pre-Lie algebras. In *Non-Commutative Geometry and Physics: Renormalisation, Motives, Index Theory*, Carey, A. (ed.). European Mathematical Society, Zurich.

Mencattini, I., Quesney, A., Silva, P. (2019). Post-symmetric braces and integration of post-Lie algebras. arXiv:1907.12081.

Milnor, J. and Moore, J. (1965). On the structure of Hopf algebras. *Ann. Math., Second Ser.*, 81, 211.

Newman, K. and Radford, D.E. (1979). The Cofree irreducible Hopf algebra on an algebra. *Am. J. Math.*, 101(5), 1025.

Novelli, J.-C. and Thibon, J.-Y. (2009). A one-parameter family of dendriform identities. *J. Combin. Theory, Ser. A*, 116, 864.

Oudom, J.-M. and Guin, D. (2008). On the Lie enveloping algebra of a pre-Lie algebra. *J. K-Theory*, 2(1), 147.

Patras, F. (1991). Construction géométrique des idempotents eulériens. Filtration des groupes de polytopes et des groupes d'homologie de hochschild. *Bull. Soc. Math. France*, 119, 173.

Patras, F. (1994). L'algèbre des descentes d'une bigébre gradué. *J. Algebra*, 170(2), 547.

Patras, F. (1997). Adams operations, iterated integrals and free loop spaces cohomology. *J. Pure Appl. Algebra*, 114, 273.

Radford, D. (2012). *Hopf Algebras*. World Scientific, Hackensack, NJ.

Ree, R. (1958). Lie elements and an algebra associated with shuffles. *Ann. Math., Second Series*, 68, 210.

Reutenauer, C. (1993). *Free Lie Algebras*. Oxford University Press, Oxford.

Rota, G.-C. (1969). Baxter algebras and combinatorial identities I, II. *Bull. Am. Math. Soc.*, 75, 325, 330.

Rota, G.-C. (1995). Baxter operators, an introduction. In *Gian-Carlo Rota on Combinatorics, Introductory Papers and Commentaries*, Kung, J.P.S. (ed.). Birkhäuser, Boston, MA.

Rota, G.-C. (1998). Ten mathematics problems I will never solve. *DMV Mittellungen Heft*, 2, 45–52.

Rota, G.-C. and Smith, D. (1972). Fluctuation theory and Baxter algebras. *Istituto Nazionale di Alta Matematica*, IX, 179.

Schützenberger, M.P. (1958/59). Sur une propriété combinatoire des algèbres de Lie libres pouvant être utilisée dans un problème de mathématiques appliquées. Technical report, Séminaire Dubreil–Jacotin Pisot (Algèbre et théorie des nombres), Paris.

Semenov-Tian-Shansky, M.A. (1983). What is a classical r-matrix? *Funct. Ana. Appl.*, 17, 254.

Solomon, L. (1976). A Mackey formula in the group algebra of a finite coxeter group. *J. Algebra*, 41, 255.

Spitzer, F. (1956). A combinatorial lemma and its application to probability theory. *Trans. Am. Math. Soc.*, 82, 323.

Wilson, K. (1975). The renormalization group: Critical phenomena and the Kondo problem. *Rev. Mod. Phys.*, 47(4), 773.

3

Noncommutative Symmetric Functions, Lie Series and Descent Algebras

Jean-Yves THIBON

Gustave Eiffel University, Marne-la-Vallée, France

3.1. Introduction

Classical Hopf algebra structures on the algebra of symmetric functions describe the groups of formal power series in one variable under multiplication and composition. Similarly, noncommutative symmetric functions can be applied to the study of formal power series with coefficients in a noncommutative algebra, in particular to the Lie series. The main tool is identification of the Hopf algebra of noncommutative symmetric functions with the direct sum of the descent algebras of all symmetric groups. Classical formulas, such as the Magnus expansion, the Baker–Campbell–Hausdorff series and its continuous version or the Fer–Zassenhaus expansion, can then be interpreted in terms of idempotents of the descent algebras. These idempotents admit continuous deformations which can interpolate between the various expansions.

Algebra and Applications 2,
coordinated by Abdenacer MAKHLOUF.

3.2. Classical symmetric functions

3.2.1. *Symmetric polynomials*

The classical theory of symmetric functions is as old as algebra itself, its first manifestation being the Viète formulas ((Viète 1646), see (van der Waerden 1985, p. 78)) of the relationships between the roots and the coefficients of a polynomial:

$$P(x) = \prod_{i=1}^{n}(x - x_i) = \sum_{k=0}^{n}(-1)^k e_k(X)x^{n-k} \qquad [3.1]$$

where

$$e_k(X) = \sum_{i_1>i_2>\ldots>i_k} x_{i_1}x_{i_2}\cdots x_{i_k} \qquad [3.2]$$

are the *elementary symmetric functions* of the roots x_i, interpreted as functions of a *multiset* X.

At the beginning, the theory was essentially limited to the *fundamental theorem*:

> Any symmetric polynomial in x_i is a polynomial in $e_1, \ldots, e_n$, and moreover, this expression is unique.

Otherwise said, the ring of invariants $\mathbb{C}[x_1, \ldots, x_n]^{\mathfrak{S}_n}$ of the symmetric group $\mathfrak{S}_n$ in the polynomial ring $\mathbb{C}[x_1, \ldots, x_n]$ is itself a polynomial ring $\mathbb{C}[e_1, \ldots, e_n]$ (see van der Waerden 1985, section 26).

The game was then to express, in terms of the elementary symmetric polynomials e_k, or in terms of each other, various symmetric polynomials, such as the *power-sums*

$$p_k(X) = x_1^k + x_2^k + \cdots + x_n^k \qquad [3.3]$$

or the *monomial symmetric polynomials*

$$m_\lambda(X) = \sum_{\mu\in\mathfrak{S}_n(\lambda)} x^\mu \qquad [3.4]$$

where, for a *partition* $\lambda = (\lambda_1 \geq \lambda_2 \geq \ldots \geq \lambda_r > 0)$ (a non-increasing sequence of positive integers), $x^\lambda = x_1^{\lambda_1}x_2^{\lambda_2}\cdots x_r^{\lambda_r}$, and $\mathfrak{S}_n(\lambda)$ is the set of *distinct* permutations of the sequence λ completed by zeros (if $r < n$)[1]. Another family of interest is the sequence of *complete homogeneous symmetric polynomials* h_k, defined as the sum of all monomials of degree k.

1. This sum is traditionally denoted by Σx^λ.

Examples of such expressions are

$$E(t;X) = \prod_{i=1}^{n}(1+tx_i) = \sum_{k=0}^{n} t^k e_k(X)k, \quad [3.5]$$

$$H(t;X) = \prod_{i=1}^{n}(1-tx_i)^{-1} = \sum_{k\geq 0} t^k h_k(X) = E(-t;X)^{-1}, \quad [3.6]$$

$$h_k = h_{k-1}e_1 - h_{k-2}e_2 + \cdots + (-1)^{k-1}e_k, \quad [3.7]$$

$$E(t;X) = \exp\left[\sum_{k\geq 1}(-1)^{k-1}\frac{t^k}{k}p_k(X)\right], \quad [3.8]$$

$$H(t;X) = \exp\left[\sum_{k\geq 1}\frac{t^k}{k}p_k(X)\right], \quad [3.9]$$

$$kh_k = h_{k-1}p_1 + h_{k-2}p_2 + \cdots + h_0 p_k \quad (h_0 = 1). \quad [3.10]$$

The theory is not limited to the study of algebraic equations. For example, the advent of linear algebra and matrices allowed for a fresh interpretation of the Viète formulas. If $P(x)$ is the characteristic polynomial of a matrix M, x_i are its eigenvalues and $e_k(X) = \mathrm{tr}\,\Lambda^k(M)$, where $\Lambda^k(M)$ is the kth exterior power of M, i.e. the matrix whose coefficients are the minors of order k of M.

At this stage, it is more convenient to assume that x_i are the reciprocals of the eigenvalues, so that

$$|I - xM| = \prod_{i=1}^{n}(1-x_i x) = \sum_{k=0}^{n} e_k(X)(-x)^k \quad [3.11]$$

is invertible as a formal power series, and its inverse

$$|I - xM|^{-1} = \prod_{i=1}^{n}(1-x_i x)^{-1} = \sum_{k\geq 0} h_k(X)x^k \quad [3.12]$$

has the complete homogeneous symmetric functions $h_k(X)$ as coefficients, which can be interpreted as the traces of the symmetric powers $S^k(M)$, the matrix whose coefficients are the minors of order k of the permanent of M. This last statement is essentially MacMahon's "Master Theorem" (MacMahon 2004, vol. 1, p. 93).

Exterior and symmetric powers are the simplest examples of irreducible representations of the group $GL(n,\mathbb{C})$. The trace of the representation matrix $\rho(g)$, expressed as a symmetric function of the eigenvalues x_i of the matrix g, is called the *character* of the representation ρ. As found by I. Schur (1901), there are other irreducible polynomial representations of $GL(n, \mathbb{C})$, one for each partition in at most n parts. Their characters are the so-called *Schur functions* $s_\lambda(X)$ (described below).

Exercise 3.1.– *MacMahon's Master Theorem.* Let $A = (a_{ij})_{m\times m}$ be a complex matrix, and let $x_1, \dots, x_m$ be formal variables. Consider the coefficients

$$G(k_1, \dots, k_m) = \left[x_1^{k_1} \cdots x_m^{k_m}\right] \prod_{i=1}^{m} \left(a_{i1}x_1 + \cdots + a_{im}x_m\right)^{k_i}.$$

(Here, the notation $[f]g$ means the coefficient of f in g). Let $t_1, \dots, t_m$ be another set of formal variables, and let $T = (\delta_{ij}t_i)_{m\times m}$ be a diagonal matrix. Then

$$\sum_{(k_1,\dots,k_m)} G(k_1, \dots, k_m)\, t_1^{k_1} \cdots t_m^{k_m} = \frac{1}{\det(I_m - TA)},$$

where the sum runs over all non-negative integer vectors $K = (k_1, \dots, k_m)$ and I_m denotes the identity matrix of size m.

Hint:

$$\frac{1}{\det(I_m - TA)} = \sum_{n\geq 0} \operatorname{tr} S^n(TA)$$

The symmetric power $S^n(TA)$ acts naturally on $\mathbb{K}[x_1, \dots, x_m]$, and, denoting by $\langle\cdot|.\cdot\rangle$ its natural scalar product, we have

$$\begin{aligned}\operatorname{tr} S^n(TA) &= \sum_{|K|=n} \langle x^K | S^n(TA)x^K\rangle \\ &= \sum_{|K|=n} \left\langle x^K \Big| \prod_{i=1}^{m} (t_i a_{i1}x_1 + \cdots + t_i a_{im}x_m)^{k_i} \right\rangle.\end{aligned}$$

Application: prove Dixon's identity

$$\sum_{k=0}^{2n} (-1)^k \binom{2n}{k}^3 = (-1)^n \binom{3n}{n, n, n}.$$

3.2.2. *The Hopf algebra of symmetric functions*

Symmetric "functions" are symmetric polynomials in an infinite set of indeterminates X (the term "function" is traditional here). We assume that the coefficients are in some field $\mathbb{K}$ of characteristic 0.

The sequence of elementary symmetric functions is now infinite, with generating series[2]

$$\lambda_t(X) \text{ or } E(t;X) = \prod_{i\geq 1}(1+tx_i) = \sum_{n\geq 0} e_n(X)t^n\,. \qquad [3.13]$$

The fundamental theorem (Macdonald 2015, equation (2.4), p. 20) then states that $e_n(X)$ are algebraically independent, and the algebra of symmetric functions is

$$Sym(X) = \mathbb{K}[e_1, e_2, \dots]\,, \qquad [3.14]$$

naturally graded by the degree in X.

The complete homogeneous functions are given by

$$\sigma_t(X) \text{ or } H(t;X) = \prod_{i\geq 1}(1-tx_i)^{-1} \qquad [3.15]$$

and we also have

$$Sym(X) = \mathbb{K}[h_1, h_2, \dots]\,. \qquad [3.16]$$

The h_n are on the same footing as the e_n, and are often preferred as generators of Sym.

Now, let $Y = \{y_i | i \geq 1\}$ be a second set of indeterminates and denote by $X + Y$ the disjoint union of X and Y. For $f \in Sym$, set

$$\Delta f = f(X+Y)\,. \qquad [3.17]$$

It is clear that $\lambda_t(X+Y) = \lambda_t(X)\lambda_t(Y)$, so that

$$\Delta e_n = \sum_{i=0}^{n} e_i(X)e_{n-i}(Y)\,, \qquad [3.18]$$

which determines Δf for any f, thanks to the fundamental theorem [3.14]. Similarly,

$$\Delta h_n = \sum_{i=0}^{n} h_i(X)h_{n-i}(Y)\,, \qquad [3.19]$$

so that the involutive automorphism ω exchanging h_n and e_n commutes with Δ.

2. $\lambda_t(X)$ is Grothendieck's notation for the generating series of exterior powers in a λ-ring, and indeed, Sym is the free λ-ring on one generator; see Knutson (1973) and section 3.2.3 below.

Identifying a product $u(X)v(Y)$ with the element $u \otimes v$ in the tensor product of algebras $Sym \otimes Sym$, we can interpret Δ as a *comultiplication*, i.e. a linear map $Sym \to Sym \otimes Sym$. If Z is a third set of variables, the obvious equality

$$f((X+Y)+Z) = f(X+(Y+Z)) \qquad [3.20]$$

means that Δ is *coassociative*. The no less obvious fact that $\Delta(fg) = \Delta(f)\Delta(g)$ means that Δ is a homomorphism of algebras, so Δ endows Sym with the structure of a *bialgebra*. Being graded and *connected* (this just means that the degree 0 component is one-dimensional), it is automatically a Hopf algebra (it has an *antipode*), but we do not need to resort to this argument: the antipode is the map $S : f(X) \mapsto f(-X)$, where symmetric functions of $-X$ are defined by

$$\sigma_t(-X) := \sigma_t(X)^{-1} = \lambda_{-t}(X), \qquad [3.21]$$

so that $f(X - X + Y) = f(Y)$, as expected. In particular, $f(X - X) = f(\emptyset)$ is the constant term of f, which we may denote by $\epsilon(f)$. Denoting by $\mu : f \otimes g \mapsto fg$ the multiplication map, and by $u : 1_{\mathbb{K}} \mapsto 1_{Sym}$, we can rewrite this as

$$\mu \circ (I \otimes S) \circ \Delta = u \circ \epsilon\,. \qquad [3.22]$$

This is the usual definition of the antipode: the inverse of the identity map I for the *convolution* $\star$ of endomorphisms, $F \star G = \mu \circ (F \otimes G) \circ \Delta$, for which the projection onto the scalar component $u \circ \epsilon$ is the neutral element. The linear form ϵ is called the *co-unit*.

Recall that, in a coalgebra, an element x such that $\Delta x = x \otimes 1 + 1 \otimes x$ is said to be *primitive*. Now, note that $p_n(X+Y) = p_n(X) + p_n(Y)$, so that the p_n are primitive elements of Sym, and that their relationship to the other generators is given by the identity (Macdonald 2015, equation (2.10), p. 23)

$$\sigma_t(X) = \exp\left[\sum_{m\geq 1} p_m(X)\frac{t^m}{m}\right]. \qquad [3.23]$$

3.2.3. *The λ-ring notation*

The modern point of view interprets symmetric functions as operators. For our concerns, it will be sufficient to regard them as acting on polynomial rings. The simplest operators are the power-sums p_n, which act as algebra morphisms. On variables, p_n acts by $p_n(x) = x^n$, and on scalars by $p_n(\alpha) = \alpha p_n(1) = \alpha$.

We can identify our alphabet X with the formal sum of its elements:

$$X = \sum_i x_i \Rightarrow p_n(X) = \sum_i x_i^n \qquad [3.24]$$

This provides a convenient notation for the powers of the generating series:

$$\sigma_t(X)^\alpha = \exp\left(\alpha \sum_{k\geq 1} p_k(X)\frac{t^k}{k}\right) = \sigma_t(\alpha X)\,. \qquad [3.25]$$

In particular, $\sigma_t(-X) = \sigma_t(X)^{-1}$, which coincides with the previous definition.

The operators can be composed, for example, $p_n \circ p_m = p_{mn}$. The symmetric function $f \circ g$ is called the *plethysm* of g by f, and is sometimes denoted by $f[g]$.

The term λ-*ring* comes from the usual notation Λ^k for the exterior powers. What we have defined is instead called a ψ-ring, but over a field of characteristic 0, both notions are equivalent. In the original definition, the exterior powers correspond to elementary symmetric functions. A variable is called an *element of rank* 1, since $e_k(x) = x$ for $k = 1$ and is 0 for $k > 1$, as are the exterior powers of a one-dimensional vector space. A scalar α is called a *binomial element* since its exterior powers are $e_k(\alpha) = \binom{\alpha}{k}$. In this context, the power-sums are known as *Adams operations*, and are generally denoted by ψ_n (Knutson 1973; Lascoux 2003).

3.2.4. *Symmetric functions and formal power series*

The Hopf structure of Sym is a typical example of a Hopf algebra associated with a group. If we denote by G the multiplicative group of formal power series with constant term 1,

$$G = 1 + t\mathbb{K}[[t]] = \{a(t) = 1 + a_1 t + a_2 t^2 + \cdots\}, \qquad [3.26]$$

we can interpret h_n as the coordinate function

$$h_n(a(t)) = a_n\,, \qquad [3.27]$$

and Sym as the algebra of polynomial functions on G. With this interpretation,

$$h_n(x(t)y(t)) = \sum_{i=0}^{n} x_i y_{n-i} = \sum_{i=0}^{n} (h_i \otimes h_{n-i})(x(t) \otimes y(t)), \qquad [3.28]$$

The standard coproduct for functions on a group (see (Abe 1980)) turns a function $f(x)$ into a function of two variables $\Delta f(x, y) = f(xy)$, and the antipode is $Sf(x) = f(x^{-1})$. This is precisely what the coproduct of Sym does, although this structure has other interpretations.

We can also consider h_n as a coordinate function on the group of formal diffeomorphisms of $\mathbb{R}$ tangent to the identity

$$G_1 = \{A(t) = ta(t) | a(t) \in G\}\,. \qquad [3.29]$$

This leads to a second Hopf algebra structure on Sym, which is known as the *Faà di Bruno algebra* $\mathcal{F}$ (see, for example (Figueroa and Gracia-Bondía 2005)). The coproduct $\Delta_1 h_n$ of h_n is then the function picking the coefficient of t^{n+1} in the composition $A(B(t)) = tb(t)a(tb(t))$. Setting $a_n = h_n(X)$ and $b_n = h_n(Y)$, we obtain

$$A(B(t)) = t\sigma_t(Y)\sum_{k\geq 0} t^k h_k(X)\sigma_t(Y)^k = \sum_{k\geq 0} t^{k+1} h_k(X)\sigma_t((k+1)Y) \qquad [3.30]$$

so that the coefficient of t^{n+1} is

$$\Delta_1(h_n) = \sum_{k=0}^{n} h_k(X)h_{n-k}((k+1)Y). \qquad [3.31]$$

The antipode $S_1 h_n = h_n^*$ of h_n is the function associating with $A(t)$ the coefficient of t^{n+1} in its functional inverse $B(t) = A(t)^{\langle -1\rangle}$ ($A(B(t)) = t$). Setting $a_n = h_n(X)$ and $b_n = h_n^*(X)$, the relationship between $u = t\sigma_t^*(X)$ and $\sigma_t(X)$ is

$$u = t\sigma_t(X) \quad \text{or} \quad t = u\sigma_u^*(X) \qquad [3.32]$$

whose solution is given by the Lagrange inversion formula (Macdonald 2015, Example 24, p. 35):

$$h_n^*(X) = \frac{1}{n+1}h_n(-(n+1)X). \qquad [3.33]$$

3.2.5. *Duality*

A graded bialgebra has a graded dual, the direct sum of the duals of its homogeneous components, endowed with the product dual to its coproduct, and with the coproduct dual to its product. We can show that Sym is *self-dual*, i.e. that it is isomorphic to its graded dual. This is equivalent to the existence of a scalar product satisfying

$$\langle f\cdot g\,,\,h\rangle = \langle f\otimes g\,,\,\Delta h\rangle\,. \qquad [3.34]$$

This scalar product has a very explicit description. To define it, we need some more bases of Sym.

From e_n or h_n, we can build bases of Sym, which are labeled by integer partitions $\lambda = (\lambda_1 \geq \lambda_2 \geq \ldots \geq \lambda_r > 0)$:

$$e_\lambda = e_{\lambda_1}e_{\lambda_2}\cdots e_{\lambda_r} \text{ and } h_\lambda = h_{\lambda_1}h_{\lambda_2}\cdots h_{\lambda_r}\,. \qquad [3.35]$$

These are examples of *multiplicative bases*. There is also an obvious (non-multiplicative) basis, the *monomial symmetric functions* (sums over orbits)

$$m_\lambda = \sum_{\mu \in \mathfrak{S}(\lambda)} x^\mu \,, \qquad [3.36]$$

where $\mathfrak{S}(\lambda)$ denotes the set of distinct permutations of λ. The dimension of the homogeneous component Sym_n is thus the number of partitions of n,

$1, 1, 2, 3, 5, 7, 11, 15, 22, 30, 42, 56, 77, 101, 135, 176, 231, 297, 385, 490, 627, 792, 1002 \ldots$

a quite non-trivial combinatorial sequence.

The identification of an alphabet with the formal sum of its elements allowed us to consistently define sums and differences of alphabets. We can also take products, so that

$$XY = \sum_{i,j} x_i y_j, \qquad [3.37]$$

and the power-sums behave nicely with respect to this operation:

$$p_n(XY) = \sum_{i,j} x_i^n y_j^n = p_n(X)p_n(Y). \qquad [3.38]$$

The behavior of the complete functions is also not too bad:

$$\sigma_1(XY) = \prod_{i,j \geq 1} (1 - x_i y_j)^{-1} = \prod_{i \geq 1} \sigma_{x_i}(Y) = \sum_\lambda m_\lambda(X) h_\lambda(Y) \,. \qquad [3.39]$$

Recall that the identity map of a vector space V with basis (b_i) and dual basis (b_i^*) can be written $id_V = \sum_i b_i \otimes b_i^*$ (such an expression is often called a resolution of the identity). Equation [3.39] will acquire this meaning if we introduce *Hall's scalar product*, defined by

$$\langle h_\lambda \,,\, m_\mu \rangle = \delta_{\lambda\mu} \qquad [3.40]$$

which, moreover, satisfies [3.34]. Equation [3.39] is a so-called *Cauchy-type* identity, which is satisfied by any pair (u, v) of mutually adjoint bases:

$$\sigma_1(XY) = \sum_\lambda u_\lambda(X) v_\lambda(Y) \quad \text{iff} \quad \langle u_\lambda \,,\, v_\mu \rangle = \delta_{\lambda\mu} \,, \qquad [3.41]$$

(see Macdonald (2015, equation (4.6), p. 63)).

Exercise 3.2.– Show that the dual basis p_λ^* of $p_\lambda := p_{\lambda_1} \cdots p_{\lambda_r}$ is given by

$$p_\lambda^* = \frac{p_\lambda}{z_\lambda}, \qquad [3.42]$$

where for $\lambda = 1^{m_1} 2^{m_2} \cdots n^{m_n}$,

$$z_\lambda = 1^{m_1} m_1! \cdots 2^{m_2} m_2! \cdots n^{m_n} m_n! \,. \tag{3.43}$$

The original Cauchy identity (a classical exercise on determinants) expresses precisely this for the *Schur functions*, of which a possible definition (not the original one) is

$$s_\lambda = \det(h_{\lambda_i + j - i}) \,. \tag{3.44}$$

It reads (Macdonald 2015, equation (4.8), p. 64)

$$\sigma_1(XY) = \sum_\lambda s_\lambda(X) s_\lambda(Y) \tag{3.45}$$

which is equivalent to the fact that the Schur functions form an orthonormal basis for the Hall scalar product.

The original definition of Schur functions (which were actually defined by Jacobi) proceeds by antisymmetrization. Start with n variables and define the antisymmetrization of a monomial x^I by

$$A_n(x^I) = A_n(x_1^{i_1} x_2^{i_2} \cdots x_n^{i_n}) = \sum_{\sigma \in \mathfrak{S}_n} \varepsilon(\sigma) x_{\sigma(1)}^{i_1} x_{\sigma(2)}^{i_2} \cdots x_{\sigma(n)}^{i_n} \tag{3.46}$$

This is an antisymmetric polynomial, which is zero if the exponents are not all distinct, and $A_n(x^J) = \pm A_n(x^I)$ is J is a permutation of I. A basis for the image of this operator is thus formed by the images of x^α with $\alpha_1 > \alpha_2 > \ldots > \alpha_n \geq 0$. We can write $\alpha = \lambda + \rho$, where $\rho = (n-1, n-2, \ldots, 1, 0)$ and λ is a partition. The polynomial $A_n(x^{\lambda+\rho})$ is divisible by the product of the differences

$$\Delta_n = \prod_{i<j} (x_i - x_j) \tag{3.47}$$

and the quotient

$$s_\lambda(x_1, \ldots, x_n) = \frac{1}{\Delta} A_n(x^{\lambda+\rho}) \tag{3.48}$$

is the Schur function indexed by the partition λ. It has a well-defined limit in Sym for $n \to \infty$, which can be shown to be given by [3.44], known as the *Jacobi–Trudi* formula.

The importance of the Schur functions comes from the fact that they encode the irreducible characters of the symmetric groups. This fundamental result, due to Frobenius, as well as the character theory of finite groups in general, reads

$$\chi_\mu^\lambda = \langle s_\lambda \,,\, p_\mu \rangle \tag{3.49}$$

where χ^λ_μ denotes the value of the irreducible character χ^λ on an element of the conjugacy class μ, and p_n are the *power-sums*:

$$p_n(X) = \sum_{i\geq 1} x_i^n , \quad p_\mu = p_{\mu_1} \cdots p_{\mu_r} . \qquad [3.50]$$

Thus, the homogeneous component Sym_n of Sym can be identified with the character ring of $\mathfrak{S}_n$ via the *Frobenius characteristic map* $\mathrm{ch}(\chi^\lambda) = s_\lambda$. This leads to the definition of a second product, denoted by $*$, on Sym_n:

$$s_\lambda * s_\mu = ch(\chi^\lambda) * \mathrm{ch}(\chi^\mu) := \mathrm{ch}(\chi^\lambda \chi^\mu) . \qquad [3.51]$$

We can then prove that

$$h_\lambda * h_\mu = \sum_{M\in \mathrm{Mat}(\lambda,\mu)} h_M \qquad [3.52]$$

where $\mathrm{Mat}(\lambda, \mu)$ is the set of non-negative integer matrices with row-sum vector λ and column-sum vector μ, and $h_M = \prod_{i,j} h_{m_{ij}}$.

This so-called *internal product* $*$ has a more elementary definition, in fact, which is more appropriate for our purposes: the map δ, defined by $\delta f = f(XY)$, induces a coproduct of each Sym_n, since obviously $\delta p_\lambda = p_\lambda(X)p_\lambda(Y)$. The dual product is precisely $*$.

Exercise 3.3.– Check this, and use it to prove [3.52].

Finally, note that the Faà di Bruno Hopf algebra is not self-dual. It is clearly not cocommutative, and its graded dual is the universal enveloping algebra $U(\mathfrak{g}_1)$ of the Lie algebra $\mathfrak{g}_1$ of G_1. It is easy to see that $\mathfrak{g}_1$ is spanned by the formal vector fields $d_n = -t^{n+1}\frac{d}{dt}$, which satisfy the relationships

$$[d_m, d_n] = (m - n)d_{m+n}. \qquad [3.53]$$

3.3. Noncommutative symmetric functions

3.3.1. *Basic definitions*

The Hopf algebra **Sym** of noncommutative symmetric functions has a very simple definition: replace the complete symmetric functions h_n by non-commuting indeterminates S_n, and keep the coproduct formula

$$\Delta S_n = \sum_{k=0}^{n} S_k \otimes S_{n-k} . \qquad [3.54]$$

That is, **Sym** is the free associative algebra $\mathbb{K}\langle S_1, S_2, \ldots\rangle$ over an infinite sequence S_n, with the grading $\deg S_n = n$ and the coproduct as above. It can be realized in terms of polynomials over an auxiliary set $A = \{a_i | i \geq 1\}$ of non-commuting variables, endowed with a total order:

$$\sigma_t(A) = \overrightarrow{\prod_{i\geq 1}}(1 - ta_i)^{-1} = \sum_{n\geq 0} S_n(A)t^n \quad (\mapsto h_n \text{ for } a_i \mapsto x_i), \tag{3.55}$$

so that

$$\lambda_t(A) = \overleftarrow{\prod_{1\leq i}}(1 + ta_i) = \sum_{n\geq 0} \Lambda_n(A)t^n \quad (\mapsto e_n \text{ for } a_i \mapsto x_i). \tag{3.56}$$

Equivalently,

$$S_n(A) = \sum_{i_1\leq i_2\leq \ldots \leq i_n} a_{i_1}a_{i_2}\cdots a_{i_n}, \tag{3.57}$$

$$\Lambda_n(A) = \sum_{i_1 > i_2 \ldots > i_n} a_{i_1}a_{i_2}\cdots a_{i_n}, \tag{3.58}$$

Then, the coproduct is again

$$\Delta F = F(A + B) \tag{3.59}$$

where $A + B$ is now interpreted as the *ordinal sum* of the ordered sets A and B (the disjoint union, ordered by $a < b$ for $a \in A$ and $b \in B$, so that this sum is not commutative) and where A commutes with B for the multiplication (so that the terms in A and B can be separated).

This structure again has an obvious interpretation in terms of the multiplicative group of formal power series over a noncommutative algebra $\mathcal{A}$:

$$a(t) = \sum_{n\geq 0} a_n t^n, \quad a_n \in \mathcal{A},\ a_0 = 1_{\mathcal{A}}. \tag{3.60}$$

We can identify S_n with the function $a(t) \mapsto a_n$.

Although such series do not form a group under composition, the noncommutative version of the Faà di Bruno algebra does exist, and has applications in quantum field theory (Brouder *et al.* 2006). Indeed, picking the coefficient of t^{n+1} in the right composition of $A(t) = ta(t)$ and $B(t) = tb(t)$ again defines a coassociative coproduct

$$\Delta_1 S_n = \sum_{k=0}^{n} S_k(A) \otimes S_{n-k}((k + 1)A) \tag{3.61}$$

where $\sigma_t(nA) = \sigma_t(A)^n$, as above. The antipode of this algebra is given by a version of the noncommutative Lagrange inversion formula; see (Brouder *et al.* 2006; Novelli and Thibon 2008).

Calling $\mathbf{Sym}$ *noncommutative symmetric functions* implies looking at it in a special way. In particular, looking for analogues of the classical families of symmetric functions, of the numerous operations (internal products, plethysms) existing on Sym and of the various interpretations of Sym. For example, in representation theory, $\mathbf{Sym}$ is to the tower of 0-Hecke algebras what Sym is to the sequence of symmetric groups[3] (Krob and Thibon 1997).

As a Hopf algebra, $\mathbf{Sym}$ is not self-dual. This is clear from the definition of the coproduct, which is obviously cocommutative. Its (graded) dual is the commutative algebra $QSym$ of *quasi-symmetric functions* (Gessel 1984, pp. 289–317; Gelfand *et al.* 1995).

3.3.2. *Generators and linear bases*

From the generators S_n, we can form a linear basis

$$S^I = S_{i_1} S_{i_2} \cdots S_{i_r} \qquad [3.62]$$

of the homogeneous component $\mathbf{Sym}_n$, parametrized by *compositions* of n, i.e. finite ordered sequences $I = (i_1, \ldots, i_r)$ of positive integers summing to n. We often write $I \models n$ to mean that I is a composition of n. The dimension of $\mathbf{Sym}_n$ is 2^{n-1} for $n \geq 1$.

Similarly, from Λ_n, we can build a basis Λ^I. We can also look for analogues of the power-sum symmetric functions. It is here that the non-trivial questions arise. Indeed, in the commutative case, the power-sum p_n is, up to a scalar factor, the unique primitive element of degree n (this is easily seen by writing an arbitrary primitive element as a polynomial in p_k). Here, the primitive elements form a free Lie algebra (more on this later), and it is not immediately obvious to identify the ones which should deserve the name "noncommutative power-sums".

However, we can at least give one example: the series σ_t being group-like ($\Delta\sigma_t = \sigma_t \otimes \sigma_t$), its logarithm is primitive, and writing

$$\log \sigma_t = \sum_{n\geq 1} \Phi_n \frac{t^n}{n}, \qquad [3.63]$$

3. With the difference that these algebras are not semi-simple. To be precise, $\mathbf{Sym}_n$ is $K_0(H_n(0))$ (Grothendieck group of projective modules).

we can reasonably interpret Φ_n as a noncommutative power-sum (of course, it is not $\sum_i a_i^n$, this element is not even in $\mathbf{Sym}$).

Now recall from [3.57] and [3.58] that, in terms of our auxiliary ordered alphabet, $A = \{a_1, a_2, \ldots\}$ S_n is the sum of non-decreasing words, and Λ_n is the sum of strictly decreasing words.

Let us say that a word $w = a_{i_1} a_{i_2} \cdots a_{i_n}$ has a *descent* at k if $i_k > i_{k+1}$, and denote by $\mathrm{Des}(w)$ the set of such k (the *descent set* of w).

Thus, $S_n(A)$ is the sum of words of length n with no descent and $\Lambda_n(A)$ is the sum of words with descents at all possible places. Now, obviously,

$$S^I = S_{i_1} S_{i_2} \cdots S_{i_r} \qquad [3.64]$$

is the sum of words of length $i_1 + i_2 + \cdots + i_r$ whose descent set is contained in

$$\{i_1, i_1 + i_2, \ldots, i_1 + i_2 + \cdots i_{r-1}\}. \qquad [3.65]$$

Let us denote this set by $\mathrm{Des}(I)$ and call it the *descent set of the composition I*. Symmetrically, we call I the *descent composition* and write $I = C(w)$, of any word of length n with $\mathrm{Des}(I)$ as its descent set.

Exercise 3.4.– Let $\varphi : \mathbf{Sym} \to \mathbb{K}$ be the algebra morphism defined by $\varphi(S_n) = \frac{1}{n!}$. Show that, for a composition I of n, $n!\varphi(S^I)$ is the number of permutations $\sigma \in \mathfrak{S}_n$ such that $\mathrm{Des}(\sigma) \subseteq \mathrm{Des}(I)$.

It seems now natural to introduce the *noncommutative ribbon Schur functions*[4]

$$R_I(A) = \sum_{C(w)=I} w \qquad [3.66]$$

so that we have

$$S^I = \sum_{J \preceq I} R_J \qquad [3.67]$$

where $J \preceq I$ is the *reverse refinement order*, which means that $\mathrm{Des}(J) \subseteq \mathrm{Des}(I)$.

Note that some bases, such as R_I, have lower indices, while others, like S^I, have upper ones. This convention allows us to distinguish between *multiplicative bases*, which are products of generators, and those which are not.

4. Their commutative images are indeed the skew Schur functions indexed by ribbon diagrams.

Exercise 3.5.– Prove that

$$R_I R_J = R_{I\cdot J} + R_{I\triangleright J} \qquad [3.68]$$

where for $I = (i_1, \dots, i_r)$ and $J = (j_1, \dots, j_s)$, $I \cdot J = (i_1, \dots, i_r, j_1, \dots, j_s)$ and $I \triangleright J = (i_1, \dots, i_{r-1}, i_r + j_1, j_2, \dots, j_s)$.

Exercise 3.6.– Prove that, for a composition I of n, $n!\varphi(R_I)$ is the number of permutations $\sigma \in \mathfrak{S}_n$ such that $\mathrm{Des}(\sigma) = \mathrm{Des}(I)$.

Exercise 3.7.– Prove that

$$\left(\sum_{n\geq 0}(-1)^n S_{2n}\right)^{-1} = 1 + R_2 + R_{22} + R_{222} + \cdots = \sum_{n\geq 0} R_{(2^n)} \qquad [3.69]$$

and that

$$\left(\sum_{n\geq 0}(-1)^n S_{2n}\right)^{-1} \cdot \left(\sum_{n\geq 0}(-1)^n S_{2n+1}\right)$$

$$= R_1 + R_{21} + R_{221} + \cdots = \sum_{n\geq 0} R_{(2^n 1)}\,. \qquad [3.70]$$

Deduce from these identities the following classical result of D. André: if we define the *Euler numbers* by

$$\sec x = \sum_{n\geq 0} E_{2n}\frac{x^{2n}}{(2n)!} \text{ and } \tan x = \sum_{n\geq 0} E_{2n+1}\frac{x^{2n+1}}{(2n+1)!} \qquad [3.71]$$

then E_n is the number of *alternating* permutations in $\mathfrak{S}_n$, i.e. permutations with descents at each even position.

The commutative ribbon Schur functions were investigated by MacMahon before 1915 in connection with combinatorial problems. We will see that the noncommutative version has an interesting interpretation.

3.3.3. *Duality*

Since the coproduct of $\mathbf{Sym}$ is cocommutative, we may attempt a realization of its graded dual in terms of commutative polynomials in an infinite set of variables. Since Sym is a quotient of $\mathbf{Sym}$, the dual should contain $Sym^* = Sym$.

Let $X = \{x_1 < x_2 < x_3 < \dots\}$ be an infinite totally ordered set of *commuting* indeterminates, and let $\mathbb{K}[X]$ denote the algebra of formal series of bounded degree[5] over X.

A (generalized) polynomial f is said to be *quasi-symmetric* (Gessel 1984, pp. 289–317) if, for any two finite sequences $y_1 < y_2 < \dots < y_k$ and $z_1 < z_2 < \dots < z_k$ of elements of X, and any exponents $i_1, i_2, \dots, i_k \in \mathbb{N}$, the monomials $y_1^{i_1} y_2^{i_2} \cdots y_k^{i_k}$ and $z_1^{i_1} z_2^{i_2} \cdots z_k^{i_k}$ have the same coefficient in f.

The quasi-symmetric polynomials form a subalgebra of $\mathbb{K}[X]$, denoted by $QSym(X)$, or just $QSym$ if X is understood. An obvious basis of this subalgebra is provided by the *quasi-monomial functions* (Gessel 1984, pp. 289–317)

$$M_I = \sum_{y_1 < y_2 < \dots < y_k} y_1^{i_1} y_2^{i_2} \cdots y_k^{i_k}$$

where $I = (i_1, \dots, i_k)$ is any composition of n. In particular, the dimension of $QSym_n$ is $2^{n-1} = \dim \mathbf{Sym}_n$. Another convenient basis, also introduced in Gessel (1984, pp. 289–317), is constituted by

$$F_I = \sum_{J \succeq I} M_J$$

which are called *quasi-ribbons*.

Let Y be a second infinite totally ordered set of indeterminates and denote by $X + Y$ the ordered sum of X and Y, i.e. their disjoint union ordered by $x < y$ for $x \in X$ and $y \in Y$, and by the previous orderings on X and Y. We can define a coproduct γ on $QSym$ by the usual trick

$$\gamma(f) = f(X + Y). \qquad [3.72]$$

Indeed, it is easily seen from the definition that

$$\gamma(M_I) = \sum_{J \cdot K = I} M_J \otimes M_K \qquad [3.73]$$

where $J \cdot K = (j_1, \dots, j_r, k_1, \dots, k_s)$ denotes the concatenation of compositions. This suggests that we identify M_I with the dual basis of some multiplicative basis of $\mathbf{Sym}$, but we need to determine which one.

5. These are therefore not *polynomials* in the usual sense, but rather elements of a projective limit of polynomial rings, and might thus be called *prolynomials*. We will however refrain from introducing this terminology and continue to refer to these objects as (generalized) polynomials.

This can be settled by means of an analogue of the classical Cauchy formula of the commutative theory. Let A be an ordered noncommutative alphabet and X be a totally ordered commutative alphabet, as above. We can define the symmetric functions of the noncommutative alphabet XA by means of the generating series

$$\sigma_1(XA) = \overrightarrow{\prod_{x \in X}} \sigma_x(A) , \qquad [3.74]$$

the above product being taken with respect to the total order of X. The expansion of this infinite product leads at once to the relationship

$$\sigma_1(XA) = \sum_I M_I(X)\, S^I(A) . \qquad [3.75]$$

Exercise 3.8.– Check this.

Expanding each S^I on the ribbon functions, we obtain the following identities:

$$\sigma_1(XA) = \sum_I M_I(X) \sum_{J \preceq I} R_J(A) = \sum_J \sum_{I \succeq J} M_I(X) R_J(A)$$
$$= \sum_J F_J(X) R_J(A). \qquad [3.76]$$

Thus, if we identify M_I^* with S^I, then we will have, as in the commutative case that for any basis (U_I) of $\mathbf{Sym}$ with dual basis (V_I) in $QSym$,

$$\sigma_1(XA) = \sum_I V_I(X) U_I(A). \qquad [3.77]$$

In particular, F_I will be identified with the dual basis of R_I.

The pairing

$$\langle M_I, S^J \rangle = \delta_{IJ} \qquad [3.78]$$

induces an isomorphism of Hopf algebras between the graded dual $\mathbf{Sym}^*$ of $\mathbf{Sym}$ and the Hopf algebra $QSym$. More precisely, we have, for $f, g \in QSym$ and $P, Q \in \mathbf{Sym}$

$$\langle f \,,\, PQ \rangle = \langle \gamma f \,,\, P \otimes Q \rangle \qquad [3.79]$$

$$\langle fg \,,\, P \rangle = \langle f \otimes g \,,\, \Delta P \rangle. \qquad [3.80]$$

The following property is useful. If $g \in QSym$ is a symmetric function, then

$$\langle g\,,\,R_I\rangle_{Qsym\times\mathbf{Sym}} = \langle g\,,\,r_I\rangle_{Sym\times Sym} \qquad [3.81]$$

where, in the right-hand side, r_I is the commutative ribbon Schur function. In other words, $g = \sum_I \langle g\,,\,r_I\rangle F_I$.

Exercise 3.9.– Define the virtual alphabet $X_q = (1-q)$ by setting

$$\sigma_t((1-q)A) = \lambda_{-qt}(A)\sigma_t(A). \qquad [3.82]$$

Show that

$$S_n((1-q)A) = (1-q)\sum_{k=0}^{n-1}(-q)^k R_{(1^k,n-k)}(A). \qquad [3.83]$$

Deduce from this the specializations $F_I(1-q)$ and $M_I(1-q)$.

Exercise 3.10.– The quasi-symmetric functions of the ordered alphabet $Y_q = \{\dots < q^3 < q^2 < q < 1\}$ can be computed by expanding

$$\sigma_t(Y_qA) = \prod_{k\geq 0}^{\leftarrow} \sigma_{qt^k}(A) = \cdots\sigma_{tq^2}(A)\sigma_{tq}(A)\sigma_t(A). \qquad [3.84]$$

Show that

$$M_I(Y_q) = \frac{q^{\mathrm{maj}\,(I)}}{(1-q^{i_1})(1-q^{i_1+i_2})\cdots(1-q^{i_1+i_2+\cdots+i_r})} \qquad [3.85]$$

where

$$\mathrm{maj}\,(I) = \sum_{d\in\mathrm{Des}(I)} d \qquad [3.86]$$

and that, for a composition I of n,

$$F_I(Y_q) = \frac{q^{\mathrm{maj}\,(I)}}{(1-q)(1-q^2)\cdots(1-q^n)}. \qquad [3.87]$$

3.3.4. *Solomon's descent algebras*

Apart from the already mentioned relationship with 0-Hecke algebras, which will not be covered here (see (Krob and Thibon 1997)), there is another noncommutative analogue of the relationship between symmetric functions and characters of symmetric groups. The story goes back to a discovery by Solomon (1976). His construction, which provides a noncommutative lift in the group algebra of Mackey's formula for a product of induced characters, is valid in general for finite Coxeter groups, but we will only need the case of symmetric groups.

Let (W, S) be a Coxeter system (see (Humphreys 1990)). We say that $w \in W$ has a descent at $s \in S$ if w has a reduced word ending by s. For $W = \mathfrak{S}_n$ and $s_i = (i, i+1)$, this means that $w(i) > w(i+1)$, whence the terminology. In this case, we instead say that i is a descent of w. Let $\mathrm{Des}(w)$ denote the descent set of w, and for a subset $E \subseteq S$, set

$$D_E = \sum_{\mathrm{Des}(w)=E} w \ \in \mathbb{Z}W \,. \qquad [3.88]$$

Solomon has shown that D_E spans a $\mathbb{Z}$-subalgebra $\Sigma(W)$ of $\mathbb{Z}W$. Moreover

$$D_{E'} D_{E''} = \sum_{E} c^{E}_{E' E''} D_E \qquad [3.89]$$

where the coefficients $c^{E}_{E' E''}$ are non-negative integers.

In the case of $W = \mathfrak{S}_n$, we encode descent sets by compositions of n as explained above. If $E = \{d_1, \ldots, d_{r-1}\}$, then we set $d_0 = 0$, $d_r = n$ and $I = C(E) = (i_1, \ldots, i_r)$, where $i_k = d_k - d_{k-1}$. From now on, we will write D_I instead of D_E.

DEFINITION 3.1.– *We denote by Σ_n the descent algebra of $\mathfrak{S}_n$ with scalars extended to the same ground field $\mathbb{K}$ as* $\mathbf{Sym}$.

Thus, Σ_n has the same dimension as $\mathbf{Sym}_n$, and both have bases labeled by compositions of n. It is therefore interesting to look for a good isomorphism between them.

DEFINITION 3.2.– *Let α be the linear map $\Sigma_n \to \mathbf{Sym}_n$ defined by*

$$\alpha : D_I \to R_I. \qquad [3.90]$$

Remembering what has been said about descents, α may appear as a natural choice for a correspondence $\Sigma_n \to \mathbf{Sym}_n$.

This choice is not only natural but also *canonical*. Indeed, Solomon proved that the structure constants of his descent algebra were the same as the decomposition coefficients of certain tensor products of representations of $\mathfrak{S}_n$. Precisely, if we set

$$B^I = \sum_{J \preceq I} D_J \ \in \Sigma_n \qquad [3.91]$$

and

$$B^I B^J = \sum_K b_K^{IJ} B^K , \qquad [3.92]$$

then, the product of the characters β^I of $\mathfrak{S}_n$, induced by the trivial representations of the parabolic subgroups $\mathfrak{S}_{i_1} \times \cdots \times \mathfrak{S}_{i_r}$, decomposes as

$$\beta^I \beta^J = \sum_K b_K^{IJ} \beta^K . \qquad [3.93]$$

Such products of induced characters can be calculated by Mackey's formula, and therefore the title of Solomon's paper (Solomon 1976).

Why is the above correspondence canonical? This is because it is compatible with the Frobenius characteristic map, from $\mathfrak{S}_n$-characters to symmetric functions. Indeed, $\mathrm{ch}(\beta^I) = h_I$, the commutative image of S^I.

Recall that the Frobenius characteristic map allows us to define the internal product $*$ on symmetric functions by equation [3.51]. Comparing equation [3.52] with Solomon's formula for the b_K^{IJ}, we can see that

$$h_\lambda * h_\mu = \mathrm{ch}(\beta^\lambda \beta^\mu). \qquad [3.94]$$

This equation could have been used as a definition of the internal product on Sym. We can use this remark to define an internal product on noncommutative symmetric functions using the descent algebras instead of the character rings.

For technical reasons, which will soon become clear, we want our correspondence to be an *anti-isomorphism*. We set

$$S^I * S^J = \sum_K b_K^{JI} S^K . \qquad [3.95]$$

This is because we want to interpret permutations as endomorphisms of tensor algebras: if $f_\sigma(\mathbf{w}) = \mathbf{w}\sigma$, then $f_\sigma \circ f_\tau = f_{\tau\sigma}$ (see below).

As already shown in Gessel (1984, pp. 289–317), the other coproduct δ of Sym can also be extended to $QSym$. Define $X \times Y$ to be XY endowed with the lexicographic ordering, i.e. $x_1 y_1 < x_2 y_2$ iff $x_1 < x_2$ or $x_1 = x_2$ and $y_1 < y_2$. The extension of δ to $QSym$ is then defined by setting

$$\delta(f) = f(X \times Y) \qquad [3.96]$$

for $f \in QSym$. The dual operation in **Sym** coincides with the internal product defined as above. Note that $F * G = 0$ if F and G are homogeneous of different degrees. The internal product is explicitly given by

$$S^I * S^J = \sum_{M \in \mathrm{Mat}(I,J)} S^M , \qquad [3.97]$$

where S^M is the product of $S^{m_{ij}}$ in the row-reading order, an analogue of the commutative formula for $h_\lambda * h_\mu$. From this, we can deduce, by linearity, the *splitting formula*

$$(F_1 \cdots F_r) * G = \mu_r[(F_1 \otimes \cdots \otimes F_r) *_r \Delta^r G] \qquad [3.98]$$

where μ_r is the r-fold multiplication and Δ^r is the iterated coproduct with values in $\mathbf{Sym}^{\otimes r}$.

Exercise 3.11.– Show that the maps $\theta_q : F \mapsto F * \sigma_1(X_q A)$ and $\kappa_q : F \mapsto F * \sigma_1(Y_q A)$, where $X_q = (1-q)$ and $Y_q = \{\cdots < q^2 < q < 1\}$, as above, are bialgebra morphisms and inverse to each other. For this reason, Y_q is often identified with the geometric series $(1-q)^{-1}$.

3.4. Lie series and Lie idempotents

3.4.1. *Permutational operators on tensor spaces*

Let V be a vector space over some field $\mathbb{K}$ of characteristic 0. Let $T(V)$ be its tensor algebra, and $L(V)$ the free Lie algebra generated by V (Reutenauer 1993). We denote by $L_n(V) = L(V) \cap V^{\otimes n}$ its homogeneous component of degree n. The group algebra $\mathbb{K}\mathfrak{S}_n$ acts on the right on $V^{\otimes n}$ by

$$(v_1 \otimes v_2 \otimes \cdots \otimes v_n) \cdot \sigma = v_{\sigma(1)} \otimes v_{\sigma(2)} \otimes \cdots \otimes v_{\sigma(n)} \,. \qquad [3.99]$$

This action commutes with the left action of $GL(V)$, and when $\dim V \geq n$, which we will usually assume, these actions are the commutant of each other (Schur–Weyl duality).

Any $GL(V)$-equivariant projector $\Pi_n : V^{\otimes n} \to L_n(V)$ can therefore be regarded as an idempotent π_n of $\mathbb{K}\mathfrak{S}_n$: $\Pi_n(\mathbf{v}) = \mathbf{v} \cdot \pi_n$. By definition (see (Reutenauer 1993)), such an element is called a *Lie idempotent* whenever its image is $L_n(V)$. Then, a homogeneous element $P_n \in V^{\otimes n}$ is in $L_n(V)$ if and only if $P_n \pi_n = P_n$.

From now on, we fix a basis $A = \{a_1, a_2, \ldots\}$ of V. We identify $T(V)$ with the free associative algebra $\mathbb{K}\langle A \rangle$ and $L(V)$ with the free Lie algebra $L(A)$.

3.4.2. *The Hausdorff series*

Lie idempotents arise naturally in the investigation of the Hausdorff series

$$H(a_1, a_2, \ldots, a_N) = \log\left(e^{a_1} e^{a_2} \cdots e^{a_N}\right) = \sum_{n \geq 0} H_n(A) \qquad [3.100]$$

which is known to be a *Lie series*, i.e. each homogeneous component $H_n(A) \in L_n(A)$. This is known as the Baker–Campbell–Hausdorff (BCH) "formula". It follows immediately from the characterization of $L(A)$ as the space of primitive elements of the standard comultiplication of $\mathbb{K}\langle A\rangle$ (Friedrichs' criterion[6]).

The BCH formula is not actually a formula at all, only actually a (very important) property, which is the basis of the correspondence between Lie groups and Lie algebras. However, various applications (including real-world ones) require the explicit calculation of the polynomials $H_n(A)$. This is where Lie idempotents come into play.

For small values of n and N, this can be done by hand, and we obtain, for example

$$\begin{aligned}H_3(a_1,a_2,a_3) = {} & \frac{1}{12}a_1a_1a_2 + \frac{1}{12}a_1a_1a_3 - \frac{1}{6}a_1a_2a_1 + \frac{1}{12}a_1a_2a_2 + \frac{1}{3}a_1a_2a_3 \\ & - \frac{1}{6}a_1a_3a_1 + \frac{1}{12}a_1a_3a_3 + \frac{1}{12}a_2a_1a_1 - \frac{1}{6}a_2a_1a_2 \\ & - \frac{1}{6}a_2a_1a_3 + \frac{1}{12}a_2a_2a_1 + \frac{1}{12}a_2a_2a_3 - \frac{1}{6}a_2a_3a_1 - \frac{1}{6}a_2a_3a_2 \\ & + \frac{1}{12}a_2a_3a_3 + \frac{1}{12}a_3a_1a_1 - \frac{1}{6}a_3a_1a_2 - \frac{1}{6}a_3a_1a_3 + \frac{1}{3}a_3a_2a_1 \\ & + \frac{1}{12}a_3a_2a_2 - \frac{1}{6}a_3a_2a_3 + \frac{1}{12}a_3a_3a_1 + \frac{1}{12}a_3a_3a_2\end{aligned}$$

Already, it might not be obvious at first sight that this is indeed a Lie polynomial, which can be rewritten in the form

$$\begin{aligned}H_3(a_1,a_2,a_3) = {} & \frac{1}{12}[a_1,[a_1,a_2]] + \frac{1}{12}[[a_1,a_2],a_2] + \frac{1}{12}[a_1,[a_1,a_3]] \\ & + \frac{1}{12}[[a_1,a_3],a_3] + \frac{1}{12}[a_2,[a_2,a_3]] + \frac{1}{12}[[a_2,a_3],a_3] \\ & + \frac{1}{6}[a_1,[a_2,a_3]] + \frac{1}{6}[[a_1,a_2],a_3]\,.\end{aligned}$$

The problem of finding a systematic procedure for expressing the Hausdorff series as a linear combination of commutators was raised at the Gelfand seminar in the 1940s, and Dynkin (1947) came up with the following solution (also discovered independently by Specht (1948) and Wever (1949)):

6. This criterion, which appears only as a footnote (Friedrichs 1953, p. 203) in a text on quantum field theory, is, for us, together with its converse (the Milnor–Moore theorem (Milnor and Moore 1965), due to Cartier (1957)), the "founding act" of combinatorial Hopf algebra theory, which actually has little to do with Hopf's original motivations.

THEOREM 3.1.– *The element of the group algebra of* $\mathfrak{S}_n$ *defined by*[7]

$$\theta_n = \frac{1}{n}[\dots[[[1,2],3],\dots,],n] \qquad [3.101]$$

is a Lie idempotent, called the Dynkin idempotent. *Therefore, expanding* H_n *as a linear combination of words, and writing* $H_n = H_n\theta_n$*, gives the required expression.*

Exercise 3.12.– For $n = 2$, $ab\theta_2 = \frac{1}{2}(ab - ba)$ and $ab\theta_2^2 = \frac{1}{2}(\frac{1}{2}(ab - ba) - \frac{1}{2}(ba - ab)) = \frac{1}{2}(ab - ba)$. Check that $abc\theta_3^2 = abc\theta_3$.

In order to apply this recipe, we need a reasonably efficient way to find the expansion on words

$$H_n(A) = \sum_{w \in A^n} c_w w\,. \qquad [3.102]$$

We can show that $H_n(A)$ is the image of the homogeneous component $E_n(A)$ of the product of exponentials

$$E(A) = e^{a_1}e^{a_2}\cdots e^{a_N} = \sum_{n\geq 0} E_n(A)\,. \qquad [3.103]$$

under an element of $\mathbb{K}\mathfrak{S}_n$

$$H_n(A) = E_n(A)\cdot\phi_n\,, \qquad [3.104]$$

where

$$\phi_n = \frac{1}{n}\sum_{\sigma\in\mathfrak{S}_n} \frac{(-1)^{d(\sigma)}}{\binom{n-1}{d(\sigma)}}\sigma \qquad (d(\sigma) = |\operatorname{Des}(\sigma)|)\,. \qquad [3.105]$$

Here, $\operatorname{Des}(\sigma) = \{i|\sigma(i) > \sigma(i+1)\}$ denotes the *descent set* of σ. This formula is due to Solomon (1968) and, independently, to Bialynicki-Birula (1969). It can be shown, although none of these properties are clearly apparent on the expression [3.105], that ϕ_n is a Lie idempotent. It has been rediscovered many times, and is also known as the (first) Eulerian idempotent.

We can now write the Hausdorff polynomials

$$H_n(A) = E_n(A)\phi_n\theta_n = \sum_{w\in A^n} c_w \cdot w\theta_n \qquad [3.106]$$

as linear combinations of commutators $w\theta_n$. However, these commutators are far from being linearly independent, and we would be interested in an expansion of H_n on the *basis* of the free Lie algebra.

7. In this expression, permutations are regarded as words over the alphabet of positive integers.

One way to achieve this is to use *Klyachko's basis* of the free Lie algebra. This little known basis is obtained from a third Lie idempotent, discovered by Klyachko (1974), and originally introduced as the solution of a different problem.

This problem was the following: the character of $GL(V)$ on $L_n(V)$ is known (Witt has given formulas for the dimensions of its weight spaces, see (Reutenauer 1993)), and its expression shows that as a $GL(V)$-module, $L_n(V)$ is isomorphic to the space $\Gamma_n(V)$, image of the idempotent

$$\gamma_n = \frac{1}{n}\sum_{k=0}^{n-1} \omega^k c^k , \qquad [3.107]$$

where $c = (12\ldots n)$ is an n-cycle of $\mathfrak{S}_n$ and ω is a primitive nth root of unity. Klyachko's idempotent κ_n is an intertwiner between these two isomorphic representations. This means that for any word $w \in A^n$, (i) $w\kappa_n = 0$ if w is not primitive (i.e. $w = v^d$ for some non-trivial divisor d of n) and (ii) if w is primitive, $wc\kappa_n = \omega w\kappa_n$.

Hence, applying κ_n to some set of representatives of circular classes of primitive words, for example, to Lyndon words (defined as the primitive words which are lexicographically minimal among their circular shifts), we obtain a basis[8] of $L_n(A)$.

There is a closed formula:

$$\kappa_n = \frac{1}{n}\sum_{\sigma\in\mathfrak{S}_n} \omega^{\mathrm{maj}\,(\sigma)}\sigma \qquad [3.108]$$

where $\omega = e^{2i\pi/n}$ and $\mathrm{maj}\,(\sigma) = \sum_{j\in\mathrm{Des}(\sigma)} j$ is the *major index* of σ.

Now, applying κ_n to the Lyndon words appearing in the expansion of $E_n(A)$ obtained from the Eulerian idempotent gives an expansion of $H_n(A)$ on the basis of $L_n(A)$, which can be further expanded in terms of commutators by means of Dynkin's idempotent if needed.

At this point, we are facing three idempotents, which, admittedly, are given by rather different formulas.

It may therefore come as a surprise that the element[9]

$$\varphi_n(q) = \frac{1}{n}\sum_{\sigma\in\mathfrak{S}_n} \frac{(-1)^{d(\sigma)}}{\begin{bmatrix} n-1 \\ d(\sigma)\end{bmatrix}_q} q^{\mathrm{maj}\,(\sigma) - \binom{d(\sigma)+1}{2}}\sigma \qquad [3.109]$$

is a Lie idempotent, interpolating between our three examples (Krob *et al.* 1997).

8. Lyndon words can also be used to define the so-called Lyndon basis, which has nothing to do with the present one; see Reutenauer (1993).

9. See Chapoton (2009) for a recent interpretation of this element.

Apart from the fact that $\varphi_n(1) = \phi_n$, none of these properties are evident. Nevertheless, we can show that $\varphi_n(0) = \theta_n$ and $\varphi_n(\omega) = \kappa_n$ (Krob *et al.* 1997). The explanation of these strange facts starts with the observation that the three idempotents do have something in common: they all belong to the descent algebra.

3.4.3. *Lie idempotents in the descent algebra*

It is clear from the definitions that ϕ_n and κ_n are in the descent algebra of $\mathfrak{S}_n$. For θ_n, this is also quite easy to see. Writing permutations into ribbon diagrams, i.e. writing σ_{i+1} to the right of σ_i if $\sigma_{i+1} > \sigma_i$ and below σ_i otherwise (i.e. if i is a descent of σ), we can observe the following pattern:

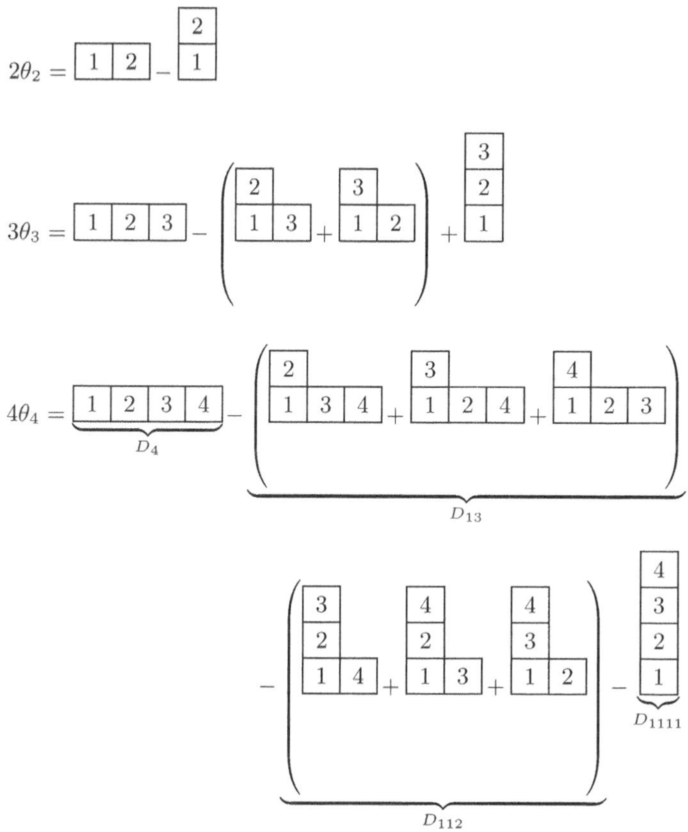

We remark that the descent composition of a permutation consists of the lengths of the rows of its ribbon diagram, read from top to bottom. Thus,

$$n\theta_n = \sum_{k=0}^{n-1} (-1)^k D_{1^k\, n-k}\,. \qquad [3.110]$$

3.5. Lie idempotents as noncommutative symmetric functions

3.5.1. *Noncommutative power-sums*

As already mentioned, the first really interesting question about noncommutative symmetric functions is perhaps "what are the noncommutative power-sums?". The answer to this question is far from being unique.

If we start from the classical expression

$$\sigma_t(X) = \sum_{n\geq 0} h_n(X)t^n = \exp\left\{\sum_{k\geq 1} p_k \frac{t^k}{k}\right\}, \qquad [3.111]$$

then we can choose to define noncommutative power-sums Φ_k by the same formula

$$\sigma_t(A) = \sum_{n\geq 0} S_n(A)t^n = \exp\left\{\sum_{k\geq 1} \Phi_k \frac{t^k}{k}\right\}, \qquad [3.112]$$

but a noncommutative version of the Newton formulas

$$nh_n = h_{n-1}p_1 + h_{n-2}p_2 + \cdots + p_n \qquad [3.113]$$

which are derived by taking the logarithmic derivative of [3.111], leads to different noncommutative power-sums Ψ_k, inductively defined by

$$nS_n = S_{n-1}\Psi_1 + S_{n-2}\Psi_2 + \cdots + \Psi_n\,, \qquad [3.114]$$

or, equivalently, by

$$\psi(t) := \sum_{n\geq 1} t^{n-1}\Psi_n = \lambda_{-t}\frac{d\sigma_t}{dt}\,. \qquad [3.115]$$

A bit of computation then reveals that

$$\Psi_n = R_n - R_{1,n-1} + R_{1,1,n-2} - \cdots = \sum_{k=0}^{n-1}(-1)^k R_{1^k,n-k}\,, \qquad [3.116]$$

which is analogous to the classical expression of p_n as the alternating sum of hook Schur functions. Therefore, in the descent algebra, Ψ_n corresponds to Dynkin's element, $n\theta_n$.

Exercise 3.13.– Using [3.115] and [3.98], show directly that $\Psi_n * \Psi_n = n\Psi_n$.

Hint: compute $\psi(x) * \psi(y)$.

The Φ_n can also be expressed on the ribbon basis without much difficulty, and we find

$$\Phi_n = \sum_{|I|=n} \frac{(-1)^{l(I)-1}}{\binom{n-1}{l(I)-1}} R_I \qquad [3.117]$$

so that Φ_n corresponds to $n\phi_n$.

Exercise 3.14.– Prove directly that $\Phi_n * \Phi_n = n\Phi_n$.

Since $\sigma_t' = \sigma_t\psi(t)$, we can express the solution of the differential equation in terms of iterated integrals

$$\begin{aligned}\sigma_t = 1 &+ \int_0^t dt_1\, \psi(t_1) + \int_0^t dt_1 \int_0^{t_1} dt_2\, \psi(t_2)\psi(t_1) \\ &+ \int_0^t dt_1 \int_0^{t_1} dt_2 \int_0^{t_2} dt_3\, \psi(t_3)\psi(t_2)\psi(t_1) + \cdots \\ &= T\exp\left\{\int_0^t \psi(s)ds\right\}, \qquad [3.118]\end{aligned}$$

or as Dyson's T-exponential.

Define a nonlinear operator $h(t) \mapsto E_q[h(t)]$, acting on noncommutative power series $h(t) = \sum_{n\geq 1} H_n t^{n-1}$ by

$$E_q[h(t)] = \sum_I c_I(q) H^I t^{|I|} \text{ where } \sigma_t = \sum_I t^{|I|} c_I(q)\varphi^I(q) \qquad [3.119]$$

we have

$$E_1[h(t)] = \exp \int_0^t h(s)ds \qquad [3.120]$$

while E_0 is Dyson's chronological exponential, and

$$\begin{aligned}E_0[h(t)] &= T\exp \int_0^t h(s)ds \\ &= 1 + \int_0^t dt_1 h(t_1) + \int_0^t dt_1 \int_0^{t_1} dt_2 h(t_2)h(t_1) + \cdots \qquad [3.121]\end{aligned}$$

Thus, the family $\varphi_n(q)$ provides an interpolation between the chronological exponential and the true exponential.

3.5.2. *The Magnus expansion*

The problem of expressing Φ_n as a linear combination of the Ψ^I is a classical question known as the problem of the *continuous Baker–Campbell–Hausdorff (BCH) exponents* (Magnus 1954; Wilcox 1967; Bialynicki-Birula 1969; Mielnik and Plebański 1970). The usual formulation of this problem is as follows: let $H(t)$ be a differentiable function with values in some operator algebra, and consider the solution $E(t; t_0)$ (the *evolution operator*) of the Cauchy problem

$$\begin{cases} \frac{\partial}{\partial t}E(t;t_0) = E(t;t_0)H(t) \\ \qquad E(t_0;t_0) = 1 \end{cases} \qquad [3.122]$$

The problem is to express the *continuous BCH exponent* $\Omega(t; t_0)$, defined by $E(t; t_0) = \exp \Omega(t; t_0)$ in terms of $H(t)$.

We can set $E(t) = \sigma_t$, $H(t) = \psi(t)$, $\Omega(t) = \Phi(t)$ (see equation [3.115]).

The first answer to this question has been given by Magnus (1954). His formula provides an implicit relationship between $\Phi(t)$ and $\psi(t)$, allowing the recursive computation of the coefficients Φ_n. From our point of view, it gives the explicit expression of Ψ_n in terms of Φ^I. Following Wilcox (1967), we will first derive this formula from an identity of independent interest:

LEMMA 3.1.– *The derivative of $\sigma_t = \exp \Phi(t)$ is given by*

$$\sigma_t' = \frac{d}{dt}e^{\Phi(t)} = \int_0^1 e^{(1-u)\Phi(t)}\frac{d\Phi(t)}{dt}e^{u\Phi(t)}du\,. \qquad [3.123]$$

PROOF.– Expanding $\sigma_t = e^{\Phi(t)}$ in powers of $\Phi(t)$, we have

$$\sigma_t' = \frac{d}{dt}\sum_{n\geq 0}\frac{1}{n!}\Phi(t)^n = \sum_{r,s\geq 0}\frac{\Phi(t)^r\Phi'(t)\Phi(t)^s}{(r+s+1)!}$$

$$= \sum_{r,s\geq 0}\frac{r!s!}{(r+s+1)!}\frac{\Phi(t)^r}{r!}\Phi'(t)\frac{\Phi(t)^s}{s!}\,,$$

and, using the integral representation

$$\frac{r!s!}{(r+s+1)!} = \mathrm{B}(r+1,s+1) = \int_0^1 (1-u)^r u^s du$$

we find

$$\sigma_t' = \sum_{r,s\geq 0} \int_0^1 \frac{[(1-u)\Phi(t)]^r}{r!} \Phi'(t) \frac{[u\Phi(t)]^s}{s!} du$$

$$= \int_0^1 e^{(1-u)\Phi(t)} \Phi'(t) e^{u\Phi(t)} du \,.$$

■

To obtain an expression for Ψ_n, we observe that Lemma 3.1 can be rewritten

$$\sigma_t' = \sigma_t \int_0^1 e^{-u\Phi(t)} \Phi'(t) e^{u\Phi(t)} du \,,$$

so that

$$\psi(t) = \int_0^1 e^{-u\Phi(t)} \Phi'(t) e^{u\Phi(t)} du \qquad [3.124]$$

$$= \int_0^1 \sum_{r\geq 0} \frac{(-u)^r}{r!} \Phi(t)^r \Phi'(t) \sum_{s\geq 0} \frac{u^s}{s!} \Phi(t)^s \, du$$

$$= \sum_{r,s\geq 0} \frac{(-1)^r}{(r+s+1)!} \binom{r+s}{r} \Phi(t)^r \Phi'(t) \Phi(t)^s \,.$$

Extracting the coefficient of t^{n-1}, we finally obtain:

PROPOSITION 3.1.– *The expansion of Ψ_n in the basis (Φ^K) is given by*

$$\Psi_n = \sum_{|K|=n} \left[\sum_{i=1}^{\ell(K)} (-1)^{i-1} \binom{\ell(K)-1}{i-1} k_i \right] \frac{\Phi^K}{\ell(K)!\pi(K)} \qquad [3.125]$$

where, for $K = (k_1, \ldots, k_r)$, $\ell(K) := r$ and $\pi(K) = k_1 k_2 \cdots k_r$.

■

Using the symbolic notation

$$\{\Phi_{i_1} \cdots \Phi_{i_r} \,,\, F\} = \operatorname{ad} \Phi_{i_1} \operatorname{ad} \Phi_{i_2} \cdots \operatorname{ad} \Phi_{i_r}(F)$$
$$= [\Phi_{i_1}, [\Phi_{i_2}, [\ldots [\Phi_{i_r} \,,\, F] \ldots]]]$$

and the classical identity

$$e^a b e^{-a} = \sum_{n\geq 0} \frac{(\operatorname{ad} a)^n}{n!} b = \{e^a \,,\, b\} \,,$$

we can rewrite [3.124] as

$$\psi(t) = \sum_{n\geq 0} \frac{(-1)^n}{(n+1)!}\{\Phi(t)^n \,,\, \Phi'(t)\} = \left\{\frac{1-e^{-\Phi(t)}}{\Phi(t)} \,,\, \Phi'(t)\right\}$$

which, by inversion, gives the Magnus formula:

$$\Phi'(t) = \left\{\frac{\Phi(t)}{1-e^{-\Phi(t)}} \,,\, \psi(t)\right\} = \sum_{n\geq 0} \frac{B_n}{n!}(\mathrm{ad}\,\Phi(t))^n\,\psi(t) \qquad [3.126]$$

with B_n being the Bernoulli numbers.

3.5.3. *The continuous BCH expansion*

As already mentioned, formula [3.126] allows the recursive computation of Φ_n in terms of iterated integrals. There exists, however, an explicit expression of this type, which is due to Bialynicki-Birula (1969) (see also (Mielnik and Plebański 1970)) and rediscovered by several authors (see, for example (Strichartz 1987; Vinokurov 1991)).

We denote by $d(\sigma)$ the number of descents of the permutation σ.

THEOREM 3.2.– *(Continuous BCH formula) The expansion of* $\Phi(t)$ *in the basis* (Ψ^I) *is given by the series*

$$\Phi(t) = \sum_{r\geq 1}\int_0^t dt_1 \cdots \int_0^{t_{r-1}} dt_r \sum_{\tau\in\mathfrak{S}_r} \frac{(-1)^{d(\tau)}}{r}\binom{r-1}{d(\tau)}^{-1} \psi(t_{\tau(r)})\cdots\psi(t_{\tau(1)})\,. \qquad [3.127]$$

Thus, the coefficient of $\Psi^I = \Psi_{i_1}\cdots\Psi_{i_r}$ *in the expansion of* Φ_n *is equal to*

$$n\int_0^1 dt_1\cdots\int_0^{t_{r-1}} dt_r \sum_{\tau\in\mathfrak{S}_r}\frac{(-1)^{d(\tau)}}{r}\binom{r-1}{d(\tau)}^{-1} t_{\tau(r)}^{i_1-1}\cdots t_{\tau(1)}^{i_r-1}\,.$$

Note that this formula has been recently exploited in Ebrahimi-Fard *et al.* (2009) to solve Bogoliubov's recursion in quantum field theory.

Let us prove it within the formalism of noncommutative symmetric functions. The starting point is to write

$$\Phi(t) = \Phi(1) * \sigma_t \qquad [3.128]$$

using σ_t as a reproducing kernel and expressing it in the form

$$\sigma_t = 1 + \sum_{r\geq 1}\int_0^t dt_1\cdots\int_0^{t_{r-1}} dt_r\,\psi(t_r)\cdots\psi(t_1)\,.$$

The problem is thus reduced to the computation of $\Phi(1) * \psi(t_r) \cdots \psi(t_1)$, which is given by the following lemma:

LEMMA 3.2.– *Let $F_1, \dots, F_r$ be primitive for Δ. Then,*

$$\Phi(1) * (F_1 \cdots F_r) = \sum_{\tau \in \mathfrak{S}_r} \frac{(-1)^r}{r} \binom{r-1}{d(\tau)}^{-1} F_{\tau(1)} \cdots F_{\tau(r)} \,.$$

That is, if $\phi_r = \alpha^{-1}(\Phi_r/r)$, identified with an element of the descent algebra,

$$\Phi(1) * (F_1 \cdots F_r) = (F_1 \cdots F_r) \cdot \phi_r \,,$$

then the right action of a permutation $\tau \in \mathfrak{S}_r$ on $F_1 \cdots F_r$ being, as usual, $(F_1 \cdots F_r) \cdot \tau = F_{\tau(1)} \cdots F_{\tau(r)}$.

PROOF.– Since $\Phi(1) = \log(1 + (\sigma_1 - 1))$,

$$\Phi(1) * (F_1 \cdots F_r) = \sum_{k \geq 1} \frac{(-1)^k}{k} (\sigma_1 - 1)^k * (F_1 \cdots F_r) \,,$$

and by the splitting formula,

$$(\sigma_1 - 1)^k * (F_1 \cdots F_r)$$

$$= \mu_k \left[(\sigma_1 - 1)^{\otimes k} * \sum_{A_1, \dots, A_k} F_{a_1^1} \cdots F_{a_{m_1}^1} \otimes \cdots \otimes F_{a_1^k} \cdots F_{a_{m_k}^k} \right]$$

(recall that μ_k denotes the k-fold multiplication $\mathbf{Sym}^{\otimes k} \to \mathbf{Sym}$) where the sum runs over all decompositions $\{1, \dots, r\} = A_1 \cup \cdots \cup A_k$ into disjoint subsets $A_i = \{a_1^i < \dots < a_{m_i}^i\}$. This expression is equal to

$$\sum_{d(\tau) \leq k-1} F_{\tau(1)} \cdots F_{\tau(r)} \,,$$

so that

$$\Phi(1) * (F_1 \cdots F_r) = (F_1 \cdots F_r) \cdot \sum_{k \geq 1} \frac{(-1)^{k-1}}{k} \sum_{d(\tau) \leq k-1} \tau = (F_1 \cdots F_r) \cdot \phi_r \,.$$

■

Using the step function

$$\theta(t) = \begin{cases} 1 & t > 0 \\ 0 & t < 0 \end{cases}$$

and the notations

$$\theta_{i,j} = \theta(t_i - t_j), \quad \Theta_n = \theta_{1,2} + \theta_{2,3} + \cdots + \theta_{n-1,n} ,$$

formula [3.127] can be rewritten in the form of Mielnik and Plebański (1970)

$$\Phi(t) = \sum_{r\geq 1} \int_0^t dt_1 \cdots \int_0^{t_{r-1}} dt_r \frac{(-1)^{\Theta_r}}{r} \binom{r-1}{\Theta_r}^{-1} \psi(t_r) \cdots \psi(t_1). \quad [3.129]$$

3.5.4. *Another proof of the Magnus expansion*

The above property of Φ_n (Lemma 3.2) has an analogue for Ψ_n. It is proved in Krob *et al.* (1997) that, for any sequence $F = (F_n)_{n\geq 1}$ of primitive elements such that F_n belongs to $\mathbf{Sym}_n$ for every $n \geq 1$,

$$\Psi_n * F^I = i_1 \left[[\ldots [F_{i_1}, F_{i_2}], \ldots,], F_{i_r} \right] = i_1 (F_{i_1} \cdots F_{i_r}) \cdot \theta_n . \quad [3.130]$$

This implies the Magnus formula. Indeed, setting $F(t) = \sum_{n\geq 1} F_n t^n$, formula [3.130] can be rewritten as

$$\psi(1) * (F(t_1)F(t_2) \cdots F(t_r)) = \{t_1 F'(t_1) , F(t_2) \cdots F(t_r)\} \quad [3.131]$$

where $\{a_1 , a_2 \cdots a_r\} = [\ldots [a_1 , a_2] , a_3], \ldots , a_r]$. Thus,

$$t\psi(t) = \sum_{n\geq 1} \Psi_n t^n = \psi(1) * \sigma_t = \psi(1) * e^{\Phi(t)}$$

$$= \sum_{n\geq 0} \frac{1}{n!} \psi(1) * \Phi(t)^n = \sum_{n\geq 1} \frac{1}{n!} \{t\Phi'(t) , \Phi(t)^{n-1}\}$$

$$= \left\{ t\Phi'(t) , \frac{e^{\Phi(t)} - 1}{\Phi(t)} \right\}$$

which is the required expression.

3.5.5. *The* $(1-q)$*-transform*

The case of Klyachko's idempotent is even more interesting, but to explain it, we will need the $(1-q)$-transform, to be defined below.

In its classical (commutative) version, the $(1-q)$-transform ϑ_q is the algebra endomorphism of Sym defined on the power-sums by $\vartheta_q(p_n) = (1-q^n)p_n$. In λ-ring notation, which is particularly convenient for dealing with such transformations, it reads $f(X) \mapsto f((1-q)X)$. We have to pay attention to the abuse of notation in using the same minus sign for the λ-ring and for scalars, though these operations are quite different. That is, ϑ_{-1} maps p_n to $2p_n$ if n is odd, and to 0 otherwise. Thus, $\vartheta_{-1}(f(X)) = f((1-q)X)_{q=-1}$ is not the same as $f((1+1)X) = f(2X)$.

In Krob *et al.* (1997), a consistent definition of $\vartheta_q(F) = F((1-q)A)$ has been introduced as follows: we first define the complete symmetric functions $S_n((1-q)A)$ via their generating series (Krob *et al.* 1997, Definition 5.1)

$$\sigma_t((1-q)A) := \sum_{n\geq 0} t^n S_n((1-q)A) = \sigma_{-qt}(A)^{-1}\sigma_t(A)\,, \qquad [3.132]$$

and then ϑ_q is defined as the ring homomorphism such that

$$\vartheta_q(S_n) = S_n((1-q)A). \qquad [3.133]$$

It can then be shown (Krob *et al.* 1997, Theorem 4.17) that

$$F((1-q)A) = F(A) * \sigma_1((1-q)A)\,. \qquad [3.134]$$

For generic q, ϑ_q is an automorphism, and its inverse is the $1/(1-q)$-transform

$$\sigma_t\left(\frac{A}{1-q}\right) = \overleftarrow{\prod_{n\geq 0}} \sigma_{tq^n}(A)\,. \qquad [3.135]$$

Computing the image of S_n, we arrive at

$$K_n(q) := (q)_n S_n\left(\frac{A}{1-q}\right) = \sum_{|I|=n} q^{\mathrm{maj}\,(I)} R_I(A)\,. \qquad [3.136]$$

Hence, Klyachko's element $n\kappa_n = K(\omega)$ is the specialization of this expression at $q = \omega$. This is puzzling: the commutative image of [3.136] is a Hall–Littlewood function ($\tilde{Q}'_{(1^n)}$, precisely), and the specialization of such functions at roots of unity are known. In this case, we get the power-sum p_n (Macdonald 2015, Example 7, p. 234).

3.5.6. *Hopf algebras enter the scene*

At this point, we can see that the commutative images of our three Lie idempotents are the same: $\frac{1}{n}p_n$. The symmetric functions $\frac{1}{n}p_n$ have two significant properties: (i) they are idempotent for the internal product and (ii) they are primitive elements for the coproduct of Sym.

We may therefore suspect that our Lie idempotents might be primitive elements of **Sym**. That this is true can be verified directly for each of them, but we have the following much stronger result (Krob *et al.* 1997). Let us say that an element e of an algebra is *quasi-idempotent* if $e^2 = c \cdot e$ for some (possibly 0) scalar c.

THEOREM 3.3.– *Let* $F = \alpha(\pi)$ *be an element of* $\mathbf{Sym}_n$, *where* $\pi \in \Sigma_n$ *(recall definitions 3.1 and 3.2).*

(i) The following assertions are equivalent:

1. π *is a Lie quasi-idempotent;*

2. F *is a primitive element for* Δ;

3. F *belongs to the Lie algebra* $L(\Psi)$ *generated by the* Ψ_n.

(ii) Moreover, π *is a Lie idempotent iff* $F - \frac{1}{n}\Psi_n$ *is in the Lie ideal* $[L(\Psi), L(\Psi)]$.

Thus, Lie idempotents are essentially the same thing as "noncommutative power-sums" (up to a factor n), and we will, from now on, identify both notions: a Lie idempotent in $\mathbf{Sym}_n$ is a primitive element whose commutative image is p_n/n.

3.5.7. *A one-parameter family of Lie idempotents*

Theorem 3.3 suggests a recipe for constructing new examples. Start from a known family, for example, Dynkin elements, and take its image by a bialgebra automorphism, for example, ϑ_q^{-1}. The result is then automatically a sequence of Lie idempotents. In the case under consideration, we get

$$\varphi_n(q) = \frac{1-q^n}{n}\Psi_n\left(\frac{A}{1-q}\right) = \frac{1}{n}\sum_{|I|=n} \frac{(-1)^{\ell(I)-1}}{\begin{bmatrix} n-1 \\ \ell(I)-1 \end{bmatrix}_q} q^{\mathrm{maj}\,(I)-\binom{\ell(I)}{2}} R_I(A), \qquad [3.137]$$

i.e. [3.109]. Obtaining the closed form on the r.h.s. requires a fair amount of calculation, but the fact that $\varphi_n(q)$ is a Lie idempotent is automatic. This being granted, it is not difficult to show that

$$\varphi_n(0) = \frac{1}{n}\,\Psi_n\,,\ \varphi_n(\omega) = \frac{1}{n}K_n(\omega)\,,\ \varphi_n(1) = \frac{1}{n}\,\Phi_n\,. \qquad [3.138]$$

Other Hopf automorphisms, like the noncommutative analogues of the transformation

$$f(X) \longmapsto f\left(\frac{1-t}{1-q}X\right) \qquad [3.139]$$

used in the theory of Macdonald polynomials, lead to other families of Lie idempotents. It is not always possible, however, to obtain such a clean closed form for them.

3.5.8. *The iterated q-bracketing and its diagonalization*

There is another one-parameter family of Lie idempotents, for which no closed expression is known, but which is of fundamental importance.

The reproducing kernel of ϑ_q, $S_n((1-q)A)$ is easily seen to be the image under α of the iterated q-bracketing operator

$$S_n((1-q)A) = (1-q)\alpha\left([[\cdots[1,2]_q,3]_q,\ldots,n]_q\right), \qquad [3.140]$$

a natural q-analogue of Dynkin's idempotent. For generic q, this is not an idempotent at all, but an automorphism.

The most important property of ϑ_q is its diagonalization (Krob *et al.* 1997, Theorem 5.14): there is a unique family of Lie idempotents $\pi_n(q)$ with the property

$$\vartheta_q(\pi_n(q)) = (1-q^n)\pi_n(q). \qquad [3.141]$$

Moreover, ϑ_q is semi-simple, and its eigenvalues in $\mathbf{Sym}_n$, the nth homogeneous component of $\mathbf{Sym}$, are $p_\lambda(1-q) = \prod_i(1-q^{\lambda_i})$, where λ runs over the partitions of n. The projectors on the corresponding eigenspaces are the maps $F \mapsto F * \pi^I(q)$ (Krob *et al.* 1997, section 3.4).

Here are the first values of $\pi_n(q)$:

$$\pi_1(q) = \Psi_1\,,\quad \pi_2(q) = \frac{\Psi_2}{2}\,,\quad \pi_3(q) = \frac{\Psi_3}{3} + \frac{1}{6}\,\frac{1-q}{(1+2q)}\,[\Psi_2,\Psi_1]\,,$$

$$\pi_4(q) = \frac{\Psi_4}{4} + \frac{1}{12}\,\frac{(1-q)(2q+1)}{(1+q+2q^2)}\,[\Psi_3,\Psi_1] + \frac{1}{24}\,\frac{(1-q)^2}{(1+q+2q^2)}\,[[\Psi_2,\Psi_1],\Psi_1]\,,$$

$$\pi_5(q) = \frac{\Psi_5}{5} + \frac{1}{20}\,\frac{(1-q)(3q^2+2q+1)}{(2q^3+q^2+q+1)}\,[\Psi_4,\Psi_1] + \frac{1}{30}\,\frac{(1-q)(q+2)}{(2q^2+2q+1)}\,[\Psi_3,\Psi_2]$$

$$+\frac{1}{60}\,\frac{(1-q)^2(4q^3+7q^2+7q+2)}{(2q^2+q+2)(2q^3+q^2+q+1)}\,[[\Psi_3,\Psi_1],\Psi_1]$$

$$-\frac{1}{120}\,\frac{(1-q)^2(4q^2+9q+7)}{(q^2+3q+1)(2q^2+1+2q)}\,[[\Psi_1,\Psi_2],\Psi_2]$$

$$+\frac{1}{120}\,\frac{(1-q)^3(2q^5+2q^4+q^3+5q^2+9q+6)}{(2q^3-q^2+q+3)(2q^2+q+2)(2q^3+q^2+q+1)}\,[[[\Psi_2,\Psi_1],\Psi_1],\Psi_1]\,.$$

The idempotents $\pi_n(q)$ have interesting specializations. The easiest one is $q = 1$:

$$\pi_n(1) = \frac{\Psi_n}{n} \,. \qquad [3.142]$$

This has the strange consequence that, for *any* Lie idempotent $F_n \in \mathbf{Sym}_n$,

$$\lim_{q\to 1} \frac{F_n((1-q)A)}{1-q^n} = \frac{\Psi_n}{n} \,. \qquad [3.143]$$

Next, we have, for ω, a primitive n-th root of unity,

$$\pi_n(\omega) = \kappa_n \,. \qquad [3.144]$$

Again, a curious consequence is that

$$\lim_{q\to\omega} (1-q^n)\, F_n(\frac{A}{1-q}) = \kappa_n \,, \qquad [3.145]$$

for any Lie idempotent $F_n \in \mathbf{Sym}_n$.

To describe the next specialization, we need to introduce a new family of noncommutative power-sums. The *noncommutative power-sums* of the *third kind* Z_n are defined by

$$\sigma_t(A) = \exp(Z_1\, t)\, \exp(\frac{Z_2}{2}\, t^2) \,\ldots\, \exp(\frac{Z_n}{n}\, t^n) \,\ldots \qquad [3.146]$$

The Fer–Zassenhauss formula (see (Wilcox 1967)) shows that every Z_n is a Lie element. It is also clear that the commutative image of Z_n is p_n.

The first values of Z_n are

$$Z_1 = \Psi_1 \,,\; Z_2 = \Psi_2 \,,\; Z_3 = \Psi_3 + \frac{1}{2}\,[\Psi_2, \Psi_1] \,,$$

$$Z_4 = \Psi_4 + \frac{1}{3}\,[\Psi_3, \Psi_1] + \frac{1}{6}\,[[\Psi_2, \Psi_1], \Psi_1] \,,$$

$$Z_5 = \Psi_5 + \frac{1}{4}\,[\Psi_4, \Psi_1] + \frac{1}{3}\,[\Psi_3, \Psi_2] + \frac{1}{12}\,[[\Psi_3, \Psi_1], \Psi_1]$$

$$-\frac{7}{24}\,[\Psi_2, [\Psi_2, \Psi_1]] + \frac{1}{24}\,[[[\Psi_2, \Psi_1], \Psi_1], \Psi_1] \,.$$

A recent application of the Z_n to quantum field theory can be found in (Ebrahimi-Fard and Patras 2008).

Then, we have (Krob *et al.* 1997; Duchamp *et al.* 2000)

$$\pi_n(0) = \frac{Z_n}{n} \,. \qquad [3.147]$$

Thus, the nonlinear operator associated with this sequence of idempotents by equation [3.119] interpolates between the chronological exponential and the Fer–Zassenhaus expansion.

3.6. Decompositions of the descent algebras

3.6.1. *Complete families of minimal orthogonal idempotents*

Let $F = (F_n)$ be a sequence of Lie quasi-idempotents with non-zero normalization factors, F_n belonging to $\mathbf{Sym}_n$. We define, as usual

$$F^I = F_{i_1}\,F_{i_2}\,\ldots\,F_{i_r}$$

for a composition $I = (i_1, i_2, \ldots, i_r)$. The family (F^I) is then a basis of $\mathbf{Sym}$. It follows that

$$S_n = \sum_{|I|=n} c_I\,F^I$$

for some scalar coefficients (c_I). Then, we define

$$E_\lambda(F) = \sum_{I\downarrow=\lambda} c_I\,F^I$$

where $I\downarrow$ is the non-decreasing reordering of I and λ a partition of n. Hence,

$$S_n = \sum_{\lambda\vdash n} E_\lambda(F)\,.$$

For $F_n = \Phi_n$, $E_\lambda(\Phi)$ are the Garsia–Reutenauer idempotents (Garsia and Reutenauer 1989) (a complete set of orthogonal idempotents of the descent algebra). More generally, we have:

THEOREM 3.4.– *The family* $(E_\lambda(F))_{\lambda\vdash n}$ *is a complete family of orthogonal idempotents of* $\mathbf{Sym}_n$.

With $F_n = \Psi_n$, we find

$$E_{(3)}(\Psi) = \frac{1}{3}\Psi_3 = \frac{1}{3}\left[R_3 - R_{12} + R_{111}\right]\,,$$

$$E_{(21)}(\Psi) = \frac{1}{6}\Psi^{21} + \frac{1}{3}\Psi^{12} = \frac{1}{6}\left[3R_3 - R_{21} + R_{12} - 3R_{111}\right]\,,$$

$$E_{(111)}(\Psi) = \frac{1}{6}\Psi^{111} = \frac{1}{6}\left[R_3 + R_{21} + R_{12} + R_{111}\right]\,.$$

and for $n = 4$:

$$E_4(\Psi) = \frac{1}{4}\left[R_4 - R_{13} + R_{112} - R_{1111}\right]$$

$$E_{(31)}(\Psi) = \frac{1}{12}\left[4R_4 + R_{31} - 3R_{22} + 3R_{211} + 2R_{13} - R_{121} - 2R_{112} + 4R_{1111}\right] ,$$

$$E_{(22)} = \frac{1}{8}\left[R_4 - R_{31} + R_{22} - R_{211} - R_{13} + R_{121} - R_{112} + R_{1111}\right] ,$$

$$E_{(211)}(\Psi) = \frac{1}{12}\left[3R_4 + R_{22} - 2R_{211} + 2R_{13} - R_{121} - 3R_{1111}\right] ,$$

$$E_{(1111)}(\Psi) = \frac{1}{24}\left[R_4 + R_{31} + R_{22} + R_{211} + R_{13} + R_{121} + R_{112} + R_{1111}\right] .$$

3.6.2. *Eulerian idempotents*

The elements $e_n[k]$ of the descent algebra corresponding to the sums

$$E_n^{[k]} = \sum_{k \vdash n,\ \ell(\lambda)=k} E_\lambda(\Phi) \tag{3.148}$$

are known as the *Eulerian idempotents*. They span an n-dimensional commutative subalgebra of Σ_n, called the *Eulerian subalgebra.*

Although originally introduced in algebraic topology (see (Patras 1991)), the Eulerian idempotents were also discovered in Bialynicki-Birula (1969), in the context of generalized BCH expansions. To explain this, we need the results of the following exercise.

Exercise 3.15.– The noncommutative Eulerian polynomials are defined by

$$\mathcal{A}_n(t) = \sum_{k=1}^{n} t^k \Bigg(\sum_{\substack{|I|=n \\ \ell(I)=k}} R_I \Bigg) = \sum_{k=1}^{n} \mathbf{A}(n,k)\, t^k. \tag{3.149}$$

Show that the generating series of $\mathcal{A}_n(t)$ is

$$\mathcal{A}(t) := \sum_{n\geq 0} \mathcal{A}_n(t) = (1-t)\,\left(1 - t\,\sigma_{1-t}\right)^{-1} . \tag{3.150}$$

Let $\mathcal{A}_n^*(t) = (1-t)^{-n}\,\mathcal{A}_n(t)$. Show that

$$\mathcal{A}^*(t) := \sum_{n\geq 0} \mathcal{A}_n^*(t) = \sum_I \left(\frac{t}{1-t}\right)^{\ell(I)} S^I,$$

and deduce that

$$\mathcal{A}^*(t) = \sum_{k \geq 0} \left(\frac{t}{1-t} \right)^k (S_1 + S_2 + S_3 + \cdots)^k \qquad [3.151]$$

or

$$\frac{1}{1 - t\,\sigma_1} = \sum_{n \geq 0} \frac{\mathcal{A}_n(t)}{(1-t)^{n+1}} . \qquad [3.152]$$

Looking at the proof of lemma 3.2, we may observe that the argument depends only on the fact that $\Phi(1)$ is of the form $g(\sigma_1 - 1)$, where $g(t) = \sum_{n \geq 0} g_n t^n$ is a power series in one variable. More precisely, we have the following property, which allows a similar expansion of any series of the form $g(\sigma_t - 1)$.

LEMMA 3.3.– *Let $g(t) = \sum_{n \geq 0} g_n t^n \in K[[t]]$ be a formal power series in one variable, $G(t) := g(\sigma_t - 1) = \sum_{n \geq 0} G_n t^n$, $G_n \in \mathbf{Sym}_n$ and $\gamma_n := \alpha(G_n) \in \Sigma_n$. Then, if the series $F_1, \ldots, F_r$ are primitive elements for Δ,*

$$G(1) * (F_1 \cdots F_r) = (F_1 \cdots F_r) \cdot \gamma_r . \qquad ■$$

Using $\sigma(t)$ as reproducing kernel as in [3.128], we obtain:

COROLLARY 3.1.– *The expression of $G(t)$ in the basis Ψ^I is given by*

$$G(t) = \sum_{r \geq 0} \int_0^t dt_1 \cdots \int_0^{t_{r-1}} dt_r (\psi(t_r) \cdots \psi(t_1)) \cdot \gamma_r . \qquad ■$$

In particular, with $g(t) = (1+t)^x$, we find

$$\sigma_t^x = 1 + \sum_{r \geq 1} \int_0^t dt_1 \cdots \int_0^{t_{r-1}} dt_r \sum_{k \geq 0} x^k \left(\psi(t_r) \cdots \psi(t_1)\right) \cdot e_r^{[k]} \qquad [3.153]$$

where the $e_r^{[k]}$ are the Eulerian idempotents. Using the expression of Eulerian idempotents on the basis of descent classes and changing the variables in the multiple integrals, we can, as above, get rid of the summations over permutations by introducing an appropriate kernel.

As explained in Bialynicki-Birula (1969) and Mielnik and Plebański (1970), such a transformation is always possible, and to find the general expression of the kernel associated with an arbitrary analytic function $g(t)$, we just have to find it for the particular series

$$f_z(t) = \frac{1}{z-t} := \sum_{n \geq 0} \frac{1}{z^{n+1}} t^n$$

since by the Cauchy integral formula

$$G(t) = \frac{1}{2\pi i}\oint_{z=0} \frac{g(z)}{z-(\sigma(t)-1)}dz\,.$$

Using corollary 3.1, we have

$$F_z(t) = \sum_{r\geq 0}\int_0^t dt_1 \cdots \int_0^{t_{r-1}} dt_r(\psi(t_r)\cdots\psi(t_1))\cdot\gamma_r$$

where $\gamma_r = \alpha(F_{z,r})$ are given by

$$\sum_{r\geq 0} F_{z,r} = \frac{1}{z-(\sigma(1)-1)} = \frac{1}{z+1}\cdot\frac{1}{1-(z+1)^{-1}\sigma(1)} = \sum_{n\geq 0}\frac{\mathcal{A}_n\left(\frac{1}{z+1}\right)}{[1-(z+1)^{-1}]^{n+1}}$$

$$= \sum_{n\geq 0}\left(\frac{z+1}{z}\right)^{n+1}\mathcal{A}_n\left(\frac{1}{z+1}\right) = \sum_I \frac{(z+1)^{|I|-(\ell(I)-1)}}{z^{|I|+1}}R_I$$

(by formula [3.152]). Thus,

$$\gamma_r = \frac{1}{z^{r+1}}\sum_{\sigma\in\mathfrak{S}_r}(z+1)^{r-d(\sigma)}\sigma \qquad [3.154]$$

and

$$F_z(t) = \sum_{r\geq 0}\int_0^t\cdots\int_0^t dt_1\cdots dt_r\,\frac{1}{z^{r+1}}(z+1)^{\Theta_r}\,\psi(t_r)\cdots\psi(t_1)$$

so that

$$G(t) = \frac{1}{2\pi i}\oint_{z=0} g(z)F_z(t)dz$$

$$= \sum_{r\geq 0}\int_0^t\cdots\int_0^t dt_1\cdots dt_r\, K_r[g](t_1,\ldots,t_r)\,\psi(t_r)\cdots\psi(t_1)\,,$$

the kernels being given by

$$K_r[g](t_1,\ldots,t_r) = \frac{1}{2\pi i}\oint_{z=0}\frac{g(z)}{z^{r+1}}(z+1)^{\Theta_r}dz\,. \qquad [3.155]$$

We see from equation [3.154] that γ_r is always in the Eulerian subalgebra, so that the direct sum $\mathcal{E} = \bigoplus \mathcal{E}_r$ appears as a natural operator algebra acting on a series of iterated integrals (often called Chen's series). This is Mielnik and Plebański's (1970) algebra of *permutors*.

Exercise 3.16.– Let

$$S_n^{[p]} = \sum_{|I|=n,\ell(I)\leq p} S^I \,,$$

we have

$$S_n^{[p]} = pE_n^{[1]} + p^2E_n^{[2]} + \cdots + p^nE_n^{[n]}. \qquad [3.156]$$

Expanding the factors $(1-t)^{-(n+1)}$ on the right-hand side of equation [3.152] by the binomial theorem, and taking the coefficient of t^k in terms of weight n in both sides, we obtain

$$S_n^{[k]} = \sum_{i=0}^{k} \binom{n+i}{i} \mathbf{A}(n, k-i) \,, \qquad [3.157]$$

an identity due to Loday, which shows in particular that the noncommutative Eulerian functions also span the Eulerian subalgebra. Similarly, we have

$$\frac{\mathcal{A}_n(t)}{(1-t)^{n+1}} = \sum_{k\geq 0} t^k\, S_n^{[k]} \,,$$

so that

$$\mathbf{A}(n,p) = \sum_{i=0}^{p} (-1)^i \binom{n+i}{i} S_n^{[p-i]} \,. \qquad [3.158]$$

Another natural basis of the Eulerian subalgebra $\mathcal{E}_n$ is constituted by the elements

$$M_n^{[k]} := \sum_{|I|=n,\ell(I)=k} S^I \,. \qquad [3.159]$$

Indeed, putting $x = t/(1-t)$, we have

$$\sum_{k\geq 0} x^k\,(S_1+S_2+\cdots)^k = \sum_{n\geq 0} \mathcal{A}_n^*\left(\frac{x}{1+x}\right) = \sum_{n\geq 0} (1+x)^n\, \mathcal{A}_n\left(\frac{x}{1+x}\right),$$

so that

$$\sum_{k=1}^{n} x^k\, M_n^{[k]} = \sum_{j=1}^{n} x^j\,(1+x)^{n-j}\,\mathbf{A}(n,j)\,. \qquad [3.160]$$

Another kind of generating function can be obtained by writing

$$\sigma_1^x = [1 + (S_1 + S_2 + \cdots)]^x = \sum_{k\geq 0} \binom{x}{k} M^{[k]} \, .$$

Comparing this with $\sigma_1^x = \exp{(x\, E^{[1]})}$, it follows that

$$\sum_{k=1}^{n} x^k\, E_n^{[k]} = \sum_{k=1}^{n} \binom{x}{k} M_n^{[k]} \, . \qquad [3.161]$$

Similarly we obtain the following expansion of $E_n^{[k]}$ on the basis $\mathbf{A}(n,i)$, which is a noncommutative analogue of Worpitzky's identity (see Garsia (1990, pp. 309–382) or Loday (1989)):

$$\sum_{k=1}^{n} x^k\, E_n^{[k]} = \sum_{i=1}^{n} \binom{x-i+1}{n} \mathbf{A}(n,i) \, . \qquad [3.162]$$

3.6.3. *Generalized Eulerian idempotents*

We can define analogues of the Eulerian idempotents associated with any series (F_n) of Lie idempotents by the formula

$$E_n^{[k]}(F) = \sum_{|\lambda|=n,\, \ell(\lambda)=k} E_\lambda(F) \, . \qquad [3.163]$$

In all cases, these idempotents generate an n-dimensional commutative subalgebra of Σ_n. The original Eulerian idempotents are obtained for $F_n = \Phi_n$.

3.7. Decompositions of the tensor algebra

Any complete set $(e_\lambda)_{\lambda\vdash n}$ of orthogonal idempotents of the descent algebra Σ_n determines a decomposition

$$T^n(V) = \bigoplus_{\lambda\vdash n} V_\lambda \qquad [3.164]$$

where $V_\lambda = T^n(V)e_\lambda$.

In the case where e_λ are the Garsia–Reutenauer idempotents $e_\lambda = \alpha^{-1}(E_\lambda(\Phi))$, the characters L_λ of the $GL(V)$-modules V_λ are known. Indeed, if $\lambda = (1^{m_1}2^{m_2}\dots n^{m_n})$, then V_λ is isomorphic to a product of symmetric powers

$$S^{m_1}(L^1(V)) \otimes S^{m_2}(L^2(V)) \otimes \cdots \otimes S^{m_n}(L^n(V))$$

of homogeneous components $L^i(V)$ of the free Lie algebra $L(V)$ over V. Thus, as a symmetric function, L_λ is given by a product of plethysms

$$L_\lambda = h_{m_1}[\ell_1]h_{m_2}[\ell_2]\cdots h_{m_n}[\ell_n] \qquad [3.165]$$

where $\ell_n = \frac{1}{n}\sum_{d|n}\mu(d)p_d^{n/d}$ is the character of $GL(V)$ in $L^n(V)$.

The character of V_λ is always the same, whenever e_λ are defined by the above construction. In fact, this character depends only on the commutative image of the noncommutative symmetric function $E_\lambda = \alpha(e_\lambda)$.

THEOREM 3.5.– *(Krob* et al. *1997). Let e'_λ and e''_λ be two idempotents of Σ_n such that the associated noncommutative symmetric functions $E'_\lambda = \alpha(e_\lambda)$ and $E''_\lambda = \alpha(e''_\lambda)$ both have as commutative image the normalized product of power-sums p_λ/z_λ. Set $V'_\lambda = T^n(V)e'_\lambda$ and $V''_\lambda = T^n(V)e''_\lambda$. Then, the endomorphism ϕ of $T^n(V)$ defined by*

$$\phi(v_1\otimes\cdots\otimes v_n) = (v_1\otimes\cdots\otimes v_n)(id - e'_\lambda - e''_\lambda)$$

induces an isomorphism of $GL(V)$-modules between V'_λ and V''_λ.

Hence, the $GL(V)$-modules $V^{\otimes n}e_\lambda$ constructed from the complete sets of orthogonal idempotents associated with any sequence of Lie idempotents are all isomorphic. It is possible to describe them more precisely (Duchamp *et al.* 1997):

PROPOSITION 3.2.– *Let $\pi = (\pi_n)_{n\geq 1}$ be a sequence of Lie idempotents, with $\pi_n \in$ $\mathbf{Sym}_n$, let $\lambda = (\lambda_1, \lambda_2, \ldots, \lambda_r)$ be a partition of n and let*

$$E_\lambda(\pi) = \sum_{I\downarrow=\lambda} c_I\,\pi^I$$

be the idempotent associated with λ as above. The image $T^n(V)e_\lambda$ of this idempotent is spanned by the polynomials

$$\sum_{\sigma\in\mathfrak{S}_r} c_{\lambda\cdot\sigma}\,P_{l_{\sigma(1)}}\,P_{l_{\sigma(2)}}\,\cdots\,P_{l_{\sigma(r)}}$$

where P_{l_k} is a homogeneous Lie polynomial of degree l_k.

The proposition is a special case of the following lemma:

LEMMA 3.4.– *With the same hypotheses, let*

$$E = \sum_{I\downarrow=\lambda} c_I\,\pi^I$$

be an arbitrary linear combination of all π^I indexed by a permutation of λ. The image of this element is then spanned by the polynomials

$$\sum_{\sigma \in \mathfrak{S}_r} c_{\lambda \cdot \sigma} P_{l_{\sigma(1)}} P_{l_{\sigma(2)}} \cdots P_{l_{\sigma(r)}}$$

where, as above, P_{l_k} is a homogeneous Lie polynomial of degree l_k.

For each partition $\lambda = (\lambda_1, \ldots, \lambda_r)$, there then exists a "probability distribution" c_λ on $\mathfrak{S}_r$ with the following properties:

– $c_\lambda(\sigma) = c_\lambda(\tau)$ whenever the compositions $\lambda \cdot \sigma$ and $\lambda \cdot \tau$ are equal;

– the image of the projector $E_\lambda(\pi)$ is the module generated by all c_λ-symmetrized products

$$(P_1, \ldots, P_r)_{c_\lambda} = \sum_{\sigma \in \mathfrak{S}_r} c_\lambda(\sigma) P_{\sigma(1)} \cdots P_{\sigma(r)}$$

where each P_i is a homogeneous Lie polynomial of degree λ_i.

The family $\Phi = (\Phi_n)_{n \geq 1}$ of Solomon idempotents is characterized by the fact that the associated probability distributions are uniform.

The distributions associated with the Dynkin idempotents are given by:

$$c_\lambda(\sigma) = \frac{\lambda_1 \ldots \lambda_r}{\lambda_{\sigma(1)} (\lambda_{\sigma(1)} + \lambda_{\sigma(2)}) \cdots (\lambda_{\sigma(1)} + \cdots + \lambda_{\sigma(r)})} \qquad [3.166]$$

More examples can be found in Duchamp *et al.* (1997).

3.8. General deformations

Let X be a totally ordered commutative alphabet (which may be "virtual"), and let π_n be a Lie quasi-idempotent in $\mathbf{Sym}_n$. Then, it is easy to show (Duchamp *et al.* 1997) that

$$S_n(X\,A) * \pi_n(A) = p_n(X)\, \pi_n(A)\,. \qquad [3.167]$$

On the other hand, $\pi_n(XA) = \pi_n(A) * S_n(XA)$ is not, in general, proportional to $\pi_n(A)$. If π_n is a Lie idempotent in $\mathbf{Sym}_n$, then $\pi_n(XA)$ is another one. Actually, we can show the following:

PROPOSITION 3.3.– *For any two sequences (Π_n) and (π_n) of Lie idempotents, there exists a virtual alphabet X (i.e. a specialization of quasi-symmetric functions) such that*

$$\Pi_n(A) = \frac{\pi_n(X\,A)}{p_n(X)}\,.$$

We can again show that the operator $S_n(X\,A)$ is diagonalizable with $p_\lambda(X)$ as eigenvalues (where λ is a partition of n) and that the corresponding eigenspaces can still be constructed by the above method, based on the unique family of Lie idempotents $\pi_{X,n}$ such that

$$\pi_{X,n}(X\,A) = p_\lambda(X)\,\pi_{X,n}(A)\,. \qquad [3.168]$$

Here are the first $\pi_{X,n}$. Their coefficients are rational quasi-symmetric functions of X.

$$\pi_{X,1} = \Psi_1\,,\quad \pi_{X,2} = \frac{\Psi_2}{2}\,,$$

$$\pi_{X,3} = \frac{\Psi_3}{3} + \frac{\frac{1}{6}\,M_{12} - \frac{1}{3}\,M_{21}}{M_{12} + M_{21}}\,[\Psi_2, \Psi_1]\,.$$

For $n = 3$, the spectral projectors are

$$E_3 = \pi_{X,3}\,,\ E_{21} = \frac{M_{21}\,\pi_X^{21} + M_{12}\,\pi_X^{12}}{M_{12} + M_{21}}\,,\ E_{111} = \frac{\Psi^{111}}{6}\,.$$

3.9. Lie quasi-idempotents as Lie polynomials

The Lie elements of the descent algebra are in particular elements of the multilinear component $L_{(1^n)}([n])$ of the free Lie algebra over the alphabet $[n] = \{1, \ldots, n\}$. In this section, we give explicit expansions on the basis of multilinear alternants for some of these elements.

3.9.1. *The left derivative*

The *left derivative* Λ_1^{-1} is the linear operator on $\mathbf{Sym}$ defined on the basis Λ^I by the relationships

$$\Lambda_1^{-1} \cdot \Lambda^{i_1,\ldots,i_r} = \begin{cases} \Lambda^{i_2,\ldots,i_r} & \text{if } i_1 = 1 \\ 0 & \text{if } i_1 \neq 1 \end{cases}\,. \qquad [3.169]$$

It is characterized by the following three properties:

1) Λ_1^{-1} is an additive morphism of $\mathbf{Sym}$;

2) Λ_1^{-1} satisfies the property

$$\Lambda_1^{-1} \cdot (FG) = (\Lambda_1^{-1} \cdot F)\,G \,+\, (F|1)\,\Lambda_1^{-1} \cdot G \qquad [3.170]$$

for all $F, G \in Sym$, where $(F|1)$ denotes the constant term of F;

3) $\Lambda_1^{-1} \cdot \Lambda_n = \begin{cases} 1 & \text{if } n = 1 \\ 0 & \text{if } n \neq 1 \end{cases}\,.$ [3.171]

We can describe the action of Λ_1^{-1} on the usual bases:

PROPOSITION 3.4.– *For $n \geq 1$ and for any composition $I = (i_1, \ldots, i_r)$,*

$$\Lambda_1^{-1} \cdot S_n = S_{n-1} \, , \; \Lambda_1^{-1} \cdot \Psi_n = S_{n-1} = \sum_{|I|=n-1} \frac{1}{\pi_u(I)} \, \Psi^I \, , \qquad [3.172]$$

$$\Lambda_1^{-1} \cdot R_I = \begin{cases} R_{i_1-1, i_2, \ldots, i_r} & \textit{if } i_1 \neq 1 \\ 0 & \textit{if } i_1 = 1 \end{cases} . \qquad [3.173]$$

PROOF.– The expression of $\Lambda_1^{-1} \cdot R_I$ (and hence of $\Lambda_1^{-1} \cdot S_n$) follows from

$$M(R, \Lambda)_n = \begin{pmatrix} -1 & 1 \\ 1 & 0 \end{pmatrix}^{\otimes(n-1)} .$$

In particular,

$$\Lambda_1^{-1} \cdot \sigma_t = \sum_{n \geq 1} t^n \Lambda_1^{-1} S_n = t \, \sigma_t \, . \qquad [3.174]$$

On the other hand, using equation [3.170], we obtain

$$\Lambda_1^{-1} \cdot \psi(t) = \Lambda_1^{-1} \cdot (\lambda_{-t} \frac{d}{dt} \sigma_t) = (\Lambda_1^{-1} \cdot \lambda_{-t}) \frac{d}{dt} \sigma_t) + (\lambda_{-t} | 1) \, \Lambda_1^{-1} \cdot \frac{d}{dt} \sigma_t$$

$$= (-t) \frac{d}{dt} \sigma_t + \Lambda_1^{-1} \cdot \frac{d}{dt} \sigma_t = (-t) \frac{d}{dt} \sigma_t + \frac{d}{dt} (\Lambda_1^{-1} \cdot \sigma_t) .$$

Now using relationship [3.174], we get

$$\Lambda_1^{-1} \cdot \psi(t) = (-t) \frac{d}{dt} \sigma_t + \frac{d}{dt} (t \, \sigma_t) = \sigma_t \, .$$

■

3.9.2. *Multilinear Lie polynomials*

Let π be a Lie quasi-idempotent of $\mathbf{Sym}_n$. Let π' denote its inverse image by α in Σ_n. By hypothesis, π' is a Lie projector (up to a normalization) or a Lie nilpotent. This means that the image $\ell_n(\pi)$ of the standard word $1\,2\,\ldots\,n$ by π' is a standard Lie polynomial. That is,

$$\ell_n(\pi) = 1\,2\,\ldots\,n \cdot \pi' \;\; \in L_{1^n}(\{1, 2, \ldots, n\}) \, , \qquad [3.175]$$

where $L_{1^n}(\{1, 2, \ldots, n\})$ denotes the multihomogeneous component of multidegree (1^n) of the free Lie algebra over the alphabet $\{1, 2, \ldots, n\}$. Let us now give an explicit description of the Lie element $\ell_n(\pi)$.

Recall first that Witt's formulas imply that $L_{1^n}(\{1,2,\ldots,n\})$ is a free module of dimension $(n-1)!$. It follows from this property that the family

$$[\sigma] := [[\ldots[[1,\sigma(2)],\sigma(3)],\ldots],\sigma(n)]\,, \qquad [3.176]$$

where σ runs over permutations of $\{2,3,\ldots,n\}$, in a basis of $L_{1^n}(\{1,2,\ldots,n\})$. In particular, every element P of $L_{1^n}(\{1,2,\ldots,n\})$ may be decomposed in this basis and therefore encoded by an element of the $\mathbb{Z}$-algebra of the symmetric group $\mathfrak{S}_{\{2,\ldots,n\}}$, or, equivalently, of $\mathbb{Z}[\mathfrak{S}_{n-1}]$, we can define a natural mapping τ from $L_{1^n}(\{1,2,\ldots,n\})$ into $\mathbb{Z}[\mathfrak{S}_{n-1}]$ as follows: for every Lie polynomial P of $L_{1^n}(\{1,2,\ldots,n\})$ uniquely decomposed as

$$P = \sum_{\sigma\in\mathfrak{S}_{\{2,\ldots,n\}}} P_\sigma\,[\sigma]\,,$$

we set

$$\tau(P) = \sum_{\sigma\in\mathfrak{S}_{n-1}} P_{\sigma'}\,\sigma$$

where σ' is the permutation of $\mathfrak{S}_{\{2,\ldots,n\}}$ defined by

$$\sigma'(i) = \sigma(i)+1$$

for $i\in[1,n-1]$. We can now describe $\ell_n(\pi)$ for every Lie element π of $\mathbf{Sym}_n$.

THEOREM 3.6.– *Let π be a Lie quasi-idempotent of $\mathbf{Sym}_n$. Then, the element $\tau(l_n(\pi))$ belongs to Σ_{n-1}. We can associate with it the element $\tau_n(\pi)$ of $\mathbf{Sym}_{n-1}$ defined by*

$$\tau_n(\pi) = \alpha(\tau(l_n(\pi)))\,.$$

Then, in $\mathbf{Sym}$

$$\tau_n(\pi) = \Lambda_1^{-1}\cdot\pi\,. \qquad [3.177]$$

PROOF.– Let us first introduce some notations. For $x\in\mathbb{Z}[\mathfrak{S}_{n-1}]$, we denote by $i^n_{n-1}(x)$ the element of $\mathbb{Z}[\mathfrak{S}_n]$ corresponding to x in the identification

$$\mathfrak{S}_{n-1} = Fix(1)\subset\mathfrak{S}_n\,,$$

where $Fix(1)$ is the set of permutations fixing 1. For $x\in\mathbb{Z}[\mathfrak{S}_n]$, we denote by $x\cap Fix(1)$ the element defined by

$$x\cap Fix(1) = \sum_{\sigma\in Fix(1)} (x|\sigma)\,\sigma\,.$$

Let now π be a Lie quasi-idempotent of $\mathbf{Sym}_n$ and let

$$\pi = \sum_{i\geq 2} a_{iI}\, R_{iI} \;+ \sum_{I} a_{1I}\, R_{1I}$$

be its decomposition on the ribbon basis. According to proposition 3.4,

$$\Lambda_1^{-1} \cdot \pi = \sum_{i\geq 2} a_{iI}\, R_{i-1,I} \;.$$

Let us now identify the elements of $\mathbf{Sym}$ with their inverse images under α in $\mathbb{Z}[\mathfrak{S}_n]$. It is then easy to see that

$$i_{n-1}^n(\Lambda_1^{-1} \cdot \pi) = \Big(\sum_{i\geq 2} a_{iI}\, R_{iI}\Big) \cap Fix(1) \;.$$

Since we always have $R_{1I} \cap Fix(1) = 0$, it follows from this last equality that

$$i_{n-1}^n(\Lambda_1^{-1} \cdot \pi) = \pi \cap Fix(1) \;.$$

The theorem will follow from this last formula, and from the following lemma.

LEMMA 3.5.– *Let π be a Lie element in $\mathbb{Z}[\mathfrak{S}_n]$. Then,*

$$\pi = (\pi \cap Fix(1))\theta_n \;, \qquad [3.178]$$

where θ_n denotes Dynkin's idempotent.

PROOF.– It suffices to prove the lemma when π is a Lie element of the form $\pi = [P_1, Q]$, where P_1 and Q are alternants of lower degrees, with P_1 being the alternant containing the letter 1. The proof goes by induction on the degree d of P_1. If $d = 1$, we have

$$\pi = [1, Q] = 1\,Q - Q\,1$$

In this case, the lemma reduces to

$$\pi = (1\,Q)\theta_n \;,$$

which follows from Dynkin's criterion and from an identity of Baker (see formula (1.6.5), p. 36 of Reutenauer (1993)). Now suppose that P_1 is of degree $d > 1$. Then, we have

$$\pi \cap Fix(1) = (P_1\,Q - Q\,P_1) \cap Fix(1) = (P_1 \cap Fix(1))\,Q \;.$$

Again using Baker's formula, Dynkin's criterion and the induction hypothesis, we obtain

$$(\pi \cap Fix(1))\theta_n = ((P_1 \cap Fix(1))\,Q)\theta_n = [(P_1 \cap Fix(1))\theta_d, Q] = [P_1, Q] \;.$$

■

Identifying noncommutative symmetric functions with their images under α, proposition 3.6 can be reformulated as follows:

$$\pi = \theta_n \, i^n_{n-1}(\Lambda_1^{-1} \cdot \pi)$$

for any Lie quasi-idempotent π in Σ_n.

EXAMPLE 3.1.– *Let us consider the Lie quasi-idempotent* $\pi = [\Psi_1, \Psi_2]$. *Its inverse image under* α *is*

$$\pi' = \alpha(\pi) = 2\,(\,213 + 312 - 132 - 231\,) = -2\,[[1,3],2] = -2\,[(32)]\,,$$

so that $\tau(\ell_2(\pi)) = -2\,(21)$. *From this, we get*

$$\tau_2(\pi) = -2\,\Lambda_2\,.$$

in agreement with formula [3.177]. Indeed,

$$\pi = \Psi^{12} - \Psi^{21} = -2\,\Lambda^{12} + 2\,\Lambda^{21}$$

whence $\Lambda_1^{-1} \cdot \pi = -2\,\Lambda_2$.

COROLLARY 3.2.– *The restriction of the left derivative* Λ_1^{-1} *to* $L(\Psi)$ *is injective.*

3.9.3. *Decompositions on other bases*

We have seen that for a Lie quasi-idempotent π of Σ_n,

$$\pi = i^n_{n-1}(\Lambda_1^{-1}\pi)\,\theta_n\,.$$

Thus, the encoding used for the decomposition of $\pi \in \Sigma_n$ on Dynkin's basis lies in the descent algebra Σ_{n-1} and can be expressed in terms of noncommutative symmetric functions of $\mathbf{Sym}_{n-1}$. This is a general property:

PROPOSITION 3.5.– *Let* f_n *be a Lie idempotent in* Σ_n *such that the family* $(\sigma\, f_n)_{\sigma \in Fix(1)}$ *is a basis of* $L_{1^n}(1,2,\ldots,n)$. *Let* π *be an arbitrary Lie quasi-idempotent of* Σ_n. *There then exists an element* $s(\pi, f_n)$ *of* Σ_{n-1} *such that*

$$\pi = i^n_{n-1}(s(\pi, f_n))\, f_n\,.$$

PROOF.– The decomposition of f_n on Dynkin's basis is given by

$$f_n = i^n_{n-1}(\Lambda_1^{-1} f_n)\,\theta_n\,.$$

On the other hand,

$$\theta_n = i^n_{n-1}(T(f_n))\, f_n$$

where $T(f_n)$ is some element of $K[\mathfrak{S}_{n-1}]$. The last two formulas imply that $\Lambda_1^{-1} f_n$ is necessarily invertible (for the internal product) and that $T(f_n)$ is its inverse. Hence, $T(f_n)$ belongs to Σ_{n-1}. It follows that

$$\pi = i_{n-1}^n(\Lambda_1^{-1}\pi)\,\theta_n = i_{n-1}^n((\Lambda_1^{-1}\pi) * T(f_n))\,f_n\,. \qquad \blacksquare$$

Klyachko's idempotent $\kappa_n(\zeta)$ of order n is a Lie idempotent of Σ_n, for which the family $(\sigma\,\kappa_n(\zeta))_{\sigma\in Fix(1)}$ is a basis of $L_{1^n}(1,2,\ldots,n)$ that we will call Klyachko's basis (see Reutenauer (1993)). We can therefore use the above techniques to compute the decomposition of an arbitrary Lie quasi-idempotent of Σ_n on this basis. To this purpose, we essentially need to know the decomposition of Dynkin's idempotent on Klyachko's basis, which is given by the following proposition.

PROPOSITION 3.6.– *Let ζ be a primitive n-th root of unity. Then,*

$$\theta_n = i_{n-1}^n(S_{n-1}((1-q)\,A)^*|_{q=\zeta^{-1}})\,\kappa_n(\zeta)\,. \qquad [3.179]$$

PROOF.– Going back to the definition of $\kappa_n(\zeta)$, we can write

$$\Lambda_1^{-1}\,\kappa_n(\zeta) = \frac{1}{n}\sum_{|I|=n,\,i_1\geq 2} \zeta^{\mathrm{maj}\,(I)}\,R_{i_1-1,i_2,\ldots,i_r}\,,$$

the above sum being taken over all compositions $I = (i_1,\ldots,i_r)$ of n with the first part greater than or equal to 2. Since we clearly have $\mathrm{maj}\,(i_1-1,i_2,\ldots,i_r) = \mathrm{maj}\,(i_1,\ldots,i_r) + r - 1$ when $i_1 \geq 2$, we get

$$\Lambda_1^{-1}\,\kappa_n(\zeta) = \frac{1}{n}\sum_{|I|=n-1} \zeta^{\mathrm{maj}\,(I)+l(I)-1}\,R_I\,.$$

Observe now that we always have

$$\mathrm{maj}\,(I) + \mathrm{maj}\,(\overline{I}) = (n-1)(\ell(I)-1) \equiv 1 - \ell(I)\ [n]$$

for every composition I of $n-1$. It follows that we can rewrite the last identity as

$$\begin{aligned}\Lambda_1^{-1}\,\kappa_n(\zeta) &= \frac{1}{n}\sum_{|I|=n-1} \zeta^{-\mathrm{maj}\,(\overline{I})}\,R_I \\ &= \left.\left(\frac{1}{(q)_{n-1}}\sum_{|I|=n-1} q^{\mathrm{maj}\,(I)}\,R_{\overline{I}}\right)\right|_{q=\zeta^{-1}}\,.\end{aligned}$$

This means that

$$\Lambda_1^{-1}\,\kappa_n(\zeta) = S_{n-1}(\frac{A}{1-q})^*|_{q=\zeta^{-1}}\,.$$

Hence,

$$\kappa_n(\zeta) = i_{n-1}^n(S_{n-1}(\frac{A}{1-q})^*|_{q=\zeta^{-1}})\,\theta_n\,,$$

and inverting this last equality gives the desired relationship. ■

Using the last two propositions, we can now give the explicit decomposition of any Lie quasi-idempotent π of Σ_n on Klyachko's basis:

$$\pi = i_{n-1}^n((\Lambda_1^{-1}\pi) * S_{n-1}((1-q)\,A)^*|_{q=\zeta^{-1}})\,\kappa_n(\zeta)\,. \qquad [3.180]$$

3.10. Permutations and free quasi-symmetric functions

3.10.1. *Free quasi-symmetric functions*

For a better understanding of the previous theory, we need larger algebras. We have been computing products in various algebras $\mathcal{A}$, leading to expressions such as

$$x \cdot y = \sum_z c_{xy}^z z \qquad [3.181]$$

where the structure constants c_{xy}^z turn out to be non-negative integers, enumerating certain sets of combinatorial objects. In such a situation, the next step in to search for an algebra $\mathcal{B}$, of which $\mathcal{A}$ is a quotient, and in which the product of lifts X and Y of x, y becomes multiplicity-free.

For example, it is proved in Gessel's original paper on quasi-symmetric functions (Gessel 1984, pp. 289–317) that the dual basis of (R_I), the *fundamental basis*, also called the *quasi-ribbon* basis, satisfies

$$F_J F_K = \sum_I c_{JK}^I F_I \qquad [3.182]$$

where c_{JK}^I counts certain permutations.

Our first example is an algebra based on permutations, in which the fundamental basis can be lifted so as to admit a multiplicity-free product. It is large enough to contain algebras based on binary trees or on Young tableaux. To accommodate other kinds of trees, we can imitate its construction, starting from special words generalizing permutations.

This algebra of permutations has been first investigated by Reutenauer (1993), as the convolution algebra of graded $GL(V)$-endomorphisms of a tensor algebra $T(V)$. Its Hopf algebra structure has been described in Malvenuto and Reutenauer (1995).

There is, however, a more direct and elementary approach (Duchamp *et al.* 2002). Recall that our noncommutative ribbon Schur function R_I has two interpretations:

(i) as the sum of words of shape I in the free associative algebra;

(ii) as the sum of permutations of shape I in the group algebra of the symmetric group.

So, we may ask whether is is possible to associate with each word of shape I a permutation of shape I, so as to reconcile both approaches, and interpret each permutation as a sum of words.

This is indeed possible, and the solution is given by the classical *standardization process*, familiar in combinatorics and computer science. The *standardized word* $\mathrm{Std}(w)$ of a word $w \in A^*$ is the permutation obtained by iteratively scanning w from left to right, and labeling $1, 2, \ldots$ the occurrences of its smallest letter, then numbering the occurrences of the next one and so on. Alternatively, $\sigma = \mathrm{Std}(w)^{-1}$ can be characterized as the unique permutation of minimal length such that $w\sigma$ is a non-decreasing word. For example, $\mathrm{Std}(bbacab) = 341625$.

Obviously, $\mathrm{std}(w)$ has the same descents as w. We can now define polynomials

$$\mathbf{G}_\sigma(A) := \sum_{\mathrm{std}(w)=\sigma} w\,. \qquad [3.183]$$

It is not hard to check that the linear span of these polynomials is a subalgebra, denoted by $\mathbf{FQSym}(A)$, an acronym for *free quasi-symmetric functions*.

Exercise 3.17.– Show that for $\alpha \in \mathfrak{S}_m$ and $\beta \in \mathfrak{S}_n$

$$\mathbf{G}_\alpha \mathbf{G}_\beta = \sum_{\gamma \in \alpha \star \beta} \mathbf{G}_\gamma, \qquad [3.184]$$

where $\alpha \star \beta$ is the set of permutations $\gamma \in \mathfrak{S}_{m+n}$ which, as words, are concatenations $\gamma = u \cdot v$ with $\mathrm{std}(u) = \alpha$ and $\mathrm{std}(v) = \beta$.

Since the definition of $\mathbf{G}_\sigma(A)$ involves only a totally ordered alphabet A, we can apply it to an ordinal sum $A + B$, and, as in the case of $\mathbf{Sym}$, this defines a coproduct if we assume that A commutes with B. Clearly, this coproduct is coassociative and multiplicative, so that we have a graded (and connected) bialgebra, hence, again, a Hopf algebra. By definition,

$$R_I(A) = \sum_{C(\sigma)=I} \mathbf{G}_\sigma(A) \qquad [3.185]$$

so that $\mathbf{Sym}$ embeds into $\mathbf{FQSym}$ as a Hopf subalgebra.

It is also easy to check that **FQSym** is self-dual. If we set $\mathbf{F}_\sigma = \mathbf{G}_{\sigma^{-1}}$ and $\langle \mathbf{F}_\sigma , \mathbf{G}_\tau \rangle = \delta_{\sigma,\tau}$, then $\langle FG, H \rangle = \langle F \otimes G, \Delta H \rangle$.

The graded dual of **Sym** is the (commutative) algebra $QSym$ of *quasi-symmetric functions* (see (Gessel 1984, pp. 289–317; Gelfand *et al.* 1995; Malvenuto and Reutenauer 1995)). Hence we have a surjective homomorphism $\mathbf{FQSym}^* \to QSym$. Its description is particularly simple: it consists of replacing our non-commuting variables a_i by commuting ones x_i. $\mathbf{F}_\sigma(X)$ then depends only on the descent composition $I = C(\sigma)$, and this can be taken as a definition of the quasi-symmetric function F_I (Gessel 1984, pp. 289–317).

We can embed many combinatorial Hopf algebras into **FQSym** by classifying permutations according to various features (in the case of **Sym**, the feature was the descent set). For example, we can build a Hopf algebra of standard Young tableaux by means of the Robinson–Schensted correspondence (Poirier and Reutenauer 1995). When interpreted in **FQSym**, this construction proves the famous Littlewood–Richardson rule for the multiplication of Schur functions in one stroke (Duchamp *et al.* 2002). Replacing standard tableaux by binary trees and the Robinson–Schensted insertion by the binary search tree insertion, familiar to computer scientists, we arrive (Hivert *et al.* 2005) at the Loday–Ronco Hopf algebra of binary trees (Loday and Ronco 1998).

3.11. Packed words and word quasi-symmetric functions

An optimized exposition of the previous theories, disregarding the historical order of their discoveries, could run as follows: noncommutative symmetric functions might come first. After all, it is reasonably obvious that some facts about ordinary symmetric functions, such as the scalar products $\langle h_\lambda, h_\mu \rangle$ and the internal products $h_\lambda * h_\mu$, or the combinatorics MacMahon's ribbon Schur functions, are better understood if we do not assume that h_n commute with each other. Moreover, as an abstract Hopf algebra, **Sym** occurs sporadically in various unrelated contexts such as formal groups (Ditters 1975), algebraic topology (Baker and Richter 2008), Hecke algebras (Krob and Thibon 1997) or polytopes (Bukhshtaber and Erokhovets 2011).

As we have seen, from the definition of **Sym** as the free associative algebra over the S_n, we easily arrive at the definition of the monomial quasi-symmetric functions M_I by imitating the Cauchy-type identities of the classical theory. The coproduct $X + Y$ is easily found to be dual to the product of **Sym**, and the coproduct XY is also well defined and dualizes to an internal product extending that of Sym.

At this point, we may try to resolve the multiplicities in the products $M_I M_J$. This is easily accomplished if we realize that the M_I are the abelianizations of simple noncommutative polynomials, defined by means of a slight refinement of the standardization process.

The *packed word* $u = \text{pack}\,(w)$ associated with a word $w \in A^*$ is obtained by the following process: if $b_1 < b_2 < \ldots < b_r$ are the letters occurring in w, u is the image of w by the homomorphism $b_i \mapsto a_i$.

A word u is said to be *packed* if $\text{pack}\,(u) = u$. We denote by PW the set of packed words. With such a word, we associate the "polynomial"

$$\mathbf{M}_u := \sum_{\text{pack}\,(w)=u} w\,. \qquad [3.186]$$

For example, restricting A to the first five letters a_i (written as integers i),

$$\begin{aligned}\mathbf{M}_{13132} =& 13132 + 14142 + 14143 + 24243 \\ &+ 15152 + 15153 + 25253 + 15154 + 25254 + 35354.\end{aligned}$$

Its image under the abelianization map $a_i \mapsto x_i$ is

$$\begin{aligned}&x_1^2 x_2 x_3^2 + x_1^2 x_2 x4^2 + x_1^3 x_2 x_4^2 + x_2^2 x_3 x_4^2 + x_1^2 x_2 x_5^2 \\ &\quad + x_1^2 x_2 x_5^2 + x_2^2 x_3 x_5^2 + x_1^2 x_4 x_5^2 + x_3^2 x_4 x_5^2 = M_{212}(X).\end{aligned}$$

The *evaluation* $\text{Ev}(w)$ of a word w is the sequence whose i-th term is the number of times the letter a_i occurs in w.

Under the abelianization $\chi : \mathbb{K}\langle A\rangle \to \mathbb{K}[X]$, $\mathbf{M}_u$ are mapped to the monomial quasi-symmetric functions M_I ($I = (|u|_a)_{a\in A}$ being the evaluation vector of u). These polynomials span a subalgebra of $\mathbb{K}\langle A\rangle$, called **WQSym** for word quasi-symmetric functions[10].

The coproduct $A + B$ is well defined on **WQSym** and allows us to consider its graded dual **WQSym***. Let $\mathbf{N}_u \in \mathbf{WQSym}^*$ be the dual basis of $\mathbf{M}_u$.

There is an inclusion of Hopf algebras ι : **FQSym** $\hookrightarrow$ **WQSym** given by

$$\mathbf{G}_\sigma = \sum_{\text{std}(u)=\sigma} \mathbf{M}_u, \qquad [3.187]$$

whose dual is the projection ι^* : **WQSym*** $\twoheadrightarrow$ **FQSym**

$$\mathbf{N}_u \mapsto \mathbf{F}_{\text{std}(u)}. \qquad [3.188]$$

10. It consists of the invariants of the noncommutative version of Hivert's quasi-symmetrizing action (Hivert 2000; Duchamp *et al.* 2011), which is defined by $\sigma \cdot w = w'$, where w' is such that $\text{std}(w') = \text{std}(w)$ and $\chi(w') = \sigma \cdot \chi(w)$. Hence, two words are in the same $\mathfrak{S}(A)$-orbit iff they have the same packed word.

It is easy to check that the coproduct AB is also well defined: packing with respect to the lexicographic order makes sense, and

$$\mathbf{M}_u(AB) = \sum_{\operatorname{pack}\binom{v}{w}=u} \mathbf{M}_v(A)\mathbf{M}_w(B). \qquad [3.189]$$

The dual of the coproduct AB is an internal product on each $\mathbf{WQSym}_n^*$ given by

$$\mathbf{N}_u * \mathbf{N}_v = \mathbf{N}_{\operatorname{pack}\binom{u}{v}}, \qquad [3.190]$$

where $\binom{u}{v}$ denotes the word in *biletters* $\binom{u_i}{v_i}$, lexicographically ordered with priority to the top letter.

There is also a Hopf embedding of $\mathbf{Sym}$ into $\mathbf{WQSym}^*$ given by $S_n \mapsto \hat{S}_n := \mathbf{N}_{1^n}$

$$S^I \mapsto \hat{S}^I = \sum_{\operatorname{Ev}(u)=I} \mathbf{N}_u. \qquad [3.191]$$

For example, $\hat{S}^{21} = \mathbf{N}_{112} + \mathbf{N}_{121} + \mathbf{N}_{211}$.

Under the projection ι^*,

$$\hat{S}^I \mapsto \iota^*(\mathbf{N}_{1^{i_1}} \cdots \mathbf{N}_{1^{i_r}}) = \mathbf{F}_{12\ldots i_1} \cdots \mathbf{F}_{12\ldots i_r} = S^I \qquad [3.192]$$

This projection is compatible with the internal products (recall that at this point we have defined the internal product of $\mathbf{Sym}$ as dual to XY and know nothing about the descent algebra). Indeed, by definition, it maps $f(AB)$ to $f(XY)$. Now, although the $*$ product of $\mathbf{WQSym}^*$ does not coincide with composition on permutations, we have the following compatibility.

LEMMA 3.6.– *Define a right action of $\mathfrak{S}_n$ on $\mathbf{WQSym}_n^*$ by*

$$\mathbf{N}_u \cdot \sigma := \mathbf{N}_{u\sigma}, \text{ where } u\sigma = u_{\sigma(1)}u_{\sigma(2)} \cdots u_{\sigma(n)}. \qquad [3.193]$$

Then, for any $I \vDash n$

$$\mathbf{N}_{u\sigma} * \hat{S}^I = (\mathbf{N}_u * \hat{S}^I) \cdot \sigma. \qquad [3.194]$$

PROOF. If $\hat{S}^I$ contains $\mathbf{N}_v$, it contains $\mathbf{N}_{v\tau}$ for all permutations τ, and

$$\operatorname{pack}\begin{pmatrix} u\tau \\ v \end{pmatrix} = \operatorname{pack}\begin{pmatrix} u \\ v\tau^{-1} \end{pmatrix} \cdot \tau. \qquad [3.195]$$

■

For example, with $u = 111122$, $v = 212211$, $\tau = 451623$, we have $u\tau = 121211$, $v\tau^{-1} = 211212$, $\text{pack}\binom{121211}{212211} = 232411$, $\text{pack}\binom{111122}{211212} = 211234$ and $211234\tau = 232411$.

This implies that $(f \cdot \sigma) * g = (f * g) \cdot \sigma$ for all $f \in \mathbf{WQSym}_n^*, g \in \mathbf{Sym}_n$ and $\sigma \in \mathfrak{S}_n$.

A similar argument proves the existence of the descent algebra. If we denote by $\phi = \iota^*$ the dual of the inclusion map $\mathbf{FQSym} \to \mathbf{WQSym}$, which is given by $\phi(\mathbf{N}_u) = \mathbf{F}_{\text{std}(u)}$, then, if $\sigma = \text{std}(v)$,

$$\phi(\hat{S}^I * \mathbf{N}_v) = \sum_{\text{Ev}(u)=I} \phi\left(\mathbf{N}_{\text{pack}\binom{u}{v}}\right) \quad [3.196]$$

$$= \sum_{\text{Ev}(u)=I} \mathbf{F}_{\text{std}\binom{u}{v}} \quad [3.197]$$

$$= \sum_{\text{Ev}(u)=I} \mathbf{F}_{\text{std}\binom{u\sigma^{-1}}{v\sigma^{-1}}\circ\sigma} \quad [3.198]$$

$$= \sum_{\text{Ev}(u)=I} \mathbf{F}_{\text{std}(u\sigma^{-1})\circ\sigma} = S^I \circ \mathbf{F}_\sigma \text{ (where } \mathbf{F}_\alpha \circ \mathbf{F}_\beta := \mathbf{F}_{\alpha\circ\beta}) \quad [3.199]$$

$$= \phi(\hat{S}^I) * \phi(\mathbf{N}_v). \quad [3.200]$$

This implies, in particular, that in **FQSym**,

$$S^I \circ S^J = \phi(\hat{S}^I * \hat{S}^J) = \sum_{M\in\text{Math}(I,J)} S^M \quad [3.201]$$

so that the $S^I = \sum_{C(\sigma^{-1})=I} \mathbf{F}_\sigma$ span a sub $\circ$-algebra of **FQSym** isomorphic to $(\mathbf{Sym}, *)$.

Packed words can be naturally identified with *ordered set partitions*, the letter a_i at the jth position meaning that j belongs to block i. For example,

$$u = 313144132 \leftrightarrow \Pi = (\{2,4,7\},\{9\},\{1,3,8\},\{5,6\}). \quad [3.202]$$

If we write $u = \{s'_1, \ldots, s'_k\}$ and $v = \{s''_1, \ldots, s''_l\}$ as ordered set partitions, then the packed word $w = \text{pack}\binom{u}{v}$ corresponds to the ordered set partition obtained from

$$\{s'_1 \cap s''_1, s'_1 \cap s''_2, \ldots, s'_1 \cap s''_l, s'_2 \cap s''_1, \ldots, s'_k \cap s''_l\}. \quad [3.203]$$

The corresponding algebra structure on ordered set partitions is known as the Solomon–Tits algebra (Patras and Schocker 2006). Hence, $(\mathbf{WQSym}_n^*, *)$ is isomorphic to the Solomon–Tits algebra. This structure was originally defined (again for finite Coxeter groups) by Tits in an appendix (Tits 1976) to Solomon's paper.

3.12. References

Abe, E. (1980). *Hopf Algebras*, Translated by H. Kinoshita and H. Tanaka. Cambridge University Press, Cambridge, New York.

Baker, A. and Richter, B. (2008). Quasisymmetric functions from a topological point of view. *Math. Scand.*, 103(2), 208–242.

Bialynicki-Birula, I. (1969). Explicit solution of the continuous Baker–Campbell–Hausdorff problem and a new expression for the phase operator. *Ann. Phys.*, 51, 187–200.

Brouder, C., Frabetti, A., Krattenthaler, C. (2006). Non-commutative Hopf algebra of formal diffeomorphisms. *Adv. Math.*, 200(2), 479–524.

Bukhshtaber, V.M. and Erokhovets, N.Y. (2011). Polytopes, Fibonacci numbers, Hopf algebras, and quasisymmetric functions. *Uspekhi Mat. Nauk*, 66(2), 67–162.

Cartier, P. (1957). *Séminaire "Sophus Lie" de la Faculté des Sciences de Paris, 1955–56. Hyperalgèbres et groupes de Lie formels.* Secrétariat mathématique, Paris.

Chapoton, F. (2009). A rooted-trees q-series lifting a one-parameter family of Lie idempotents. *Algebr. Number Theory*, 3(6), 611–636.

Ditters, B. (1975). *Groupes formels*. Université de Paris-Sud, Orsay.

Duchamp, G., Klyachko, A., Krob, D., Thibon, J.-Y. (1997). Noncommutative symmetric functions III. Deformations of Cauchy and convolution algebras. *Discrete Math. Theor. Comput. Sci.*, 1(1), 159–216.

Duchamp, G., Krob, D., Vassilieva, E.A. (2000). Zassenhaus Lie idempotents, q-bracketing and a new exponential/logarithm correspondence. *J. Algebr. Combin.*, 12(3), 251–277.

Duchamp, G., Hivert, F., Thibon, J.-Y. (2002). Noncommutative symmetric functions VI. Free quasi-symmetric functions and related algebras. *Int. J. Algebr. Comput.*, 12(5), 671–717.

Duchamp, G., Hivert, F., Novelli, J.-C., Thibon, J.-Y. (2011). Noncommutative symmetric functions VII: Free quasi-symmetric functions revisited. *Ann. Comb.*, 15(4), 655–673.

Dynkin, E.B. (1947). Calculation of the coefficients in the Campbell–Hausdorff formula. *Doklady Akad. Nauk SSSR (N.S.)*, 57, 323–326.

Ebrahimi-Fard, K. and Patras, F. (2008). A Zassenhaus-type algorithm solves the Bogoliubov recursion. *Bulg. J. Phys.*, 35, 303–315.

Ebrahimi-Fard, K., Manchon, D., Patras, F. (2009). A noncommutative Bohnenblust–Spitzer identity for Rota-Baxter algebras solves Bogoliubov's recursion. *J. Noncommut. Geom.*, 3(2), 181–222.

Figueroa, H. and Gracia-Bondía, J.M. (2005). Combinatorial Hopf algebras in quantum field theory I. *Rev. Math. Phys.*, 17(8), 881–976.

Friedrichs, K.O. (1953). *Mathematical Aspects of the Quantum Theory of Fields*. Interscience Publishers, Inc., New York and Interscience Publishers Ltd., London.

Garsia, A.M. (1990). Combinatorics of the free Lie algebra and the symmetric group. In *Analysis, et Cetera*, Rabinowitz, P.H. and Zehnder E. (eds). Academic Press, Boston.

Garsia, A.M. and Reutenauer, C. (1989). A decomposition of Solomon's descent algebra. *Adv. Math.*, 77(2), 189–262.

Gelfand, I.M., Krob, D., Lascoux, A., Leclerc, B., Retakh, V.S., Thibon, J.-Y. (1995). Noncommutative symmetric functions. *Adv. Math.*, 112(2), 218–348.

Gessel, I.M. (1984). Multipartite P-partitions and inner products of skew Schur functions. In *Contemporary Mathematics: Combinatorics and Algebra*, Greene, C. (ed.). American Mathematical Society, Providence, RI.

Hivert, F. (2000). Hecke algebras, difference operators, and quasi-symmetric functions. *Adv. Math.*, 155(2), 181–238.

Hivert, F., Novelli, J.-C., Thibon, J.-Y. (2005). The algebra of binary search trees. *Theor. Comput. Sci.*, 339(1), 129–165.

Humphreys, J.E. (1990). *Reflection Groups and Coxeter Groups*. Cambridge University Press, Cambridge.

Klyachko, A.A. (1974). Lie elements in a tensor algebra. *Sibirsk. Mat. Ž.*, 15, 1296–1304, 1430.

Knutson, D. (1973). *λ-Rings and the Representation Theory of the Symmetric Group*. Springer-Verlag, Berlin-New York.

Krob, D. and Thibon, J.-Y. (1997). Noncommutative symmetric functions IV. Quantum linear groups and Hecke algebras at $q = 0$. *J. Algebr. Combin.*, 6(4), 339–376.

Krob, D., Leclerc, B., Thibon, J.-Y. (1997). Noncommutative symmetric functions II. Transformations of alphabets. *Int. J. Algebra Comput.*, 7(2), 181–264.

Lascoux, A. (2003). *Symmetric Functions and Combinatorial Operators on Polynomials*. American Mathematical Society, Providence, RI.

Loday, J.-L. (1989). Opérations sur l'homologie cyclique des algèbres commutatives. *Invent. Math.*, 96(1), 205–230.

Loday, J.-L. and Ronco, M.O. (1998). Hopf algebra of the planar binary trees. *Adv. Math.*, 139(2), 293–309.

Macdonald, I.G. (2015). *Symmetric Functions and Hall Polynomials*, 2nd edition. The Clarendon Press, Oxford University Press, New York.

MacMahon, P.A. (2004). *Combinatory Analysis. Volume I & II.* Dover Phoenix Editions, Dover Publications, Inc., Mineola, New York.

Magnus, W. (1954). On the exponential solution of differential equations for a linear operator. *Comm. Pure Appl. Math.*, 7, 649–673.

Malvenuto, C. and Reutenauer, C. (1995). Duality between quasi-symmetric functions and the Solomon descent algebra. *J. Algebra*, 177(3), 967–982.

Mielnik, B. and Plebański, J. (1970). Combinatorial approach to Baker–Campbell–Hausdorff exponents. *Ann. Inst. H. Poincaré Sect. A (N.S.)*, 12, 215–254.

Milnor, J.W. and Moore, J.C. (1965). On the structure of Hopf algebras. *Ann. Math.*, 81, 211–264.

Novelli, J.-C. and Thibon, J.-Y. (2008). Noncommutative symmetric functions and Lagrange inversion. *Adv. Appl. Math.*, 40(1), 8–35.

Patras, F. (1991). Construction géométrique des idempotents eulériens. Filtration des groupes de polytopes et des groupes d'homologie de Hochschild. *Bull. Soc. Math. France*, 119(2), 173–198 [Online]. Available at: http://www.numdam.org/item?id=BSMF_1991__119_2_173_0.

Patras, F. and Schocker, M. (2006). Twisted descent algebras and the Solomon–Tits algebra. *Adv. Math.*, 199(1), 151–184.

Poirier, S. and Reutenauer, C. (1995). Algèbres de Hopf de tableaux. *Ann. Sci. Math. Québec*, 19(1), 79–90.

Reutenauer, C. (1993). *Free Lie Algebras.* The Clarendon Press, Oxford University Press, New York.

Schur, I. (1901). Über eine Klasse von Matrizen die sich einter gegeben Matrix zuordnen lassen. Inaugural-Dissertation, Berlin.

Solomon, L. (1968). On the Poincaré–Birkhoff–Witt theorem. *J. Combinatorial Theory*, 4, 363–375.

Solomon, L. (1976). A Mackey formula in the group ring of a Coxeter group. *J. Algebr.*, 41(2), 255–264.

Specht, W. (1948). Die linearen Beziehungen zwischen höheren Kommutatoren. *Math. Z.*, 51, 367–376.

Strichartz, R.S. (1987). The Campbell–Baker–Hausdorff–Dynkin formula and solutions of differential equations. *J. Funct. Anal.*, 72(2), 320–345.

Tits, J. (1976). Two properties of Coxeter complexes. *J. Algebr.*, 41(2), 265–268.

Viète, F. (1646). *Opera Mathematica*. F. van Schouten, Leiden.

Vinokurov, V.A. (1991). Logarithm of the solution of a linear differential equation, the Hausdorff formula and conservation laws. *Dokl. Akad. Nauk SSSR*, 319(4), 792–797.

van der Waerden, B.L. (1985). *A History of Algebra: From al-Khwārizmī to Emmy Noether*. Springer-Verlag, Berlin.

Wever, F. (1949). Über Invarianten in Lie'schen Ringen. *Math. Ann.*, 120, 563–580.

Wilcox, R.M. (1967). Exponential operators and parameter differentiation in quantum physics. *J. Math. Phys.*, 8, 962–982.

4

From Runge–Kutta Methods to Hopf Algebras of Rooted Trees

Ander MURUA

University of the Basque Country UPV/EHU, Donostia, Spain

4.1. Numerical integration methods for ordinary differential equations

4.1.1. *Introduction*

We are interested in numerical methods for systems of ordinary differential equations (ODEs) of the form

$$\frac{dy}{dt} = f(y), \quad f : \mathcal{U} \subset \mathbb{R}^d \to \mathbb{R}^d \qquad [4.1]$$

where t is the independent variable ("time") and $\mathcal{U}$ is a nonempty open set of $\mathbb{R}^d$. Usually, $\mathcal{U}$ is called the *phase space* of [4.1] and $f : \mathcal{U} \subset \mathbb{R}^d \to \mathbb{R}^d$ is referred to as the vector field. We will assume throughout that the vector field f is smooth, in the sense that it has continuous derivatives of all orders.

Recall that the differential equation [4.1] supplemented with the initial condition $y(t_0) = y_0$ has a unique solution $y(t)$. Numerical integration methods aim to obtain approximations $y_k \approx y(t_k)$ at the time levels $t_1, t_2, \ldots$, with small step sizes $h_k = t_k - t_{k-1}$. If a constant step size $h_k = h$ is considered, then $t_k = kh$ for all k.

In this chapter, we focus on *one-step methods* to integrate initial-value problems for systems of differential equations [4.1]. The simplest (one step) numerical

Algebra and Applications 2,
coordinated by Abdenacer MAKHLOUF.

method for [4.1] is the so-called (explicit or forward) Euler method: Consider the one-parameter family of maps $\psi_{h\,f} : \mathbb{R}^d \to \mathbb{R}^d$ defined as

$$\psi_{h\,f}(y) = y + hf(y). \qquad [4.2]$$

A sequence $\{y_1, y_2, \ldots\}$ of approximations to the values $y(t_k)$ of the solution of [4.1], supplemented with the initial condition $y(0) = y_0$ at times t_k $(k = 1, 2, \ldots)$, is obtained in a step-by-step manner as

$$y_k = \psi_{h_k\,f}(y_{k-1}), \quad k = 1, 2, \ldots \qquad [4.3]$$

The smaller the step size h, the more accurate the approximations. This is a consequence of the fact that, for $\tilde{y}(h) := \psi_{hf}(y_0)$, we have

$$\frac{d}{dh}\tilde{y}(t) = f(\tilde{y}(h)) + hR(y_0, h),$$

where $R(y_0, h) := \frac{1}{h}(f(y_0) - f(y_0 + hf(y_0)))$ is uniformly bounded for sufficiently small $|h|$. That is, $\tilde{y}(t)$ approximately satisfies the original differential equation [4.1] for small values of time $t = h$.

In this sense, a method is of *order* n (n a positive integer), if for each fixed y, there exists $h_0(y), C(y) > 0$, such that, for all $h \in [-h_0(y), h_0(y)]$,

$$\|\frac{\partial}{\partial h}\psi_{h\,f}(y) - f(\psi_{h\,f}(y))\| \leq C(y)h^{n+1}. \qquad [4.4]$$

From our preceding discussion, it is clear that Euler's method is of order $n = 1$.

Another simple example of a one-step integrator is the explicit trapezoidal method (due to Runge), where

$$\psi_{h\,f}(y) = y + \frac{h}{2}(f(y) + f(y + hf(y))) \qquad [4.5]$$

is used in [4.3] instead of [4.2]. It is not difficult to check that method [4.5] is of order 2 (i.e. [4.4] holds for $n = 2$).

4.1.2. *Runge–Kutta methods*

A *Runge–Kutta* (RK) method with s *stages* is specified by an *RK tableau* of real constants

$$\begin{array}{|ccc} a_{11} & \cdots & a_{1s} \\ \vdots & \ddots & \vdots \\ a_{s1} & \cdots & a_{ss} \\ \hline b_1 & \cdots & b_s \end{array}. \qquad [4.6]$$

When applied to the system [4.1], the method corresponding to [4.6] advances the numerical solution from time t_{k-1} to time $t_k = t_{k-1} + h$ through the relation $y_k = \psi_{h\,f}(y_{k-1})$, where

$$\psi_{h\,f}(y) = y + h\sum_{i=1}^{s} b_i f(Y_i), \qquad [4.7]$$

and the vectors Y_i (the so-called *internal stages*) are determined by the relations

$$Y_i = y + h\sum_{j=1}^{s} a_{ij} f(Y_j), \quad i = 1, \ldots, s. \qquad [4.8]$$

If the matrix A is strictly lower triangular (i.e. if $a_{ij} = 0$ whenever $i \leq j$), equations [4.8] provide a recursion for explicitly computing each Y_i in terms of the preceding internal stages:

$$\begin{aligned} Y_1 &= y_n, \\ Y_2 &= y_n + h\,a_{21} f(Y_1), \\ Y_3 &= y_n + h\,a_{31} f(Y_1) + h\,a_{32} f(Y_2) \\ &\vdots \end{aligned}$$

The method is then called *explicit*. The computation of one step of an explicit Runge–Kutta (RK) method thus requires s evaluations of the function f.

For general matrices A, the method is *implicit*, and [4.8] provides a coupled system of $s \times d$ algebraic equations for the $s \times d$ components of the stage vectors. By the implicit function theorem, given $y_0 \in \mathcal{U}$, there exists $h_0 > 0$ and a neighborhood $\mathcal{V}$ of $(y_0, \ldots, y_0) \in \mathbb{R}^{s\,d}$, such that the algebraic system [4.8] has a unique solution in $\mathcal{V}$ that smoothly depends on $h \in [-h_0, h_0]$.

For the tableau

$$\begin{array}{|c} 0 \\ \hline 1 \end{array}, \qquad [4.9]$$

equation [4.8] reads $Y_1 = y$ and [4.7] is

$$\psi_{h\,f} = y + hf(Y_1) = y + hf(y),$$

which is the explicit Euler formula [4.2]. The RK method given by the tableau

$$\begin{array}{|cc} 0 & 0 \\ 1 & 0 \\ \hline 1/2 & 1/2 \end{array}$$

corresponds to the explicit trapezoidal method [4.5]. The two examples of RK tableaux considered so far correspond to explicit RK methods. The RK tableau

$$\begin{array}{|cc} 0 & 0 \\ 1/2 & 1/2 \\ \hline 1/2 & 1/2 \end{array} \qquad [4.10]$$

corresponds to the implicit trapezoidal rule

$$y^* = \psi_{h\,f}(y) := y + \frac{h}{2}(f(Y_1) + f(Y_2)),$$

where $Y_1 = y$, and $Y_1 \in \mathbb{R}^d$ is implicitly defined as a function of h and y by

$$Y_2 = y + \frac{h}{2}(f(Y_1) + f(Y_2)).$$

Given two RK methods specified by two different RK tableaux determining the integration maps $\psi_{h\,f}$ and $\bar{\psi}_{h\,f}$, they are *equivalent* if, when applied to an arbitrary system [4.1] with Lipschitz continuous $f : \mathcal{U} \subset \mathbb{R}^d \to \mathbb{R}^d$, for each $y_0 \in \mathcal{U}$, there exists $h_0 > 0$, such that $\psi_{h\,f}(y_0) = \bar{\psi}_{h\,f}(y_0)$ for each $h \in (-h_0, h_0)$.

For instance, it is straightforward to check that all of the RK tableaux of the one-parameter family

$$\begin{array}{|cc} 0 & 0 \\ 0 & 0 \\ \hline \lambda & 1-\lambda \end{array}$$

are equivalent to each other, and they are equivalent to the tableau [4.9]. A less trivial example of equivalent RK tableaux is given by the following three-parameter family:

$$\begin{array}{|ccc} 0 & \lambda_1 & -\lambda_1 \\ 1/2 & \lambda_2 & 1/2-\lambda_2 \\ 1/2 & \lambda_3 & 1/2-\lambda_3 \\ \hline 1/2 & \lambda_4 & 1/2-\lambda_4 \end{array}$$

It can be shown that the tableaux in that family are all equivalent to each other and that they are actually equivalent to the tableaux [4.10], corresponding to the implicit trapezoidal rule.

4.2. Algebraic theory of Runge–Kutta methods

4.2.1. *The order conditions of RK methods*

We now aim to derive the necessary and sufficient conditions for an RK method to be of order n when applied to arbitrary (sufficiently smooth) systems of ODEs [4.1].

In the sequel, an s-stage RK method is a pair $\mu = (b, A)$, where $A = (a_{ij})_{i,j=1}^{s}$ is an $s \times s$ real matrix, and $b = (b_i)_{i=1}^{s}$ a row vector with real components, typically represented as an RK tableau [4.6]. We denote the set of RK methods as $\mathcal{RK}$.

An RK method $\mu = (b, A)$ is said to be of order n ($n \geq 1$), if [4.4] holds for the integration map $\psi_{h\,f}$ (defined by [4.7]–[4.8]) associated to each smooth vector field $f : \mathcal{U} \subset \mathbb{R}^d \to \mathbb{R}^d$.

Given an s-stage RK method $\mu = (b, A)$, for each $i = 1, \dots, s$, we denote as μ_i the RK method

$$\mu_i := \begin{array}{|ccc} a_{11} & \cdots & a_{1s} \\ \vdots & \ddots & \vdots \\ a_{s1} & \cdots & a_{ss} \\ \hline a_{i1} & \cdots & a_{is} \end{array}\,. \qquad [4.11]$$

Clearly, the integration map $\psi_{h\,f}$ associated with the method $\mu = (b, A)$ is related (for each ODE system [4.1]) to the integration maps ${}^{i}\psi_{h\,f}$ associated with the RK methods μ_i through the relation

$$\psi_{h\,f}(y) = y + h \sum_{i=1}^{s} b_i f({}^{i}\psi_{h\,f}(y)). \qquad [4.12]$$

Given $\mu = (b, A) \in \mathcal{RK}$, $n \geq 1$, and a smooth vector field $f : \mathcal{U} \subset \mathbb{R}^d \to \mathbb{R}^d$, in order to check whether [4.4] holds, we need to expand both $\psi_{h\,f}(y)$ and $f(\psi_{h\,f}(y))$ as series in powers of h. From [4.12], and since ${}^{i}\psi_{h\,f}(y) = y + \mathcal{O}(h)$, we get

$$\psi_{h\,f}(y) = y + h \left(\sum_{i=1}^{s} b_i \right) f(y) + \mathcal{O}(h^2).$$

Hence, $f(\psi_{h\,f}(y))$ can be expanded as

$$\begin{aligned} f(\psi_{h\,f}(y)) &= f(y) + f'(y)(\psi_{h\,f}(y) - y) + \mathcal{O}(h^2) \\ &= f(y) + h \left(\sum_{i=1}^{s} b_i \right) f'(y)(y) + \mathcal{O}(h^2). \end{aligned}$$

Since the latter expansion is valid for the integration map $\psi_{h\,f}(y)$ produced by an arbitrary RK method $\mu = (b, A)$, this also holds for the integration map ${}^{i}\psi_{h\,f}$ of the RK method μ_i, i.e.

$$f({}^{i}\psi_{h\,f}(y)) = f(y) + h \left(\sum_{j=1}^{s} a_{ij} \right) f'(y)(y) + \mathcal{O}(h^2).$$

By inserting that into [4.12], we in turn arrive at

$$\psi_{h\,f}(y) = y + h\left(\sum_{i=1}^{s} b_i\right) f(y) + h^2\left(\sum_{i,j=1}^{s} b_i a_{ij}\right) f'(y)f(y) + \mathcal{O}(h^3).$$

The multivariate Taylor expansion of $f(\psi_{h\,f}(y)) = f(z)$ at $z = y$ gives

$$\begin{aligned} f(\psi_{h\,f}(y)) &= f(y) + f'(y)(z-y) + \frac{1}{2} f''(y)(z-y, z-y) + \mathcal{O}(h^3) \\ &= f(y) + h\left(\sum_{i=1}^{s} b_i\right) f'(y)f(y) + h^2\left(\sum_{i,j=1}^{s} b_i a_{ij}\right) f'(y)f'(y)f(y) \\ &\quad + \frac{h^2}{2}\left(\sum_{i=1}^{s} b_i\right)^2 f''(y)(f(y), f(y)) + \mathcal{O}(h^3) \end{aligned}$$

where $f'(y)$ represents the Jacobian matrix of $f(y)$ with respect to y and $f''(y)$ represents the second Fréchet derivative at y, so that $f''(y)(f(y), f(y))$ is the vector obtained by letting $f''(y)$ act on the pair $(f(y), f(y))$. That expansion is of course valid for $f({}^{i}\psi_{h\,f}(y))$ if the coefficients of the RK method $\mu = (b, A)$ are replaced by those of μ_i. By inserting the expansions of $f({}^{i}\psi_{h\,f}(y))$ obtained in this way into [4.12], we get an expansion of $\psi_{h\,f}(y)$ with an additional power of h. In order to write that conveniently, it will be useful to introduce some notations. From now on, $\mathbb{R}^{\mathcal{RK}}$ denotes the set of functions $u : \mathcal{RK} \to \mathbb{R}$.

DEFINITION 4.1.– *Given a function $u \in \mathbb{R}^{\mathcal{RK}}$, a new function $[u] \in \mathbb{R}^{\mathcal{RK}}$ can be defined as*

$$[u](\mu) = \sum_{i=1}^{s} b_i\, u(\mu_i).$$

We denote by e the function $e \in \mathbb{R}^{\mathcal{RK}}$ defined by $e(\mu) = 1, \forall \mu \in \mathcal{RK}$. Thus, the function $[e] \in \mathbb{R}^{\mathcal{RK}}$ is defined by $[e](\mu) = \sum_{i=1}^{s} b_i$, for each $\mu = (b, A) \in \mathcal{RK}$.

Using that notation, we have

$$\begin{aligned} f(\psi_{h\,f}(y)) &= f(y) + h\,[e](\mu)\, f'(y)f(y) + h^2\,[[e]](\mu)\, f'(y)f'(y)f(y) \\ &\quad + \frac{h^2}{2}\left([e](\mu)\right)^2 f''(y)(f(y), f(y)) + \mathcal{O}(h^3), \end{aligned}$$

and hence

$$\begin{aligned} f({}^{i}\psi_{h\,f}(y)) &= f(y) + h\,[e](\mu_i)\, f'(y)f(y) + h^2\,[[e]](\mu_i)\, f'(y)f'(y)f(y) \\ &\quad + \frac{h^2}{2}\left([e](\mu_i)\right)^2 f''(y)(f(y), f(y)) + \mathcal{O}(h^3), \end{aligned}$$

which inserted into [4.12] gives

$$\begin{aligned}\psi_{h\,f}(y) &= y + h\,[e](\mu)\,f(y) + h^2\sum_{i=1}^{s} b_i[e](\mu_i)\,f'(y)f(y)\\ &\quad + h^3\sum_{i=1}^{s} b_i[[e]](\mu_i)\,f'(y)f'(y)f(y)\\ &\quad + \frac{h^3}{2}\sum_{i=1}^{s} b_i[e](\mu_i)^2\,f''(y)(f(y),f(y)) + \mathcal{O}(h^4)\\ &= y + h\,[e](\mu)\,f(y) + h^2\,[[e]](\mu)\,f'(y)f(y)\\ &\quad + h^3\,[[[e]]](\mu)\,f'(y)f'(y)f(y)\\ &\quad + \frac{h^3}{2}\,[[e]^2](\mu)\,f''(y)(f(y),f(y)) + \mathcal{O}(h^4).\end{aligned}$$

With such expansions of $\psi_{h\,f}(y)$ and $f(\psi_{h\,f}(y))$ at hand, and using the characterization [4.4] of methods of order n, we readily obtain that an RK method $\mu = (b, A) \in \mathcal{RK}$ is of order 3 if

$$[e](\mu) = 1, \quad [[e]](\mu) = \frac{1}{2}\,[e](\mu), \quad [[[e]]](\mu) = \frac{1}{3}\,[[e]](\mu),$$
$$[[e]^2](\mu) = \frac{1}{3}\,([e](\mu))^2,$$

or equivalently,

$$[e](\mu) = 1, \quad [[e]](\mu) = \frac{1}{2}, \quad [[[e]]](\mu) = \frac{1}{6}, \quad [[e]^2](\mu) = \frac{1}{3}.$$

A systematic derivation of conditions that guarantee that a method $\mu = (b, A)$ is of order n can be obtained by generalizing the above procedure. In order to do that, we consider the following sets of functions $u \in \mathbb{R}^{\mathcal{RK}}$.

DEFINITION 4.2.– *For $n \geq 1$, consider the sets $\mathcal{T}_n \subset \mathbb{R}^{\mathcal{RK}}$ recursively defined as follows: $\mathcal{T}_1 = \{[e]\}$, and for each $n \geq 2$,*

$$\mathcal{T}_n = \Big\{[u_1 \ldots u_m] \;:\; m \geq 1,\ u_i \in \mathcal{T}_{n_i} \text{ and } \sum_{i=1}^{m} n_i = n - 1\Big\}$$

We also denote $\mathcal{T} = \bigcup_{n\geq 1} \mathcal{T}_n$.

Observe that for each $u \in \mathcal{T}_n$, $u(\mu)$ is a polynomial function on the entries b_i and a_{ij} in $\mu = (b, A)$. Actually, u is a function of homogeneous degree n in the following sense: Given $\mu = (b, A) \in \mathcal{RK}$, consider for each $\lambda \in \mathbb{R}$ the new method $\lambda\mu := (\lambda b, \lambda A) \in \mathcal{RK}$.[1] A function $u \in \mathbb{R}^{\mathcal{RK}}$ is of *homogeneous degree* n if for all $\lambda \in \mathbb{R}$, $u(\lambda m) = \lambda^n u(\mu)$, and we write $|u| = n$ in that case. We will say that a function $u \in \mathbb{R}^{\mathcal{RK}}$ is *homogeneous*, if $|u| = n$ for some non-negative integer n. Obviously, $|uv| = |u| + |v|$ if u and v are functions on the set $\mathcal{RK}$.

It was observed by Butcher that the set $\mathcal{T}$ can be identified with the set of rooted trees. Indeed, we can associate the function $[e] \in \mathcal{T}_1$ with the rooted tree with one vertex, and $[u_1 \cdots u_m] \in \mathcal{T}$ (with $u_1, \ldots, u_m \in \mathcal{T}$) is associated with the rooted tree that is obtained by grafting the root of the tree corresponding to each $u_1, \ldots, u_m$ to a new root. Obviously, $|u| = n$ if the function $u \in \mathcal{T}$ is associated with a rooted tree with n vertices.

The rooted trees with up to four vertices are displayed in Table 4.1, together with their associated functions in $\bigcup_{n=1}^{4} \mathcal{T}_n \subset \mathbb{R}^{\mathcal{RK}}$.

[e]	[[e]]	[[[e]]]	[[e]^2]	[[[[e]]]]	[[[e]^2]]	[[[e]][e]]	[[e]^3]

Table 4.1. *Functions $u \in \mathcal{T}$ associated with rooted trees with up to four vertices*

We can now state Butcher's original result giving necessary and sufficient order conditions for RK methods. A proof of that result is given in section 4.2.3.

THEOREM 4.1 (Butcher).– *An RK method $\mu \in \mathcal{RK}$ is of order n if and only if*

$$u(\mu) = \frac{1}{u!} \quad \forall u \in \bigcup_{k=1}^{n} \mathcal{T}_k, \qquad [4.13]$$

where $[e]! = 1$, *and if* $u = [u_1 \cdots u_m]$, *with* $u_1, \ldots, u_m \in \mathcal{T}$,

$$u! = u_1! \cdots u_m! |u|.$$

4.2.2. *The independence of order conditions*

We will next show that, as a result of Theorem 4.2, the order conditions [4.13] given in Theorem 4.1 are all independent, in the sense that no system of

1. If $\psi_{h\,f}$ is the integration map (defined by [4.7]–[4.8]) produced by the method μ for a smooth vector field $f : \mathcal{U} \subset \mathbb{R}^d \to \mathbb{R}^d$, then the integration map associated with the method $\lambda\mu$ is precisely $\psi_{\lambda h\,f}$, i.e. the map obtained from $\psi_{h\,f}$ by rescaling the time step h as λh.

algebraic equations having the same set of solutions has fewer equations than [4.13]. Furthermore, Theorem 4.2 implies that the set of functions $\mathcal{T} \subset \mathbb{R}^{\mathcal{RK}}$ can actually be identified with the set of rooted trees, i.e. that two functions $u, v \in \mathcal{T}$ associated with two different rooted trees are also different as functions on the set $\mathcal{RK}$ of RK methods.

We first prove an auxiliary result.

LEMMA 4.1.– *Given* $u, v \in \mathbb{R}^{\mathcal{RK}}$, *if* $[u] = [v]$, *then* $u = v$.

PROOF.– This is a consequence of the following: Given $\mu = (b, A) \in \mathcal{RK}$, then the method

$$\mu' = \begin{array}{|ccccc} a_{11} & \cdots & a_{1s} & 0 \\ \vdots & \ddots & \vdots & \vdots \\ a_{s1} & \cdots & a_{ss} & 0 \\ b_1 & \cdots & b_s & 0 \\ \hline 0 & \cdots & 0 & 1 \end{array}$$

is such that

$$\forall u \in \mathbb{R}^{\mathcal{RK}}, \quad u(\mu) = [u](\mu').$$

□

THEOREM 4.2 (Butcher).– *Given arbitrary* $u_1, \ldots, u_k \in \mathcal{T}$ *associated with* k *distinct rooted trees,*

$$\forall \alpha \in \mathbb{R}^k, \quad \exists \mu \in \mathcal{RK} \quad \textit{such that} \quad \begin{pmatrix} u_1(\mu) \\ \vdots \\ u_k(\mu) \end{pmatrix} = \alpha. \qquad [4.14]$$

PROOF.– We first make the following observations:

1) Given $u_1, \ldots, u_k \in \mathcal{T}$ associated with k distinct rooted trees, [4.14] implies that the functions

$$u_1^{l_1} \cdots u_k^{l_k}$$

for different k-tuples $(l_1, \ldots, l_k)$ of non-negative integers are linearly independent;

2) The set $\mathcal{RK}$ can be endowed with a vector space structure, such that each $u \in \mathcal{T}$ is a linear function $u : \mathcal{RK} \to \mathbb{R}$. This is clearly achieved by defining, for given $(b', A'), (b'', A'') \in \mathcal{RK}$ and $\lambda', \lambda'' \in \mathbb{R}$, the linear combination $\lambda' (b', A') + \lambda'' (b'', A'')$ as the RK method $\mu = (b, A)$ given by

$$b = \left(\lambda' b' \ \lambda'' b''\right), \quad A = \begin{pmatrix} A' & 0 \\ 0 & \bar{A}'' \end{pmatrix}.$$

Since each $u_j \in \mathcal{T}$ is a linear function on the vector space $\mathcal{RK}$, [4.14] is equivalent to the linear independence of $u_1, \ldots, u_k$.

We now prove by induction on $n = \max_i |u_i|$ that the functions $u_1, \ldots, u_k \in \mathcal{T}$ associated with arbitrary k distinct rooted trees are linearly independent. The statement trivially holds for $n = 1$, as in that case $k = 1$, $u_1 = [e]$. Assume that the statement holds whenever $\max(|u_1|, \ldots, |u_k|) < n$. Hence, the functions

$$u_1^{l_1} \cdots u_k^{l_k}, \quad l_1, \ldots, l_k \geq 0,$$

are linearly independent, provided that $\max(|u_1|, \ldots, |u_k|) < n$. Assume now that there exists $v_1, \ldots, v_r \in \mathcal{T}$ associated with r distinct rooted trees with $\max(|v_1|, \ldots, |v_r|) = n$ that are linearly dependent, i.e. $\exists \lambda_1, \ldots, \lambda_r \in \mathbb{R}$, such that

$$0 = \sum_{i=1}^{r} \lambda_i v_i. \qquad [4.15]$$

By definition of the set $\mathcal{T}$, there exist $u_1, \ldots, u_k \in \mathcal{T}$, and $l_{ij} \geq 0$ for $i = 1, \ldots, r$, $j = 1, \ldots, k$, such that $v_i = [u_1^{l_{i1}} \cdots u_k^{l_{ik}}]$ for each $i = 1, \ldots, r$. By virtue of Lemma 4.1, we have

$$0 = \sum_{i=1}^{r} \lambda_i v_i = \left[\sum_{i=1}^{r} \lambda_i u_1^{l_{i1}} \cdots u_k^{l_{ik}}\right] \quad \Longrightarrow \quad 0 = \sum_{i=1}^{r} \lambda_i u_1^{l_{i1}} \cdots u_k^{l_{ik}},$$

which is in contradiction with the induction hypothesis as $\max_i |u_i| < n$. □

REMARK 4.1.– *Theorem 4.2 is also valid, as can be checked from its proof, with $\mathcal{RK}$ replaced by the set of explicit RK tableaux (i.e. the RK tableaux (b, A) with strictly lower triangular matrices A).*

4.2.3. *Proof of necessary and sufficient order conditions*

We next give some definitions and auxiliary results that we will use to prove Theorem 4.1. Recall that the set $\mathcal{T}$ can be identified with the set of rooted trees, and that for each $u \in \mathcal{T}$ with $|u| > 1$, there exists a unique $m \geq 1$ and $v_1, \ldots, v_m \in \mathcal{T}$, unique up to permutations, such that $u = [v_1 \cdots v_m]$.

DEFINITION 4.3.– *Given $f : \mathbb{R}^d \to \mathbb{R}^d$, we define for each $u \in \mathcal{T}$, a smooth map $F_u : \mathbb{R}^d \to \mathbb{R}^d$ (called the elementary differential of u) as follows. For $[e] = [e] \in \mathcal{T}_1$, $F_{[e]} = f$, and for $u = [u_1 \cdots u_m]$, where $u_1, \ldots, u_m \in \mathcal{T}$,*

$$F_u(y) = f^{(m)}(y)(F_{u_1}(y), \ldots, F_{u_m}(y)) \quad \forall y \in \mathcal{U}.$$

DEFINITION 4.4.– *We consider for* $n \geq 0$ *the sets* $\mathcal{F}_n$ *of functions on* $\mathcal{RK}$ *defined as follows:* $\mathcal{F}_0 = \{e\}$, *and for* $n \geq 1$,

$$\mathcal{F}_n = \Big\{ u_1 \dots u_m \; : \; m \geq 1, \; u_i \in \mathcal{T}_{n_i} \text{ and } \sum_{i=1}^{m} n_i = n \Big\}.$$

We also denote $\mathcal{F} = \bigcup_{n \geq 0} \mathcal{F}_n$.

Recall that $e \in \mathbb{R}^{\mathcal{RK}}$ is defined by $e(\mu) = 1$. From the proof of Theorem 4.2, we get that each $u \in \mathcal{F}$ can be uniquely factored in the form $u = u_1^{l_1} \cdots u_k^{l_k}$, where $u_1, \dots, u_k \in \mathcal{T}$ are distinct functions and $l_1, \dots, l_k \geq 1$.

It is straightforward to check that each $u \in \mathcal{F}_n$ has a homogeneous degree n. Comparing the definition of the sets $\mathcal{T}_n$ and $\mathcal{F}_{n-1}$, we see that, for each $n \geq 1$, $\mathcal{T}_n = \{[u] \; : \; u \in \mathcal{F}_{n-1}\}$. Clearly,

$$\mathcal{F} = \{e\} \cup \Big\{ u_1 \dots u_m \; : \; m \geq 1, \; u_i \in \mathcal{T} \Big\}.$$

The set of functions $\mathcal{F} \subset \mathbb{R}^{\mathcal{RK}}$ can be identified with the set of forests of rooted trees. A forest is a collection of (possibly repeated) rooted trees.

DEFINITION 4.5.– *We recursively define a positive integer for each* $u \in \mathcal{F}$, $\sigma(u)$ *(the* symmetry number *of* u*) as follows:* $\sigma(e) = 1$, $\sigma([u]) = \sigma(u)$ *if* $u \in \mathcal{F}$, *and if* $u_1, \dots, u_m \in \mathcal{T}$ *are distinct functions and* $l_1, \dots, l_m \geq 1$,

$$\sigma(u_1^{l_1} \cdots u_m^{l_m}) = l_1! \cdots l_m! \sigma(u_1)^{l_1} \cdots \sigma(u_m)^{l_m}.$$

The elementary differentials F_u, and the values of $u!$ and $\sigma(u)$ associated with rooted trees with four or less vertices are displayed in Table 4.2.

$\lvert u \rvert$	1	2	3	3	4	4	4	4
u								
F_u	f	$f'f$	$f'f'f$	$f''(f,f)$	$f'f'f'f$	$f'f''(f,f)$	$f''(f'f,f)$	$f'''(f,f,f)$
$u!$	1	2	6	3	24	12	8	4
$\sigma(u)$	1	1	1	2	1	2	1	6

Table 4.2. *Elementary differentials F_u and the values of $u!$ and $\sigma(u)$ for rooted trees u with up to four vertices*

PROPOSITION 4.1.– *Consider an ODE system [4.1] and the integration map $\psi_{h\,f}$ associated with an arbitrary RK method $\mu \in \mathcal{RK}$. Then, it holds that, for each $y \in \mathcal{U}$, $\psi_{h\,f}(y)$ can be expanded as follows:*

$$\psi_{h\,f}(y) = y + \sum_{u \in \mathcal{T}} \frac{h^{|u|}}{\sigma(u)} u(\mu)\, F_u(y), \tag{4.16}$$

$$= y + \sum_{n=1}^{\infty} h^n \sum_{u \in \mathcal{T}_n} \frac{1}{\sigma(u)} u(\mu)\, F_u(y).$$

In addition, $f(\psi_{h\,f}(y))$ can be expanded as

$$f(\psi_{h\,f}(y)) = f(y) + \sum_{v \in \mathcal{F}} \frac{h^{|v|}}{\sigma(v)} v(\mu)\, F_{[v]}(y). \tag{4.17}$$

PROOF.– We will first show that [4.16] formally implies [4.17]. In order to do so, we resort to the multivariate Taylor expansion of $f(z)$ at $z = y$,

$$f(z) = f(y) + \sum_{m \geq 1} \frac{1}{m!} f^{(m)}(y)(\overbrace{z-y, \ldots, z-y}^{m \text{ times}}). \tag{4.18}$$

Here, $f^{(m)}(y)$ is the mth Fréchet derivative of f at y. We replace z in equation [4.18] by the expansion [4.16], and taking the multilinearity of $f^{(m)}(y)$ into account, we obtain that $f(z) - f(y)$ can be expanded as

$$\sum_{m \geq 1} \sum_{u_1, \ldots, u_m \in \mathcal{T}} \frac{h^{|u_1| + \cdots + |u_m|}}{m!} \frac{u_1(\mu) \cdots u_m(\mu)}{\sigma(u_1) \cdots \sigma(u_m)} f^{(m)}(y)(F_{u_1}(y), \ldots, F_{u_m}(y))$$

and by the definition of elementary differentials F_u, we get that $f(\psi_{h\,f}(y))$ admits the expansion

$$f(y) + \sum_{m \geq 1} \sum_{u_1, \ldots, u_m \in \mathcal{T}} \frac{h^{|u_1 \cdots u_m|}}{m!} \frac{(u_1 \cdots u_m)(\mu)}{\sigma(u_1) \cdots \sigma(u_m)} F_{[u_1 \cdots u_m]}(y). \tag{4.19}$$

Collecting the repeated terms, a term per $v = u_1 \cdots u_m \in \mathcal{F}$ is obtained. If $v_1, \ldots, v_k$ are the distinct rooted trees in $\{u_1, \ldots, u_m\}$, and the number of rooted trees u_j that coincide with v_i is denoted by l_i, so that $u_1 \cdots u_m = v_1^{l_1} \cdots v_k^{l_k}$, then the term corresponding to $v = u_1 \cdots u_m$ appears $m!/(l_1! \cdots l_k!)$ times in equation [4.19]. This shows, by virtue of the definition of $\sigma(v)$ that $f(\psi_{h\,f}(y))$ can be expanded as equation [4.17].

Thus, if equation [4.16] holds up to terms of degree n in h for arbitrary RK methods $\mu \in \mathcal{RK}$, then equation [4.17] also holds up to terms of degree n in h. Then, equation [4.7] implies that equation [4.16] holds up to terms of degree $n+1$ in h. □

For each $n \geq 1$, let d_n be the cardinality of the set $\mathcal{T}^n := \cup_{k=1}^{n} \mathcal{T}_k$. Consider a total ordering $u_1 < u_2 < u_3 < \cdots$ of the set $\mathcal{T}$ such that, for each $n \geq 0$, $\mathcal{T}_{n+1} = \{u_{d_n+1}, \ldots, u_{d_{n+1}}\}$. The following result can be proven by induction on n.

LEMMA 4.2.– *Given $n \geq 1$, consider the polynomial vector field $f : \mathbb{R}^{d_n} \to \mathbb{R}^{d_n}$ defined as follows: For $y = (y^1, \ldots, y^{d_n})^T$, $f(y) = (f^1(y), \ldots, f^{d_n}(y))^T$, where $f^1(y) = 1$, and for $i \geq 2$, if $u_i = [u_{j_1}^{l_1} \cdots u_{j_m}^{l_m}]$, where $u_{j_1}, \ldots, u_{j_m} \in \mathcal{T}$ are distinct and $l_1, \ldots, l_m \geq 1$, then*

$$f^i(y) = \frac{1}{(l_1 + \cdots + l_m)!}(y^{j_1})^{l_1} \cdots (y^{j_m})^{l_m}.$$

The elementary differentials F_{u_i}, $u_i \in \mathcal{T}^n$, corresponding to f evaluated at the origin $0 \in \mathbb{R}^{d_n}$ satisfy

$$F_{u_i}^j(0) = \begin{cases} 1 & \textit{if } j = i, \\ 0 & \textit{otherwise.} \end{cases}$$

PROOF OF THEOREM 4.1.– Application of Proposition 4.1 and the characterization [4.4] of methods of order n lead to the following sufficient conditions for order n,

$$[v](\mu) = \frac{1}{k} v(\mu) \quad \forall v \in \mathcal{F}_{k-1}, \quad 1 \leq k \leq n. \qquad [4.20]$$

Lemma 4.2 implies that condition [4.20] is also necessary. Since by definition, $\mathcal{T}_k = \{[v] \ : \ \mathcal{F}_{k-1}\}$, the required result follows by observing that, by virtue of the recursive definition of $u!$, equation [4.20] is equivalent to equation [4.13]. ■

4.2.4. *Composition of RK methods, rooted trees and forests*

We next define a composition in $\mathcal{RK}$ that is compatible (for each ODE [4.1]) with the composition of integration maps $\psi_{h\,f}$ associated with RK methods.

DEFINITION 4.6.– *Given an s-stage RK method $\mu = (b, A) \in \mathcal{RK}$ and an s-stage RK method $\mu' = (b', A') \in \mathcal{RK}$, the composition $\mu\,\mu' \in \mathcal{RK}$ is the $(s + s')$-stage RK method given by the Butcher tableau*

$$\begin{array}{|cccccc}
a_{11} & \cdots & a_{1s} & 0 & \cdots & 0 \\
\vdots & \ddots & \vdots & \vdots & \ddots & \vdots \\
a_{s1} & \cdots & a_{ss} & 0 & \cdots & 0 \\
b_1 & \cdots & b_s & a'_{11} & \cdots & a'_{1s'} \\
\vdots & \ddots & \vdots & \vdots & \ddots & \vdots \\
b_1 & \cdots & b_s & a'_{s'1} & \cdots & a'_{s's'} \\
\hline
b_1 & \cdots & b_s & b'_1 & \cdots & b'_{s'}
\end{array}.$$

Clearly, if $\psi_{h\,f}$ and $\psi'_{h\,f}$ are the integration maps associated with the RK methods μ and μ', respectively, then the integration map associated to $\mu\,\mu'$ is the composition $\psi'_{h\,f} \circ \psi_{h\,f}$.

PROPOSITION 4.2.– *Given an arbitrary function* $u \in \mathbb{R}^{\mathcal{RK}}$*, for all* $\mu, \mu' \in \mathcal{RK}$,

$$[u](\mu\,\mu') = [u](\mu) + [u_\mu](\mu'), \qquad [4.21]$$

where $u_\mu \in \mathbb{R}^{\mathcal{RK}}$ *is defined as* $u_\mu(\mu') = u(\mu\,\mu')$ *for all* $\mu' \in \mathcal{RK}$.

PROOF.– We first note that, from Definition 4.6,

$$(\mu\,\mu')_i = \begin{cases} \mu_i & \text{if} \quad 1 \leq i \leq s, \\ \mu'_{i-s}\,\mu & \text{if} \quad s+1 \leq i \leq s+s'. \end{cases}$$

Given $u \in \mathbb{R}^{\mathcal{RK}}$, by applying Definition 4.1 to $[u] \in \mathbb{R}^{\mathcal{RK}}$, we obtain

$$\begin{aligned} [u](\mu\,\mu') &= \sum_{i=1}^{s} b_i u((\mu\,\mu')_i) + \sum_{i=1}^{s'} b'_i u((\mu\,\mu')_{s+i}) \\ &= \sum_{i=1}^{s} b_i u(\mu_i) + \sum_{i=1}^{s'} b'_i u(\mu\,\mu'_i) \\ &= \sum_{i=1}^{s} b_i u(\mu_i) + \sum_{i=1}^{s'} b'_i u_\mu(\mu'_i), \end{aligned}$$

and equation [4.21] follows from applying Definition 4.1 to $[u](\mu)$ and $[u_\mu](\mu')$. □

The identity [4.21] can be equivalently written as

$$[u]_\mu = [u](\mu)\,e + [u_\mu] \qquad [4.22]$$

which allows recursively writing formulas for $u(\mu\mu')$ for $u \in \mathcal{F}$, starting from $(e)_\mu = e$. For instance,

$$\begin{aligned} [e]_\mu &= [e](\mu)\,e + [e], \\ [[e]]_\mu &= [[e]](\mu)\,e + [[e](\mu)\,e + [e]] \;=\; [[e]](\mu)\,e + [e](\mu)\,[e] + [[e]], \\ ([e]^2)_\mu &= ([e](\mu)\,e + [e])^2 \;=\; [e](\mu)^2\,e + 2[e](\mu)\,[e] + [e]^2, \\ [[e]^2]_\mu &= [[e]^2](\mu)\,e + [[e](\mu)^2\,e + 2[e](\mu)\,[e] + [e]^2] \\ &= [[e]^2](\mu)\,e + [e](\mu)^2\,[e] + 2[e](\mu)\,[[e]] + [[e]^2], \\ [[[e]^2]]_\mu &= [[[e]^2]](\mu)\,e + [[[e]^2]\,e + [e](\mu)^2\,[e] + 2[e](\mu)\,[[e]] + [[e]^2]] \\ &= [[[e]^2]](\mu)\,e + [[e]^2](\mu)\,[e] + [e](\mu)^2\,[[e]] + 2[e](\mu)\,[[[e]]] + [[[e]^2]] \end{aligned}$$

or equivalently, rewriting $u_\mu(\mu')$ back as $u(\mu\,\mu')$ and representing each $u \in \mathcal{T}$ by the corresponding rooted tree,

$$
\begin{aligned}
\bullet(\mu\,\mu') &= \bullet(\mu) + \bullet(\mu'),\\
\mathbf{I}(\mu\,\mu') &= \mathbf{I}(\mu) + \bullet(\mu)\,\bullet(\mu') + \mathbf{I}(\mu'),\\
\Big(\bullet(\mu\,\mu')\Big)^2 &= \Big(\bullet(\mu)\Big)^2 + 2\bullet(\mu)\,\bullet(\mu') + \Big(\bullet(\mu')\Big)^2,\\
\mathbf{V}(\mu\,\mu') &= \mathbf{V}(\mu) + \Big(\bullet(\mu)\Big)^2\bullet(\mu') + 2\bullet(\mu)\,\mathbf{I}(\mu') + \mathbf{V}(\mu'),\\
\mathbf{Y}(\mu\,\mu') &= \mathbf{Y}(\mu) + \mathbf{V}(\mu)\,\bullet(\mu') + \Big(\bullet(\mu)\Big)^2\mathbf{I}(\mu')\\
&\quad + 2\bullet(\mu)\,\vdots(\mu') + \mathbf{Y}(\mu').
\end{aligned}
\qquad [4.23]
$$

The formulas in equation [4.23] for $u(\mu\,\mu')$ can be interpreted in terms of the rooted tree associated with u. For instance, $u(\mu\,\mu')$ for the rooted tree $u = \mathbf{Y}$ is a sum of terms of the form $v(\mu)w(\mu')$, where $(v, w) \in \mathcal{F} \times \mathcal{T}$ is obtained from pruning the original tree (where w is the rooted tree that remains after pruning and v corresponds to the pieces that are removed). In that example, the collection of such (v, w) is

$$
(\mathbf{Y}, e), (\mathbf{V}, \bullet), (\bullet\bullet, \mathbf{I}), (\bullet, \vdots), (\bullet, \vdots), (e, \mathbf{Y}).
$$

Next, we will make this precise in the general case.

The diagrams representing rooted trees in Table 4.1 can be identified with partially ordered sets U of points in the plane (the vertices of U) having only one minimal vertex (the root of U) and satisfying the following:

$$
x, y, z \in U \quad x < z, \quad y < z \quad \Longrightarrow \quad x < y \quad \text{or} \quad y < x. \qquad [4.24]
$$

Actually, a rooted tree can be defined as an isomorphism class[2] of finite partially ordered sets satisfying equation [4.24] and having only one minimal vertex (the root). In turn, a forest of rooted trees can be defined as an isomorphism class of finite partially ordered sets satisfying equation [4.24]. Given two disjoint partially ordered sets V and W representing two forests v and w, respectively, its (disjoint) union $U = V + W$ also represents a forest, which we denote by $u = vw$. Clearly, any partially ordered set U satisfying equation [4.24] and having $m > 1$ minimal vertices can be uniquely decomposed as the disjoint union of m partially ordered sets $U_1, \ldots, U_m$ representing rooted trees (i.e. satisfying equation [4.24] and having only one minimal vertex), and in that case, U represents the forest $u = u_1 \cdots u_m$.

2. An isomorphism of partially ordered sets is a bijection of the sets that preserve their partial ordering.

Given a rooted tree v, we obviously obtain a forest u by removing the root of v. We will write $v = [u]$ in that case (thus mimicking the notation we have used so far for the associated functions in $\mathbb{R}^{\mathcal{RK}}$). Any rooted tree can thus be represented as $[u]$, where u is a uniquely determined forest of rooted trees. Later on, we will make use of an additional operation on rooted trees and forests, the so-called (left) Butcher product: Given a rooted tree $v = [u]$ (u a forest) and a forest w, we denote by $w \bullet v = [uw]$, i.e. $w \bullet v$ is the rooted tree of degree $|v| + |w|$ obtained by grafting each of the labeled rooted trees in the forest w to the root of v.

Given a partially ordered set U, we will say that a pair $(x, y) \in U \times U$ is comparable in U if either $x < y$ or $y < x$. Given partially ordered subsets $V_1, \ldots, V_m$ of U, we write $(V_1 \succ \cdots \succ V_m) \subset U$ if the following three conditions are satisfied:

- each V_i is a partially ordered subset of U;
- as a set, U is the disjoint union of $V_1, \ldots, V_m$;
- if $(x, y) \in V_i \times V_{i+1}$ is comparable in U, then $x > y$.

THEOREM 4.3 (Butcher).– *Given $u \in \mathcal{F}$, let U be a partially ordered set representing the forest associated with u. Then, $\forall(\mu, \mu') \in \mathcal{RK} \times \mathcal{RK}$,*

$$u(\mu\,\mu') = \sum_{(V \succ W) \subset U} v(\mu)\, w(\mu') \qquad [4.25]$$

where the partially ordered sets V and W represent the forests associated with the functions v and w, respectively.

PROOF.– We first observe that, using the notation in Proposition 4.2, equation [4.25] can equivalently be written as

$$u_\mu = \sum_{(V \succ W) \subset U} v(\mu)\, w. \qquad [4.26]$$

We will prove by induction on n that equation [4.26] holds true for all $u \in \mathcal{F}$ with $|u| \leq n$. This trivially holds for $n = 0$, as $\mathcal{F}_0 = \{e\}$, and $e_\mu = e$, while e corresponds to the empty forest, represented by the empty partially ordered set $\emptyset$, which admits the unique decomposition $(\emptyset \succ \emptyset) \subset \emptyset$.

Assume now that for all $u \in \mathcal{F}$ with $|u| \leq n - 1$, equation [4.26] holds. Then, equation [4.22] implies that

$$[u]_\mu = [u](\mu)\, e + \sum_{(V \succ W) \subset U} v(\mu)\, [w], \qquad [4.27]$$

where we have used that, by virtue of Definition 4.1, $[\cdot] : \mathbb{R}^{\mathcal{RK}} \to \mathbb{R}^{\mathcal{RK}}$ is linear. If U' is the partially ordered set representing $[u]$, and $x \in U'$ is the root of U', then the partially ordered set $U = U' \backslash \{x\}$ represents u, and the pairs of partially ordered sets $(V', W') \neq (U, \emptyset)$, such that $(V' \succ W') \subset U'$ are in one-to-one correspondence with the pairs (V, W), such that $(V \succ W) \subset U$, with $V = V'$ and $W = W' \backslash \{x\}$. This, together with equation [4.27], implies that

$$[u]_\mu = \sum_{(V \succ W') \subset U'} v(\mu)\, w'.$$

We thus have that equation [4.26] is true for all $u \in \mathcal{T}$ with $|u| \leq n$. Given $u \in \mathcal{F} \backslash \mathcal{T}$ with $|u| = n$, consider $u', u'' \in \mathcal{F} \backslash \{e\}$, such that $u = u'u''$. If U is a partially ordered set representing u, then U must be the disjoint union of two partially ordered sets U' and U'' representing u' and u'', respectively. Then, by application of the induction hypothesis, we have

$$u_\mu = \sum_{\substack{(V' \succ W') \subset U' \\ (V'' \succ W'') \subset U''}} v'(\mu)\, v''(\mu)\, w'\, w'',$$

which finally leads to equation [4.26] due to the following property of partially ordered sets satisfying equation [4.24]: Given two disjoint partially ordered sets U' and U'' and $U = U' + U''$, if $(V \succ W) \subset U$, then there exist unique partially ordered sets V', V'', W', W'' satisfying $V = V' + V''$, $W = W' + W''$, $(V' \succ W') \subset U'$, $(V'' \succ W'') \subset U''$. □

4.2.5. *The Butcher group*

From now on, $\mathcal{T}$ and $\mathcal{F}$ will denote the set of rooted trees and the set of forests of rooted trees, respectively.

Following Butcher (1972), we will define a product $*$ on the set $\mathbb{R}^{\mathcal{F}}$ of functions

$$\begin{aligned} \alpha : \mathcal{F} &\longrightarrow \mathbb{R} \\ u &\mapsto \alpha_u \end{aligned}$$

that will endow $\mathbb{R}^{\mathcal{F}}$ with a unital (noncommutative) associative algebra structure closely related to the composition of RK methods.

Given $\mu, \mu' \in \mathcal{RK}$, consider $\alpha, \beta, \gamma \in \mathbb{R}^{\mathcal{F}}$, such that, $\forall u \in \mathcal{F}$,

$$\alpha_u = u(\mu), \quad \beta_u = u(\mu'), \quad \gamma_u = u(\mu\, \mu').$$

With that notation, equation [4.25] can be rewritten as $\gamma = \alpha * \beta$, where the product $*$ is defined next.

DEFINITION 4.7.– *Given $\alpha, \beta \in \mathbb{R}^{\mathcal{F}}$, $\alpha * \beta \in \mathbb{R}^{\mathcal{F}}$ is defined as follows: If U is a partially ordered set representing a forest $u \in \mathcal{F}$,*

$$(\alpha * \beta)_u = \sum_{(V \succ W) \subset U} \alpha_V \, \beta_W, \qquad [4.28]$$

where $\alpha_V = \alpha_v$ and $\beta_W = \beta_w$ if V and W represent the forests v and w, respectively.

The product $*$ is well defined in $\mathbb{R}^{\mathcal{F}}$ (i.e. $(\alpha * \beta)_u$ does not depend on the partially ordered set U chosen as a representative of u). Indeed, if U and U' are partially ordered sets representing the same forest $u \in \mathcal{F}$, then $(V \succ W) \subset U$ implies that $(V' \succ W') \subset U'$, where the isomorphism from U to U' sends V and W to V' and W', respectively.

The product $*$ endows the vector space $\mathbb{R}^{\mathcal{F}}$ with a unital associative algebra structure with unit $\mathbf{1} \in \mathbb{R}^{\mathcal{F}}$ defined as $\mathbf{1}(e) = 1$ and $\mathbf{1}_u = 0$, if $u \in \mathcal{F} \backslash \{e\}$. Observe that the associativity of $*$ is immediate from this definition, since

$$(\alpha * (\beta * \gamma))_u = ((\alpha * \beta) * \gamma)_u = \sum_{(V_1 \succ V_2 \succ V_3) \subset U} \alpha_{v_1} \beta_{v_2} \gamma_{v_3}.$$

An important property of the product in Definition 4.7 is that, for each $u \in \mathcal{F}$, $(\alpha * \beta - \beta_e \alpha - \alpha_e \beta)_u$ is a $\mathbb{Z}$-linear combination of terms of the form $\alpha_v \beta_w$ with $\max(|v|, |w|) < |u|$, which allow the results to be proven by induction on the number of vertices $|u|$ of the forest u.

For instance, we can easily prove that, given $\alpha \in \mathbb{R}^{\mathcal{F}}$, the equation $\alpha * \beta = \mathbf{1}$ can be uniquely solved for $\beta \in \mathbb{R}^{\mathcal{F}}$, provided that $\alpha_e \neq 0$. This proves that the subset $\{\alpha \in \mathbb{R}^{\mathcal{F}} \, : \, \alpha_e \neq 0\} \subset \mathbb{R}^{\mathcal{F}}$ is a group under the product $*$. Such a group contains an important subgroup, the *Butcher group*:

PROPOSITION 4.3 (Butcher).– *The subset of $\mathbb{R}^{\mathcal{F}}$*

$$\mathcal{G} = \{\alpha \in \mathbb{R}^{\mathcal{F}} \, : \, \alpha_e = 1, \, \forall u, v \in \mathcal{F} \; \alpha_{uv} = \alpha_u \alpha_v\}, \qquad [4.29]$$

is a group under the product $*$ with identity element $\mathbf{1}$.

PROOF.– It is sufficient to prove that $\mathcal{G}$ is a subgroup of $\{\alpha \in \mathbb{R}^{\mathcal{F}} \, : \, \alpha_e \neq 0\}$. Indeed, we can prove that $\alpha * \beta \in \mathcal{G}$ if $\alpha, \beta \in \mathcal{G}$ by following the argument used at the end of the proof of Theorem 4.3. □

Obviously, each element $\alpha \in \mathcal{G}$ is uniquely determined by its values α_u for $u \in \mathcal{T}$, so that, as a set, $\mathcal{G}$ can be identified with the set of functions $\mathbb{R}^{\mathcal{T}}$.

We end this section by stating two fundamental properties of the product [4.28]:

– P1: the restriction to $\mathcal{T}$ of $\alpha * \beta - \beta_e\alpha$ depends linearly on the restriction to $\mathcal{T}$ of β. This is a consequence, according to Definition 4.7, of the fact that, if U is a partially ordered set representing a rooted tree u, and $(V \succ W) \subset U$ with $W \neq \emptyset$, then W necessarily represents a rooted tree;

– P2: given $\beta \in \mathbb{R}^{\mathcal{F}}$ with $\beta_e = 0$ (and in particular, $\beta = \alpha - \mathbf{1}$ with $\alpha \in \mathcal{G}$), $(\beta^{*n})_u = 0$, provided that $u \in \mathcal{F}$ with $|u| < n$. Indeed, in that case,

$$(\beta^{*n})_u = \sum_{(V_1 \succ \cdots \succ V_n) \subset U} \beta_{V_1} \cdots \beta_{V_n}$$

where U is a partially ordered set representing a rooted tree u, but since $|u| < n$, some of the V_j must necessarily be the empty partially ordered set (thus representing the empty forest e), which implies that $(\beta^{*n})_u = 0$.

4.2.6. *Equivalence classes of RK methods*

Recall that two RK methods $\mu = (b, A)$, $\mu' = (b', A')$ may be equivalent (section 4.1.2) in the sense of producing, for each system [4.1], exactly the same numerical solution for sufficiently small step sizes (provided that f is Lipschitz continuous).

Proposition 4.1 implies that, if $\mu, \mu' \in \mathcal{RK}$ are equivalent, then $u(\mu) = u(\mu')$ for all $u \in \mathcal{T}$. Actually, as shown by Butcher (1972) (see also Butcher (2008)), the reverse also holds true.

Given an RK method $\mu \in \mathcal{RK}$, let us consider

$$\begin{aligned} \widehat{\mu} : \mathcal{F} &\longrightarrow \mathbb{R} \\ u &\mapsto u(\mu). \end{aligned} \qquad [4.30]$$

Then, the set of equivalence classes of RK methods can be identified with the set

$$\mathcal{G}_{RK} = \{\widehat{\mu} \, : \mu \in \mathcal{RK}\} \subset \mathbb{R}^{\mathcal{F}}. \qquad [4.31]$$

It is clear that the set $\mathcal{G}_{RK} \subset \mathbb{R}^{\mathcal{F}}$ of equivalence classes of RK methods is contained in $\mathcal{G}$. Actually, as shown in Butcher (1972), $\mathcal{G}_{RK}$ is a subgroup of $\mathcal{G}$.

Moreover, Theorem 4.2 implies that $\mathcal{G}_{RK}$ is dense in $\mathcal{G}$, with the structure of a topological group determined by the neighborhood basis $\{\mathcal{U}_n\}_{n \geq 1}$ at $\mathbf{1} \in \mathcal{G}$ given by $\mathcal{U}_n = \{\alpha \in \mathcal{G} \, : \alpha_u = 0 \text{ if } u \in \mathcal{T}_k \text{ with } k \leq n\}$ (furthermore, $\mathcal{U}_1 \subset \mathcal{U}_2 \subset \cdots$ are normal subgroups of $\mathcal{G}$).

4.2.7. *Bibliographical comments*

Essentially all of the results in section 4.2 are due to Butcher (1972). Butcher's original proof of Theorem 4.3 makes use of a different (although equivalent) recursion based on the Butcher product $w \bullet v$ of two rooted trees v and w defined in section 4.2.4 (in Butcher's original work, $v{\cdot}w$ denotes what we write as $w{\bullet}v$). The recursive formula [4.22] (written with a different notation) can be found in Chan (1998) and Butcher and Chan (2002) as an equivalent formulation of Butcher's original recursion. In Butcher's original definition of the product [4.28], subsets of a particular large partially ordered set are used instead of arbitrary partially ordered sets.

4.3. B-series and related formal expansions

In Hairer and Wanner (1974), it is observed that there are integration methods other than RK methods (for instance, multiderivative Runge–Kutta methods) that admit expansions that generalize those in Proposition 4.1: the so-called B-series (i.e. Butcher series). Working with B-series allows us to work directly with equivalence classes of Runge–Kutta methods and other one-step methods. The B-series expansion of one step of such a method is a series in powers of h that is convergent for sufficiently small h in the case of real analytic vector fields f. In addition, B-series that do not necessarily correspond to the expansion of actual integration methods (and may not converge even in the real analytic case) have proven to be very useful in the analysis of numerical methods for ODEs.

4.3.1. *B-series*

Following Hairer and Wanner (1974), we define for each $\alpha \in \mathbb{R}^{\mathcal{T}\cup\{e\}}$, a formal series as follows.

DEFINITION 4.8 (Hairer and Wanner).– *Given a vector field f of an ODE [4.1], the B-series associated with $\alpha \in \mathbb{R}^{\mathcal{T}\cup\{e\}}$ is defined as*

$$B_{hf}(\alpha, y) = \alpha_e\, y + \sum_{u \in \mathcal{T}} \frac{h^{|u|}}{\sigma(u)} \alpha_u\, F_u(y). \tag{4.32}$$

We will drop the subindex hf from $B_{hf}(\alpha, y)$ when there is no ambiguity.

With that notation, Proposition 4.1 can be restated as follows: the integration map $\psi_{hf}(y)$ of an arbitrary RK method $\mu \in \mathcal{RK}$ can be expanded as a B-series $B_{hf}(\alpha, y)$, where for each $u \in \mathcal{T} \cup \{e\}$, $\alpha_u = u(\mu)$ (and in particular, $\alpha_e = 1$), and $f(\psi_{hf}(y))$ can be expanded as a B-series $B_{hf}(\beta, y)$, where $\beta_e = 0$, and for each $u \in \mathcal{T}$, with $u = [v]$, $v \in \mathcal{T}$, $\beta_u = v(\mu)$.

Given two RK methods $\mu, \mu' \in \mathcal{RK}$ and their composition $\mu\mu' \in \mathcal{RK}$, consider their corresponding B-series $B(\alpha, y)$, $B(\beta, y)$ and $B(\gamma, y)$, i.e. for each $u \in \mathcal{T} \cup \{e\}$, $\alpha_u = u(\mu)$, $\beta_u = u(\mu')$, $\gamma_u = u(\mu\mu')$ (in particular, $\alpha_e = \beta_e = \gamma_e = 1$), then we must formally have that

$$B(\beta, B(\alpha, y)) = B(\gamma, y).$$

Theorem 4.3 states that, using the notation introduced in Definition 4.7, for each $u \in \mathcal{T} \cup \{e\}$, $\gamma_u = (\alpha * \beta)_u$, where for each forest $v \in \mathcal{F}$, $\alpha_v = v(\mu)$, i.e.

$$\alpha_{u_1 \cdots u_m} = \alpha_{u_1} \cdots \alpha_{u_m}, \quad \text{if} \quad v = u_1 \cdots u_m \in \mathcal{F}. \qquad [4.33]$$

(Recall that, according to Property P2 in section 4.2.5, the values α_v and β_w for $(v, w) \in \mathcal{F} \times (\mathcal{T} \cup \{e\})$ are required to determine all of the values $(\alpha * \beta)_u$, $u \in \mathcal{T} \cup \{e\}$.)

Furthermore, Theorem 4.2 implies that, for arbitrary $\alpha, \beta \in \mathbb{R}^{\mathcal{T} \cup \{e\}}$ with $\alpha_e = \beta_e = 1$,

$$B(\beta, B(\alpha, y)) = B(\alpha * \beta, y) \qquad [4.34]$$

where $\alpha \in \mathbb{R}^{\mathcal{T} \cup \{e\}}$ is extended to $\alpha \in \mathcal{G} \subset \mathbb{R}^{\mathcal{F}}$ by equation [4.33]. Observe that equation [4.33] defines a bijection from the set $\{\alpha \in \mathbb{R}^{\mathcal{T} \cup \{e\}} : \alpha_e = 1\}$ to $\mathcal{G}$.

Observe that $\mathbf{1} \in \mathcal{G}$ corresponds, as expected, to the B-series representing the identity map, i.e. $B(e, y) \equiv y$. Formula [4.34] shows that a B-series $B_{hf}(\alpha, y)$ (for a given vector field f) defines a homomorphism from the group $\mathcal{G}$ to the group of near-to-identity formal maps $\psi_h(y) = y + hg_1(y) + h^2 g_2(y) + \cdots$ in $\mathbb{R}^d$ (with smooth maps $g_n : \mathbb{R}^d \to \mathbb{R}^d$).

The composition formula [4.34] can also be extended to the case $\beta_e \neq 1$, leading to the following result. A direct proof (that does not make use of the results in section 4.2) was given in the original paper (Hairer and Wanner 1974). An alternative direct proof of more general results (Theorem 4.5) will be given in section 4.3.3.

THEOREM 4.4 (Hairer and Wanner).– *Given two B-series $B(\alpha, y)$ and $B(\beta, y)$, where $\alpha, \beta \in \mathbb{R}^{\mathcal{T} \cup \{e\}}$ with $\alpha_e = 1$, equation [4.34] formally holds, where $\alpha * \beta \in \mathbb{R}^{\mathcal{F}}$ is given by equations [4.28] and [4.33].*

4.3.2. *Backward error analysis, the exponential and the logarithm*

The problem of standard backward error analysis of numerical integration methods for ODEs can be stated as follows: Given $\alpha \in \mathcal{G}$ representing an integration method that, when applied to the ODE [4.1] with solution $y(t)$, provides the approximations $y_n \approx y(hn)$ at $t = nh$ in a step-by-step manner as

$$y_{n+1} = B_{hf}(\alpha, y_n), \quad n = 0, 1, 2, \ldots, \qquad [4.35]$$

determine a *modified ODE* of the form

$$\frac{d\tilde{y}}{dt} = \tilde{f}(\tilde{y}; h), \quad \tilde{y}(0) = y_0, \tag{4.36}$$

(which is expected to be a perturbation of the ODE [4.1] parameterized by the discretization parameter h), whose solution $\tilde{y}(t)$ satisfies that $\tilde{y}(nh) = y_n$ for all n, or equivalently,

$$\tilde{y}(nh) = B_{hf}(\alpha^{*n}, y_0), \quad n = 0, 1, 2, \ldots \tag{4.37}$$

Let us assume that such a modified equation exists and admits an expansion in powers of h of the form

$$\tilde{f}(\tilde{y}; h) = \tilde{f}_1(\tilde{y}) + h\tilde{f}_1(\tilde{y}) + h^2\tilde{f}_2(\tilde{y}) + \cdots$$

In that case, application of polynomial interpolation with nodes $t = nh$, $n = 0, \ldots, N$ (for an arbitrarily high positive integer N) gives an interpolant $P_N(t)$ of the solution $\tilde{y}(t)$ of the modified equation [4.36] having a B-series expansion $P_N(t) = B_{hf}(\gamma^{[N]}(t/h), y_0)$, where $\forall u \in \mathcal{T}$, $\gamma^{[N]}(\tau)_u$ is the unique polynomial of degree N in τ satisfying

$$\gamma^{[N]}(n)_u = (\alpha^{*n})_u \quad \text{for} \quad n = 0, 1, \ldots, N. \tag{4.38}$$

The application of Newton's forward finite difference formula for the interpolating polynomial gives

$$\gamma^{[N]}(\tau) = \mathbf{1} + \sum_{n \geq 1}^{N} \frac{\tau(\tau - 1)\cdots(\tau - n + 1)}{n!}(\alpha - \mathbf{1})^{*n}.$$

Standard error analysis of polynomial interpolation shows that the series in powers of h of $\tilde{y}(t)$ and $P_N(t)$ coincide up to terms of degree N in h. By considering arbitrarily high N, we get

$$\tilde{y}(t) = B_{hf}(\gamma(t/h), y_0), \tag{4.39}$$

where

$$\gamma(\tau) = \mathbf{1} + \sum_{n \geq 1} \frac{\tau(\tau - 1)\cdots(\tau - n + 1)}{n!}(\alpha - \mathbf{1})^{*n}. \tag{4.40}$$

Observe that $\gamma(\tau)$ is well defined as a function in $\mathbb{R}^{\mathcal{F}}$, because $\alpha_e = \mathbf{1}$ implies that $((\alpha_e - \mathbf{1})^{*n})_u = 0$, whenever $|u| < n$ (Property P2 in section 4.2.5).

If $\tilde{y}(t)$, given by equations [4.39]–[4.40], is the solution of an autonomous ODE [4.36], then the right-hand side of equation [4.36] can be recovered from the expansion [4.39] of its solution $\tilde{y}(t)$ as

$$\tilde{f}(y_0; h) = \left.\frac{d}{dt}\tilde{y}(t)\right|_{t=0} = \left.\frac{d}{dt}B_{hf}(\gamma(t/h), y_0)\right|_{t=0},$$

i.e.

$$\begin{aligned}\tilde{f}(y;h) &= h^{-1}B_{hf}(\beta, y) \\ &= \beta_{\bullet} f(y) + h\,\beta_{[\bullet]} f'(y)f(y) + \cdots\end{aligned} \tag{4.41}$$

where the coefficients β_u for $u \in \mathcal{T}$ are given by the formula

$$\beta = \frac{d}{d\tau}\gamma(\tau)\bigg|_{\tau=0} = \sum_{n\geq 1}\frac{(-1)^{n+1}}{n}(\alpha - \mathbf{1})^{*n}, \tag{4.42}$$

or equivalently,

$$\beta_u = \frac{d}{d\tau}\bigg|_{\tau=0}\gamma(\tau)_u = \sum_{m=1}^{|u|}\frac{(-1)^{m+1}}{m}\sum_{(V_m\succ\cdots\succ V_1)\subset U}\alpha_{V_1}\cdots\alpha_{V_m}, \tag{4.43}$$

where in the inner summation, only non-empty posets $V_1, \ldots, V_m$ are considered. For instance,

$$\begin{aligned}\beta_{\bullet} &= \alpha_{\bullet}, \\ \beta_{[\bullet]} &= \alpha_{[\bullet]} - \frac{1}{2}\alpha_{\bullet}^2, \\ \beta_{[[\bullet]]} &= \alpha_{[[\bullet]]} - \alpha_{\bullet}\alpha_{[\bullet]} + \frac{1}{3}\alpha_{\bullet}^3, \\ \beta_{[\bullet,\bullet]} &= \alpha_{[\bullet,\bullet]} - \alpha_{\bullet}\alpha_{[\bullet]} + \frac{1}{6}\alpha_{\bullet}^3.\end{aligned}$$

We thus have that $\beta = \log(\alpha)$, or equivalently, $\alpha = \exp(\beta)$, where the mutually reciprocal operators log and exp are defined as follows:

DEFINITION 4.9.– *Given $\alpha \in \mathbb{R}^{\mathcal{F}}$ with $\alpha_e = 1$, $\log(\alpha) \in \mathbb{R}^{\mathcal{F}}$ is defined as*

$$\log(\alpha) = \sum_{n\geq 1}\frac{(-1)^{n+1}}{n}(\alpha - \mathbf{1})^{*n}.$$

Given $\beta \in \mathbb{R}^{\mathcal{F}}$ with $\beta_e = 0$, $\exp(\beta) \in \mathbb{R}^{\mathcal{F}}$ is defined as

$$\exp(\beta) = \mathbf{1} + \sum_{n\geq 1}\frac{1}{n!}\beta^{*n}.$$

Though defined as infinite series, $\log(\alpha)$ and $\exp(\beta)$ are well defined as functions in $\mathbb{R}^{\mathcal{F}}$, because $\alpha_e = \mathbf{1}$ (respectively, $\beta_e = 0$) implies that $((\alpha_e - \mathbf{1})^{*n})_u = 0$ (respectively, $(\beta^{*n})_u = 0$) whenever $|u| < n$ (Property P2 in section 4.2.5).

Summing up, we have seen that if the formal solution $\tilde{y}(t)$ of a modified equation [4.36] satisfies equation [4.37], then the right-hand side of equation [4.36] is given as the B-series [4.41], where $\beta = \log(\alpha)$. We will next show that the solution $\tilde{y}(t)$ of such a modified equation actually satisfies equation [4.37]: The substitution of $\tilde{y}(t) = B_{hf}(\gamma(t/h), y_0)$ in equation [4.36] and the application of Theorem 4.4 gives

$$\frac{d}{dt}B_{hf}(\gamma(t/h), y_0) = \frac{1}{h}B_{hf}(\gamma(t/h) * \beta, y_0), \quad B_{hf}(\gamma(0), y_0) = y_0,$$

or equivalently,

$$B_{hf}(\frac{d}{d\tau}\gamma(\tau), y_0) = B_{hf}(\gamma(\tau) * \beta, y_0), \quad B_{hf}(\gamma(0), y_0) = y_0,$$

which will hold true if $\gamma(\tau)$ satisfies the ODE

$$\frac{d}{d\tau}\gamma(\tau) = \gamma(\tau) * \beta, \quad \gamma(0) = \mathbf{1}, \qquad [4.44]$$

whose solution is $\gamma(\tau) = \exp(\tau\beta)$. Hence, $\tilde{y}(t) = B_{hf}(\gamma(t/h), y_0)$ with $\gamma(\tau) = \exp(\tau \log(\alpha))$ (which can be expanded as [4.40]) is the solution of the modified equation [4.36] with equation [4.41] and $\beta = \log(\alpha)$. We finally have that [4.37] holds true as $\exp(n\log(\alpha)) = \alpha^{*n}$ for all $n \in \mathbb{Z}$.

Obviously, the right-hand side of equation [4.1] can be written as $h^{-1}B_{hf}(\delta^{\bullet}, y)$, where $\delta^{\bullet} \in \mathfrak{g}$ is defined as $\delta^{\bullet}_u = 1$ if $u = \bullet$ and $\delta^{\bullet}_u = 0$ otherwise. Thus, the exact solutions $y(t)$ of equation [4.1] admits the expansion $y(t) = B_{hf}(\gamma(t/h), y(0))$, where $\gamma(\tau) = \exp(\tau\,\delta^{\bullet})$ is the solution of the ODE [4.44] with $\beta = \delta^{\bullet}$, i.e. for each $u \in \mathcal{T}$ with $u = [v]$, $v \in \mathcal{F}$,

$$\frac{d}{d\tau}\gamma(\tau)_u = \gamma(\tau)_v, \quad \gamma(0)_u = 0,$$

which leads to $\gamma(\tau)_u = \exp(\tau\delta^{\bullet})_u = \tau^{|u|}/u!$, where the factorial $u!$ of the rooted tree u is defined as in Theorem 4.1.

The fact that $\beta_{\bullet} = \alpha_{\bullet}$ implies that equation [4.36] will be a perturbation of size $\mathcal{O}(h)$ of the original ODE [4.1], provided that $\alpha_{\bullet} = 1$. More generally, equation [4.36] will be a perturbation of size $\mathcal{O}(h^r)$ of the original ODE [4.1] if and only if $\alpha_u = \exp(\delta^{\bullet})_u = 1/u!$ for all $u \in \mathcal{T}$ with $|u| \leq r$, i.e. if $\alpha \in \mathcal{G}$ corresponds to a method of order r for the original ODE [4.1].

Unfortunately, the series in powers of h defining the right-hand side of equation [4.41] of the modified ODE [4.36] is in general divergent, even when the B-series expansion [4.35] of the integration method is convergent (for real analytic f, and sufficiently small h). For rigorous results based on modified equations, we have to consider a truncated version and estimate the differences between the numerical solution and the solution of the truncated modified ODE.

4.3.3. *Series of linear differential operators*

Let $\psi_{hf}(y)$ be one step of an integration method applied to the ODE [4.1] that admits a B-series expansion $\psi_{hf}(y) = B_{hf}(\alpha, y)$ for some $\alpha \in \mathcal{G}$. Assume that we want to expand the composition of ψ_{hf} with a smooth function $g \in C^\infty(\mathbb{R}^d)$, i.e. we want the expansion in powers of h of $g(B_{hf}(\alpha, y))$. Proceeding as in the proof of Proposition 4.1, when showing that equation [4.16] implies that $f(\psi_{hf}(y))$ admits the expansion [4.17], we readily get that $g(B_{hf}(\alpha, y))$ can be expanded as

$$g(y) + \sum_{v=u_1\cdots u_m \in \mathcal{F}\backslash\{e\}} \frac{h^{|v|}}{\sigma(v)} \alpha_v \, g^{(m)}(y)(F_{u_1}(y), \ldots, F_{u_m}(y)), \qquad [4.45]$$

where the summation goes over all non-empty forests $v \in \mathcal{F}\backslash\{e\}$, and for each such forest, $v = u_1 \cdots u_m$ is its unique decomposition in rooted trees $u_1, \ldots, u_m \in \mathcal{T}$.

This motivates us to make the following definitions.

DEFINITION 4.10.– *Given a smooth vector field f, the* elementary differential operator *X_u associated with a forest $u \in \mathcal{F}$ is an operator acting on smooth functions $g \in C^\infty(\mathbb{R}^d)$ defined as follows: If $u = e$, then $X_e \, g = g$, and if $u = u_1 \cdots u_m$, where $u_1, \ldots, u_m \in \mathcal{T}$, then the function $X_u \, g$ is defined for each $y \in \mathbb{R}^d$ as*

$$X_u \, g(y) = g^{(m)}(y)(F_{u_1}(y), \ldots, F_{u_m}(y)),$$

where the elementary differentials F_v associated with rooted trees v are given in Definition 4.3.

In particular, $X_\bullet g(y) = g'(y)f(y)$, and thus $X_\bullet$ is the Lie derivative along the smooth vector field f.

DEFINITION 4.11.– *Given a smooth vector field f, to each $\alpha \in \mathbb{R}^\mathcal{F}$ we associate the formal series $S_{hf}(\alpha)$ of differential operators acting on smooth functions $g \in C^\infty(\mathbb{R}^d)$ given as*

$$S_{hf}(\alpha) = \sum_{u \in \mathcal{F}} \frac{h^{|u|}}{\sigma(u)} \alpha_u \, X_u.$$

With Definition 4.10, we clearly have that, if $u \in \mathcal{F}$, then the ith component of the elementary differential $F_{[u]}$ associated with the rooted tree $[u]$ can be obtained as the action of X_u on the ith component of f, i.e. $F^i_{[u]} = X_u f^i$ (alternatively, $F^i_{[u]} = X_{[u]} e_i$, where e_i denotes the ith coordinate function, i.e. $e_i(y) = y^i$). We thus have that, for each $\alpha \in \mathcal{G}$,

$$B_{hf}(\alpha, y) = y + h \, S_{hf}([\alpha]^*) f(y), \qquad [4.46]$$

where $[\alpha]^* \in \mathbb{R}^{\mathcal{F}}$ is determined from the values α_u for $u \in \mathcal{T}$ as

$$([\alpha]^*)_v = \alpha_{[v]} \quad \text{for each} \quad v \in \mathcal{F}. \qquad [4.47]$$

Observe that, if $\alpha \in \mathcal{G}_{RK}$, i.e. if $\exists \mu \in \mathcal{RK}$, such that $\forall u \in \mathcal{F}$, $\alpha_u = u(\mu)$, then $([\alpha]^*)_u = [u](\mu)$.

In the following, we will assume that the vector field f is fixed, and will often drop the subindex hf from the notation. From the discussion at the beginning of this section, we have the following basic result:

PROPOSITION 4.4.– *For each $\alpha \in \mathcal{G}$ and each $g \in C^\infty(\mathbb{R}^d)$,*

$$g(B(\alpha, y)) = S(\alpha)g(y).$$

We thus have that a series of differential operators $S(\alpha)$ for $\alpha \in \mathcal{G}$ can be identified with a B-series $B(\alpha, y)$ that represents a near-to-identity map (recall that $\alpha_e = 1$ if $\alpha \in \mathcal{G}$). From Proposition 4.4, for arbitrary $g_1, g_2 \in C^\infty(\mathbb{R}^d)$, we have

$$S(\alpha)g_1g_2 = (S(\alpha)g_1)\,(S(\alpha)g_2) \quad \text{if} \quad \alpha \in \mathcal{G}. \qquad [4.48]$$

We will next see that, while in the B-series context, $\alpha * \beta$ does not make sense for arbitrary $\alpha, \beta \in \mathbb{R}^{\mathcal{F}}$, it does in terms of series of differential operators. We will first state the following auxiliary result concerning the product $*$ in Definition 4.7.

PROPOSITION 4.5.– *Given $\alpha, \beta \in \mathbb{R}^{\mathcal{F}}$,*

$$[\alpha * \beta]^* = \beta_e\,[\alpha]^* + \alpha * [\beta]^*. \qquad [4.49]$$

We omit the proof of Proposition 4.5 because of its similarity to the proof of Theorem 4.3. Observe that, in the case where $\alpha, \beta \in \mathcal{G}_{RK}$ (i.e. $\exists \mu, \mu' \in \mathcal{RK}$, such that $\forall u \in \mathcal{F}$, $\alpha_u = u(\mu)$ and $\beta_u = u(\mu)$), equation [4.49] reduces to equation [4.21] (with $(\alpha * \beta)_u = u(\mu\mu')$ for all $u \in \mathcal{F}$).

LEMMA 4.3.– *For each $\alpha \in \mathbb{R}^{\mathcal{F}}$ and each $N \geq 1$, there exist $\alpha^1, \ldots, \alpha^{m_N} \in \mathcal{G}$ and $\lambda_1, \ldots, \lambda_{m_N} \in \mathbb{R}$, such that,*

$$\alpha_u = \sum_{j=1}^{m_N} \lambda_j\, \alpha_u^j, \quad \textit{for all } u \in \mathcal{F} \textit{ with } |u| \leq N.$$

THEOREM 4.5.– *For each $\alpha, \beta \in \mathbb{R}^{\mathcal{F}}$,*

$$S(\alpha)S(\beta) = S(\alpha * \beta). \qquad [4.50]$$

PROOF.– We will prove by induction on $N \geq 1$ that, for all $(\alpha, \beta) \in \mathbb{R}^{\mathcal{F}} \times \mathbb{R}^{\mathcal{F}}$,

$$S(\alpha)S(\beta)g - S(\alpha * \beta)g = \mathcal{O}(h^{N+1}). \qquad [4.51]$$

By virtue of Lemma 4.3, and by bilinearity of both $*$ and the composition of linear differential operators, it is then enough to prove [4.51] for $\alpha, \beta \in \mathcal{G}$. According to Proposition 4.4, if $\alpha, \beta \in \mathcal{G}$, then

$$S(\alpha)S(\beta)g(y) - S(\alpha * \beta)g(y) = g(B(\beta, B(\alpha, y))) - g(B(\alpha * \beta, y)), \qquad [4.52]$$

and thus, we have to prove that, for all $(\alpha, \beta) \in \mathcal{G} \times \mathcal{G}$,

$$B(\beta, B(\alpha, y)) - B(\alpha * \beta, y) = \mathcal{O}(h^{N+1}),$$

but according to equation [4.46] and Proposition 4.4,

$$\begin{aligned} B(\beta, B(\alpha, y)) - B(\alpha * \beta, y) &= S(\alpha)B(\beta, y) - y - h\, S([\alpha * \beta]^*)f(y) \\ &= h\, S([\alpha]^*)f(y) + h\, S(\alpha)S([\beta]^*)f(y) - h\, S([\alpha * \beta]^*)f(y). \end{aligned}$$

By induction hypothesis, $S(\alpha)S([\beta]^*) = S(\alpha * [\beta]^*) + \mathcal{O}(h^N)$, and thus,

$$B(\beta, B(\alpha, y)) - B(\alpha * \beta, y) = h\, S([\alpha]^* + \alpha * [\beta]^* - [\alpha * \beta]^*)f(y) + \mathcal{O}(h^{N+1})$$

and the required result is obtained by virtue of Proposition 4.5. □

We thus have that equation [4.11] defines an algebra homomorphism from the algebra $\mathbb{R}^{\mathcal{F}}$ to the algebra $\mathcal{E}[[h]]$, where $\mathcal{E}$ denotes the algebra of endomorphisms of $C^\infty(\mathbb{R}^d)$. In particular, $S_{hf}(e)$ is the identity operator. We also have that $S_{hf}(\exp(\beta)) = \exp(S_{hf}(\beta))$ (respectively, $S_{hf}(\log(\alpha)) = \log(S_{hf}(\alpha))$) if $\beta_e = 0$ (respectively, if $\alpha_e = 1$).

The following is a consequence of equation [4.48] and Lemma 4.3.

PROPOSITION 4.6.– *For each $\alpha \in \mathbb{R}^{\mathcal{F}}$,*

$$S(\alpha)g_1g_2 = \sum_{(u,v)\in\mathcal{F}\times\mathcal{F}} \frac{h^{|u|+|v|}}{\sigma(u)\sigma(v)} \alpha(uv)\,(X_u g_1)\,(X_v g_2).$$

4.3.4. *The Lie algebra of the Butcher group*

We next introduce the Lie algebra $\mathfrak{g} = \log(\mathcal{G})$ of the Butcher group $\mathcal{G}$.

PROPOSITION 4.7.– *Given $\beta \in \mathbb{R}^{\mathcal{F}}$, the following statements are equivalent:*

– $\exp(\beta) \in \mathcal{G}$;

– *for all $u, v \in \mathcal{F}$*

$$\beta_{uv} = \mathbf{1}_u\beta_v + \beta_u\mathbf{1}_v; \qquad [4.53]$$

– *there exists a curve* $\gamma : \mathbb{R} \to \mathcal{G}$, *such that*

$$\gamma(0) = \mathbf{1}, \quad \frac{d}{dt}\gamma(t)\bigg|_{t=0} = \beta. \qquad [4.54]$$

Clearly, equation [4.53] is equivalent to

$$\beta_e = 0, \quad \beta_{u_1\cdots u_m} = 0, \text{ if } u_1, \ldots, u_m \in \mathcal{T}. \qquad [4.55]$$

Hence, as in the case of $\mathcal{G}$, each $\beta \in \mathfrak{g}$ is uniquely determined by its values β_u for $u \in \mathcal{T}$.

Given a vector field f, for each $\beta \in \mathfrak{g}$, $S_{hf}(\beta)$ is a derivation of the algebra $C^\infty(\mathbb{R}^d)$, i.e.

$$S_{hf}(\beta)g_1g_2 = (S_{hf}(\beta)g_1)\, g_2 + g_1\, (S_{hf}(\beta)g_2) \quad \text{if} \quad \beta \in \mathfrak{g}. \qquad [4.56]$$

That is, $S_{hf}(\beta)$ is (the Lie operator of) a vector field. The equality [4.56] follows from equation [4.48] by considering a curve $\gamma(\tau)$ in $\mathcal{G}$ satisfying equation [4.54].

This can also be shown directly from the definition of $S_{hf}(\beta)$, by taking into account that, if $\beta \in \mathfrak{g}$, then $\beta_u = 0$ for all $u \in \mathcal{F}\backslash\mathcal{T}$, so that

$$S_{hf}(\beta)g(y) = \frac{\partial}{\partial y}g(y)B_{hf}(\beta, y).$$

Proposition 4.7 implies that $\mathfrak{g}$ is a Lie subalgebra of the Lie algebra $\mathbb{R}^{\mathcal{F}}$ with bracket $[\alpha, \beta] = \alpha{*}\beta - \beta{*}\alpha$. Thus, the restriction of S_{hf} to $\mathfrak{g}$ defines a homomorphism from the Lie algebra $\mathfrak{g}$ to the Lie algebra of (formal series of) vector fields.

4.3.5. *The pre-Lie algebra structure on* $\mathfrak{g}$

We next show that the Lie algebra $\mathfrak{g}$ has a pre-Lie algebra structure. A nice exposition on pre-Lie algebras can be found in Chapter 1 of this book.

It will be useful to consider the projection $\pi : \mathbb{R}^{\mathcal{F}} \to \mathfrak{g}$ given by $\pi(\beta)_u = 0$ for each $u \in \mathcal{F}\backslash\mathcal{T}$, and $\pi(\beta)_u = \beta_u$ for each $u \in \mathcal{T}$. Observe that the restriction of π to $\mathcal{G}$ is (as the restriction to $\mathcal{G}$ of the logarithm) a bijection from $\mathcal{G}$ to $\mathfrak{g}$.

Property P1 in section 4.2.5 implies that $\mathfrak{g}$ has a left $\mathcal{H}^*$-module structure, with $\mathcal{H}^* \otimes \mathfrak{g} \to \mathfrak{g}$ defined as follows.

DEFINITION 4.12.– *Given* $\alpha \in \mathcal{H}^*$ *and* $\beta \in \mathfrak{g}$,

$$\alpha \cdot \beta := \pi(\alpha * \beta).$$

We use the symbol $\rhd$ *to denote the restriction to* $\mathfrak{g} \otimes \mathfrak{g}$ *of that* $\mathcal{H}^*$*-module map. Thus, given* $\beta, \beta', \beta'' \in \mathfrak{g}$,

$$\beta \rhd (\beta' \rhd \beta'') = \beta \cdot (\beta' \cdot \beta'') = (\beta * \beta') \cdot \beta''. \qquad [4.57]$$

Since $\mathfrak{g}$ *is a Lie algebra under the bracket* $[\beta, \beta'] = \beta * \beta' - \beta' * \beta$, *defined for* $\beta, \beta' \in \mathfrak{g}$, *we have* $\pi([\beta, \beta']) = [\beta, \beta']$, *and thus*

$$\forall \beta, \beta' \in \mathfrak{g}, \quad \beta \rhd \beta' - \beta' \rhd \beta = [\beta, \beta']. \qquad [4.58]$$

PROPOSITION 4.8.– *The vector space* $\mathfrak{g}$ *with the restriction to* $\mathfrak{g} \otimes \mathfrak{g}$ *of* $\rhd$ *has a left pre-Lie algebra structure.*

PROOF.– According to the definition of left pre-Lie algebra, we have to prove that, given $\beta, \beta', \beta'' \in \mathfrak{g}$,

$$(\beta \rhd \beta' - \beta' \rhd \beta) \rhd \beta'' = \beta' \rhd (\beta \rhd \beta'') - (\beta' \rhd \beta) \rhd \beta''.$$

Indeed, given $\beta, \beta', \beta'' \in \mathfrak{g}$, taking into account equations [4.58] and [4.57],

$$\begin{aligned}(\beta \rhd \beta' - \beta' \rhd \beta) \rhd \beta'' &= [\beta, \beta'] \rhd \beta'' \\ &= (\beta' * \beta - \beta * \beta') \cdot \beta'' \\ &= (\beta' * \beta) \cdot \beta'' - (\beta * \beta') \cdot \beta'' \\ &= \beta' \rhd (\beta \rhd \beta'') - \beta \rhd (\beta' \rhd \beta'').\end{aligned}$$

□

PROPOSITION 4.9.– *If* $\alpha \in \mathcal{H}^*$ *and* $\beta \in \mathfrak{g}$, *then*

$$B(\alpha \cdot \beta, y) = S(\alpha) B(\beta, y).$$

In particular, if $\alpha, \beta \in \mathfrak{g}$, *then* $B(\alpha \rhd \beta, y) = \frac{\partial}{\partial y} B(\beta, y) \cdot B(\alpha, y)$.

PROOF.– It follows from Theorem 4.5, that since for each $\gamma \in \mathcal{H}^*$, the ith component of $B(\gamma) = B(\pi(\gamma))$ coincides with the action of $S(\gamma)$ on the ith coordinate function $e_i \in C^\infty(\mathbb{R}^d)$, the required result is equivalent to

$$S(\alpha * \beta) e_i = S(\alpha) S(\beta) e_i, \quad i = 1, \ldots, d.$$

□

This shows that, given a smooth vector field f, the definition of the B-series $B_{hf}(\beta, \cdot)$ for each $\beta \in \mathfrak{g}$ determines a homomorphism from the left pre-Lie algebra $\mathfrak{g}$ to the left pre-Lie algebra of (formal series of) smooth vector fields in $\mathbb{R}^d$: Recall that the binary operation $\rhd$ endowing the vector space of smooth vector fields in $\mathbb{R}^d$ (given as maps $f : \mathbb{R}^d \to \mathbb{R}^d$) with a left pre-Lie algebra structure is defined as

$(f \triangleright g)(y) = \frac{\partial}{\partial y} g(y) \cdot f(y)$, i.e. $f \triangleright g$ is the action on g of the Lie operator associated with f.

The next result gives a characterization of Lie algebras that are obtained in this way from a left pre-Lie algebra structure.

PROPOSITION 4.10.– *Given a Lie algebra $\mathcal{L}$, consider its universal enveloping algebra $\mathcal{U}(\mathcal{L})$. The vector space $\mathcal{L}$ can be endowed with a left pre-Lie algebra operation $\triangleright : \mathcal{L} \otimes \mathcal{L} \to \mathcal{L}$, such that for all $\ell, \ell' \in \mathcal{L}$,*

$$[\ell, \ell'] = \ell \triangleright \ell' - \ell' \triangleright \ell$$

if and only if $\mathcal{L}$ has a left $\mathcal{U}(\mathcal{L})$-module structure. In that case, $\triangleright$ is the restriction to $\mathcal{L} \otimes \mathcal{L}$ of the left-module map $\mathcal{U}(\mathcal{L}) \otimes \mathcal{L} \to \mathcal{L}$.

PROOF.– Clearly, the existence of a left $\mathcal{U}(\mathcal{L})$-module structure of $\mathcal{L}$ can be replaced in the statement of Proposition 4.10 by the existence of a triplet $(\mathcal{A}, \nu, \cdot)$, where $\mathcal{A}$ is an associative algebra, $\nu : \mathcal{L} \to \mathcal{A}$ is a monomorphism of Lie algebras and $\cdot : \mathcal{A} \otimes \mathcal{L} \to \mathcal{L}$ is a left $\mathcal{A}$-module structure map. The "if" part of the statement can be proven as Proposition 4.8. The "only if" part is obtained by considering $\mathcal{A}$ as the algebra of endomorphisms of the left pre-Lie algebra $\mathcal{L}$, the canonical left $\mathcal{A}$-module structure of $\mathcal{L}$ and the injection $\nu : \mathcal{L} \to \mathcal{A}$ defined as $\nu(\ell) \cdot \ell' = \ell \triangleright \ell'$, and proving that ν is a Lie algebra morphism. Indeed, by definition of ν and the left pre-Lie algebra condition, we have

$$\begin{aligned} \nu([\ell, \ell']) \cdot \ell'' &= (\ell \triangleright \ell' - \ell' \triangleright \ell) \triangleright \ell'' \\ &= \ell \triangleright (\ell' \triangleright \ell'') - \ell' \triangleright (\ell \triangleright \ell'') \\ &= \nu(\ell) \cdot (\nu(\ell') \cdot \ell'') - \nu(\ell') \cdot (\nu(\ell) \cdot \ell'') \\ &= [\nu(\ell), \nu(\ell')] \cdot \ell''. \end{aligned}$$

□

Let us consider the vector space $\mathfrak{g}_0$ of elements in $\mathfrak{g}$ with finite support, or equivalently,

$$\mathfrak{g}_0 = \{\beta \in \mathfrak{g} \ : \ \exists N \geq 1 \text{ such that } \beta_u = 0 \text{ if } |u| \geq N\}. \qquad [4.59]$$

Clearly, $\mathfrak{g}_0$ is a left pre-Lie subalgebra of $\mathfrak{g}$. The set $\mathcal{T}$ of rooted trees can be identified with the basis $\{\delta^u \ : \ u \in \mathcal{T}\}$ of $\mathfrak{g}_0$, where for each $u \in \mathcal{T}$,

$$(\delta^u)_v = \begin{cases} \sigma(u) & \text{if } v = u \\ 0 & \text{otherwise,} \end{cases} \qquad [4.60]$$

where the normalization factor $\sigma(u)$ is determined for each $u \in \mathcal{T}$ in Definition 4.5. We thus have that, for each $\beta \in \mathfrak{g}_0$,

$$\beta = \sum_{u \in \mathcal{T}} \frac{\beta_u}{\sigma(u)} \, \delta^u.$$

A left pre-Lie algebra structure is determined for the set $\mathbb{R}\mathcal{T}$ of linear combinations or rooted trees by identifying each δ^u with the rooted tree u. It can be shown that the pre-Lie algebra $\mathfrak{g}_0$ is generated by $\delta^\bullet$ and also, that

$$\delta^u \rhd \delta^v = \sum_{i=1}^{|v|} \delta^{w_i}$$

where each w_i is the rooted tree obtained by grafting the rooted tree u to the ith vertex of v. This is precisely the left pre-Lie algebra of rooted trees as introduced by Chapoton and Livernet (2001), who proves the following fundamental result:

THEOREM 4.6.– *The left pre-Lie algebra $\mathfrak{g}_0$ of rooted trees is the free left pre-Lie algebra in one generator.*

In section 4.4.3, an alternative to the original proof in Chapoton and Livernet (2001) is given in Theorem 4.7.

4.3.6. *Bibliographical comments*

A direct proof of Theorem 4.4 based on the Fàa di Bruno formula was given in the original paper (Hairer and Wanner 1974). The construction of a modified equation for general B-series methods is originally due to Hairer (1994). Formula [4.43] for the coefficients of the modified equations is closely related to the one originally introduced in Hairer (1994). An alternative approach to recursively compute the coefficients $\beta(u)$ and $\gamma(u)$ from the ODE [4.44] supplemented by the condition $\gamma(1) = \alpha$ was considered in Calvo *et al.* (1994), which only makes use of formula [4.28] of composition of B-series. The interpolatory argument adopted in section 4.3.2 was introduced in Murua (1999), and explicit expressions for $\log(\alpha)_u$ in terms of rooted trees are given in Chartier *et al.* (2005) and Murua (2006). The linear differential operators X_u were first considered in the context of Runge–Kutta methods by Merson (2001). They can also be found later on in Grossman and Larson (1990), Murua (1999) and Murua (2006). In Chartier *et al.* (2006) and Chartier and Murua (2007), a series of such differential operators are successfully applied to study preservation properties of integration methods that can be expanded as B-series. In Chapoton and Livernet (2001), the left pre-Lie algebra structure on the linear span of rooted trees is studied, and it is shown that it is the free left pre-Lie algebra in one generator (see Theorem 4.6 below).

4.4. Hopf algebras of rooted trees

Clearly, all of the material in section 4.2.5 would be valid if the set $\mathbb{R}^{\mathcal{F}}$ of functions had been replaced by the set $\mathcal{R}^{\mathcal{F}}$ of maps from $\mathcal{F}$ to an arbitrary commutative algebra

$\mathcal{R}$. In particular, this gives a group structure $\mathbb{G}(\mathcal{R})$ on the set of maps $\alpha : \mathcal{F} \to \mathcal{R}$ satisfying $\alpha_{uv} = \alpha_u \alpha_v$. In this sense, the Butcher group is actually an *affine group scheme* $\mathbb{G}(\cdot)$, a functor from commutative algebras to groups. The category of group schemes is equivalent to the category of commutative Hopf algebras (Sweedler 1969): given a group scheme $\mathbb{G}(\cdot)$, there exists a commutative Hopf algebra $\mathcal{H}$, such that $\mathbb{G}(\mathcal{R})$ is the group of algebra maps from $\mathcal{H}$ to $\mathcal{R}$, with the convolution product as group law. The coproduct in $\mathcal{H}$ can be obtained by dualizing the group law of the affine group scheme $\mathbb{G}(\cdot)$. Applying this to the particular case of Butcher's affine group scheme gives rise to a commutative Hopf algebra $\mathcal{H}$ on the vector space of linear combinations of forests of rooted trees.

As an alternative construction of $\mathcal{H}$, the algebra $\mathbb{R}[\mathcal{T}]$ of functions on $\mathcal{RK}$ generated by $\mathcal{T}$ (each $u \in \mathcal{T}$ viewed as a function in $\mathbb{R}^{\mathcal{RK}}$, as defined in section 4.2) can be endowed with a commutative Hopf algebra structure derived from the semigroup structure of the set of RK methods. This can be done by observing that the results in section 4.2.4 imply that the algebra $\mathbb{R}[\mathcal{T}]$ is a sub-bialgebra of the bialgebra of representative functions (Hochschild 1981) of the semigroup $\mathcal{RK}$ (in particular, the coassociativity of the coproduct in $\mathcal{H}$ comes as a direct consequence of the associativity of the composition of RK methods). The existence of the antipode is guaranteed, in this context, by the conilpotency of the coproduct.

There is another important Hopf algebra associated with the Butcher group $\mathcal{G}$, the dual Hopf algebra (Sweedler 1969) $\mathcal{H}^\circ$ of $\mathcal{H}$. The Butcher group $\mathcal{G}$ coincides with the group $G(\mathcal{H}^\circ)$ of *group-like elements* of $\mathcal{H}^\circ$. The Lie algebra $\mathfrak{g}$ of the group $\mathcal{G}$ is the Lie algebra $P(\mathcal{H}^\circ)$ of *primitive elements* of $\mathcal{H}^\circ$.

4.4.1. *The commutative Hopf algebra of rooted trees*

The preceding discussion leads to a commutative Hopf algebra structure on the vector space spanned by the set $\mathcal{F}$ of forests. As an algebra, it is just the algebra $\mathbb{R}[\mathcal{T}]$ of polynomials in the commuting indeterminates $u \in \mathcal{T}$ (with the empty forest e as the unit), or equivalently, the symmetric algebra $S(\mathcal{V})$ over the vector space $\mathcal{V}$ spanned by the set of rooted trees $\mathcal{T}$. The bijection $\mathcal{F} \to \mathcal{T}$ that gives the rooted tree $[u]$ for each forest u induces an isomorphism of vector spaces $B : S(\mathcal{V}) \to \mathcal{V}$, where $B(u) = [u]$ for each $u \in \mathcal{F}$.

The coproduct

$$\begin{aligned}\Delta : \mathcal{H} &\to \mathcal{H} \otimes \mathcal{H} \\ u &\mapsto \Delta(u)\end{aligned}$$

is defined by determining $\Delta(u)$ for each $u \in \mathcal{F}$ as follows: If U is a partially ordered set representing a forest $u \in \mathcal{F}$,

$$\begin{aligned}\Delta(u) &= \sum_{(V \succ W) \subset U} v \otimes w \\ &= u \otimes e + e \otimes u + \sum_{\substack{(V \succ W) \subset U \\ V, W \neq \emptyset}} v \otimes w,\end{aligned} \qquad [4.61]$$

where V and W represent the forests v and w, respectively. The coassociativity (which can be seen as a result of the associativity of the composition of RK methods) can be directly checked from the definition of $\Delta(u)$ for $u \in \mathcal{F}$ (exactly like the associativity of the group law of $\mathcal{G}$): For each $u \in \mathcal{F}$,

$$(\Delta \otimes \mathrm{id}_{\mathcal{H}}) \circ \Delta(u) = (\mathrm{id}_{\mathcal{H}} \otimes \Delta) \circ \Delta(u) = \sum_{(V_1 \succ V_2 \succ V_3) \subset U} v_1 \otimes v_2 \otimes v_3.$$

The counit $\mathbf{1}: \mathcal{H} \to \mathbb{R}$ is defined by setting $\mathbf{1}(e) = 1$ and $\mathbf{1}(u) = 0$ for all $u \in \mathcal{F} \backslash \{e\}$. The antipode $s(u)$ can be defined for each $u \in \mathcal{H}$ by

$$\begin{aligned}s(u) &= -u - \sum_{\substack{(V \succ W) \subset U \\ V, W \neq \emptyset}} v \otimes s(w) \\ &= -u + \sum_{m \geq 2} (-1)^m \sum_{\substack{(V1 \succ V_2 \succ V_m) \subset U \\ V_1, \ldots, V_m \neq \emptyset}} v_1 \otimes \cdots \otimes v_m.\end{aligned}$$

It is obvious from the definition of the coproduct of Δ that, as a coalgebra, $\mathcal{H}$ is conilpotent (i.e. connected), i.e. for each $u \in \mathcal{H}$ such that $\mathbf{1}(u) = 0$, there exists an $n \geq 0$ such that $\overline{\Delta}^n(u) = 0$, where $\overline{\Delta}(u) = \Delta(u) - u \otimes e - e \otimes u$, and $\overline{\Delta}^{n+1} = (\overline{\Delta}^n \otimes \mathrm{id}_{\mathcal{H}}) \circ \overline{\Delta}$.

Clearly, the Hopf algebra $\mathcal{H}$ is compatible with the grading $\mathcal{H} = \bigoplus_{n \geq 0} \mathcal{H}_n$, where each homogeneous component $\mathcal{H}_n$ is spanned by the set $\mathcal{F}_n$ of forests with n vertices. Thus, $\mathcal{H}$ has a graded connected Hopf algebra structure.

Next result is a dual version of equation [4.22] (and also of equation [4.49]).

PROPOSITION 4.11 (Kreimer).– *Consider the linear map $L : \mathcal{H} \otimes \mathcal{H} \to \mathcal{H} \otimes \mathcal{V}$ defined as $L(u \otimes v) = u \otimes [v]$ for each $u, v \in \mathcal{F}$. Then, for each $u \in \mathcal{H}$,*

$$\Delta([u]) = [u] \otimes e + L(\Delta(u)) \qquad [4.62]$$

PROOF.– If U' is the partially ordered set representing $[u]$, and $x \in U'$ is the root of U', then the partially ordered set $U = U' \backslash \{x\}$ represents u, and the pairs of partially ordered sets $(V', W') \neq (U, \emptyset)$, such that $(V' \succ W') \subset U'$ are in one-to-one correspondence with the pairs (V, W), such that $(V \succ W) \subset U$, with $V = V'$ and $W = W' \backslash \{x\}$. This implies that

$$\Delta([u]) = \sum_{(V \succ W') \subset U'} v \otimes w' = [u] \otimes e + \sum_{(V \succ W) \subset U} v \otimes [w].$$

□

4.4.2. *The dual algebra* $\mathcal{H}^*$ *and the dual Hopf algebra* $\mathcal{H}^\circ$

Given a linear form $\alpha \in \mathcal{H}^*$, we denote as $\langle \alpha, u \rangle$ the value of α at $u \in \mathcal{H}$. When $u \in \mathcal{F}$, we will keep the notation $\langle \alpha, u \rangle = \alpha_u$. The coalgebra structure of $\mathcal{H}$ induces an associative unital algebra structure on the algebraic dual $\mathcal{H}^*$ of $\mathcal{H}$, with the multiplication $*$ given by

$$\langle \alpha * \beta, u \rangle = \langle \alpha \otimes \beta, \Delta(u) \rangle,$$

where, as usual, $\mathcal{H}^* \otimes \mathcal{H}^*$ is considered as a subspace of $(\mathcal{H} \otimes \mathcal{H})^*$ (i.e. given $\alpha, \beta \in \mathcal{H}^*$, $\alpha \otimes \beta \in (\mathcal{H} \otimes \mathcal{H})^*$ is defined as $\langle \alpha \otimes \beta, u \otimes v \rangle = \langle \alpha, u \rangle \langle \beta, v \rangle$).

Clearly, the algebra $\mathcal{H}^*$ coincides with the algebra structure on $\mathbb{R}^{\mathcal{F}}$ defined in section 4.2.5.

Given $\alpha \in \mathcal{H}^*$, consider $\Delta(\alpha) \in (\mathcal{H} \otimes \mathcal{H})^*$.

$$\langle \Delta(\alpha), u \otimes v \rangle = \langle \alpha, uv \rangle.$$

Let $\mathcal{H}^\circ$ be the subalgebra of $\mathcal{H}^*$ consisting of $\alpha \in \mathcal{H}^*$, such that $\Delta(\alpha) \in \mathcal{H}^* \otimes \mathcal{H}^*$. Following this (Sweedler 1969), $\mathcal{H}^\circ$ has a cocommutative Hopf algebra structure. The group of group-like elements

$$G(\mathcal{H}^\circ) = \{\alpha \in \mathcal{H}^\circ \ : \ \Delta(\alpha) = \alpha \otimes \alpha\}$$

of $\mathcal{H}^\circ$ (or equivalently, the group of *characters* of $\mathcal{H}$) coincides with the Butcher group $\mathcal{G}$. The Lie algebra of primitive elements

$$P(\mathcal{H}^\circ) = \{\beta \in \mathcal{H}^\circ \ : \ \Delta(\beta) = \beta \otimes \mathbf{1} + \mathbf{1} \otimes \beta\}$$

of $\mathcal{H}^\circ$ (or equivalently, the Lie algebra of *infinitesimal characters* of $\mathcal{H}$) coincides with the Lie algebra $\mathfrak{g}$ of the Butcher group, which as shown in section 4.3.5, actually has a left pre-Lie algebra structure.

Observe that the notation introduced in equation [4.47] corresponds to the dualization $[\cdot]^* : \mathcal{H}^* \to \mathcal{H}^*$ of $[\cdot] : \mathcal{H} \to \mathcal{H}$ given by

$$\langle [\alpha]^*, u \rangle = \langle \alpha, [u] \rangle, \quad \text{for each} \quad u \in \mathcal{H}.$$

The Lie (and pre-Lie) subalgebra $\mathfrak{g}_0$ of $\mathfrak{g}$ considered in section 4.3.5 is the Lie algebra of primitive elements of the graded dual Hopf algebra $\mathcal{H}^{\text{gr}*} \subset \mathcal{H}^\circ$ of $\mathcal{H}$. The dual basis of the basis $\{\frac{1}{\sigma(u)} u \ : \ u \in \mathcal{F}\}$ of $\mathcal{H}$ is $\{\delta^u \in \mathcal{H}^* \ : \ u \in \mathcal{H}\}$, where each δ^u is defined by [4.60]. Clearly, we have

$$\forall u \in \mathcal{F}, \quad [\delta^{[u]}]^* = \delta^u.$$

The following result can be proven from the definitions of δ^u, the product $*$ and the $\mathcal{H}^*$-module map $\cdot$.

PROPOSITION 4.12.– *For each $u \in \mathcal{F}$ and $v \in \mathcal{T}$,*

$$\delta^u \cdot \delta^\bullet = \delta^{[u]},$$

$$\delta^v * \delta^u = \delta^{vu} + \sum_{i=1}^{|u|} \delta^{w_i},$$

where the forests w_i are obtained by grafting the rooted tree v to each of the vertices of the forest u.

By the Cartier–Milnor–Moore theorem, the Hopf algebra $\mathcal{H}^{\text{gr}*}$ is isomorphic to the universal enveloping Hopf algebra $\mathcal{U}(\mathfrak{g}_0)$ of the Lie algebra $\mathfrak{g}_0$ (defined in equation [4.59]).

4.4.3. *B-series and series of differential operators revisited*

Recall that, given a left pre-Lie algebra $\mathcal{P}$ with pre-Lie operation $\rhd$, Proposition 4.10 guarantees the existence of a left $\mathcal{U}(\mathcal{P})$-module structure on $\mathcal{P}$, such that, for each $p, p' \in \mathcal{P}$, $p \rhd p' = \nu(p) \cdot p'$, where ν is the canonical monomorphism of Lie algebras from $\mathcal{P}$ to $\mathcal{U}(\mathcal{P})$.

THEOREM 4.7.– *Given a left pre-Lie algebra $\mathcal{P}$, for each element $p \in \mathcal{P}$, there exists a unique pre-Lie algebra morphism $B_p : \mathfrak{g}_0 \to \mathcal{P}$ and a unique Hopf algebra morphism $S_p : \mathcal{U}(\mathfrak{g}_0) \to \mathcal{U}(\mathcal{P})$, such that $B_p(\delta^\bullet) = p$ and for each $\alpha \in \mathcal{H}^{\text{gr}*}$, $\beta \in \mathfrak{g}_0$,*

$$S_p(\alpha) \cdot B_p(\beta) = B_p(\alpha \cdot \beta), \quad \forall (\alpha, \beta) \in \mathcal{H}^{\text{gr}*} \times \mathfrak{g}_0 \tag{4.63}$$

$$S_p(\beta) = \nu(B_p(\beta)), \quad \forall \beta \in \mathfrak{g}_0. \tag{4.64}$$

Furthermore, if $\mathcal{P}$ is a graded left pre-Lie algebra $\mathcal{P} = \bigoplus_{n \geq 1} \mathcal{P}_n$ (which induces a grading of the Hopf algebra $\mathcal{U}(\mathcal{P})$) and $p \in \mathcal{P}_1$, then S_p and B_p are compatible with the grading of $\mathcal{H}^{\text{gr}*}$, $\mathcal{P}$ and $\mathcal{U}(\mathcal{P})$.

PROOF.– If there exists a pre-Lie algebra morphism $B_p : \mathfrak{g}_0 \to \mathcal{P}$, such that $B_p(\delta^\bullet) = p$, then by the universal property of $\mathcal{H}^{\text{gr}*} = \mathcal{U}(\mathfrak{g}_0)$, there exists a unique algebra morphism $S_p : \mathcal{U}(\mathfrak{g}_0) \to \mathcal{U}(\mathcal{P})$, such that equation [4.64]. In order to prove equation [4.63], it is clearly enough to check it for $\alpha = \beta_1 * \cdots * \beta_r$, where $\beta_1, \ldots, \beta_r \in \mathfrak{g}_0$:

$$
\begin{aligned}
S_p(\alpha) \cdot B_p(\beta) &= S_p(\beta_1) \cdot (S_p(\beta_2) \cdots (S_p(\beta_r) \cdot B_p(\beta)) \cdots) \\
&= B_p(\beta_1) \rhd (B_p(\beta_2) \rhd \cdots (B_p(\beta_r) \rhd B_p(\beta)) \cdots) \\
&= B_p(\beta_1 \rhd (\beta_2 \rhd \cdots (\beta_r \rhd \beta) \cdots)) \\
&= B_p(\beta_1 \cdot (\beta_2 \cdot \cdots (\beta_r \cdot \beta) \cdots)) \\
&= B_p(\alpha \cdot \beta).
\end{aligned}
$$

If such B_p and S_p exist, then according to Proposition 4.12, for each $u \in \mathcal{F}$ and $v \in \mathcal{T}$,

$$B_p(\delta^{[u]}) = B_p(\delta^u \cdot \delta^\bullet) \;=\; S_p(\delta^u) \cdot p, \qquad [4.65]$$

$$S_p(\delta^{vu}) = S_p(\delta^v) * S_p(\delta^u) - \sum_{i=1}^{|u|} S_p(\delta^{w_i}), \qquad [4.66]$$

where the forests w_i (determined as in Proposition 4.12) have the same number m of connected components as the forest u. This allows us to prove by induction on $m + |u|$ (where $u = u_1 \cdots u_m$, $u_1, \ldots, u_m \in \mathcal{T}$) that $S_p(\delta^u)$, $u \in \mathcal{F}$ are uniquely determined from equations [4.64]–[4.66]. In turn, equation [4.65] uniquely determines $B_p(\delta^v)$, $v \in \mathcal{T}$. We thus have the uniqueness of S_p and B_p. We next show that S_p and B_p uniquely determined by equations [4.64]–[4.66] actually satisfy the required conditions.

Proposition 4.12 and equation [4.66] imply that S_p is an algebra morphism. This, together with equation [4.65], implies that equation [4.63] holds true. The map B_p being a pre-Lie algebra morphism is a consequence of [4.65], since for $\alpha, \beta \in \mathfrak{g}_0$,

$$
\begin{aligned}
B_p(\alpha \rhd \beta) &= B_p(\alpha \cdot \beta) \\
&= S_p(\alpha) \cdot B_p(\beta) \\
&= \nu(B_p(\alpha)) \cdot B_p(\beta) \\
&= B_p(\alpha) \rhd B_p(\beta).
\end{aligned}
$$

In order to prove that S_p is a Hopf algebra morphism, we have to check that $\mathcal{H}^{\mathrm{gr}*} \xrightarrow{S_p} \mathcal{U}(\mathcal{P}) \xrightarrow{\Delta} \mathcal{U}(\mathcal{P}) \otimes \mathcal{U}(\mathcal{P})$ and $\mathcal{H}^{\mathrm{gr}*} \xrightarrow{\Delta} \mathcal{H}^{\mathrm{gr}*} \otimes \mathcal{H}^{\mathrm{gr}*} \xrightarrow{S_p \otimes S_p} \mathcal{U}(\mathcal{P}) \otimes \mathcal{U}(\mathcal{P})$ coincide. But both are algebra morphisms from $\mathcal{H}^{\mathrm{gr}*}$ to $\mathcal{U}(\mathcal{P}) \otimes \mathcal{U}(\mathcal{P})$, whose restriction to $\mathfrak{g}_0$ coincide, and thus the universal property of $\mathcal{H}^{\mathrm{gr}*} = \mathcal{U}(\mathfrak{g}_0)$ gives the required result. It is straightforward to check that, in case $\mathcal{P}$ is a graded Lie algebra, B_p and S_p are compatible with the gradings of $\mathcal{H}^{\mathrm{gr}*}$, $\mathcal{P}$ and $\mathcal{U}(\mathcal{P})$. $\square$

THEOREM 4.8.– *Given a left pre-Lie algebra $\mathcal{P}$ that is compatible with a decreasing filtration and complete with respect to it, let $\bar{\mathcal{U}}(\mathcal{P})$ be the complete universal enveloping algebra of the Lie algebra $\mathcal{P}$. Then, for each element $p \in \mathcal{P}$, there exists a unique filtered pre-Lie algebra morphism $B_p : \mathfrak{g} \to \mathcal{P}$ and a unique filtered algebra morphism $S_p : \mathcal{H}^* \to \bar{\mathcal{U}}(\mathcal{P})$, such that $B_p(\delta^\bullet) = p$, and equations [4.63]–[4.64] hold for each $\alpha \in \mathcal{H}^*$, $\beta \in \mathfrak{g}$.*

As a result of the last statement in Theorem 4.8, the restriction to $\mathcal{G} = G(\mathcal{H}^\circ)$ of S_p defines a morphism of topological groups from $\mathcal{G} = \exp(\mathfrak{g})$ to $\exp(\mathcal{P})$ (with the topology induced by their decreasing filtrations).

Consider the left pre-Lie algebra $\mathcal{P}$ of series of smooth vector fields in $\mathbb{R}^d$ of the form $hf_1 + h^2 f_2 + \cdots$, where each f_j is a smooth vector field in $\mathbb{R}^d$. The complete universal enveloping algebra $\bar{\mathcal{U}}(\mathcal{P})$ can be realized as the completion of the graded algebra of endomorphisms parametrized by h generated by derivations of the form $h^n\, g(y)\, \frac{\partial}{\partial y^j}$, where $n \geq 1$, $j \in \{1, \ldots, d\}$, and $g \in C^\infty(\mathbb{R}^d)$. We have seen in section 4.3.5 that, given a smooth vector field f, the definition of B-series $B_{hf}(\beta) := B_{hf}(\beta, \cdot)$ for each $\beta \in \mathfrak{g}$ determines a morphism from the left pre-Lie algebra $\mathfrak{g}$ to the left pre-Lie algebra $\mathcal{P}$ satisfying that $B_{h\,f}(\delta^\bullet) = hf$. Theorem 4.8 implies that such a pre-Lie algebra morphism $B_{hf} : \mathfrak{g} \to \mathcal{P}$ satisfying $B_{hf}(\delta^\bullet) = hf$ is unique. We have also seen that the map $S_{hf} : \mathcal{H}^* \to \bar{U}(\mathcal{P})$ defined in section 4.3.3 is an algebra morphism and satisfies equations [4.63]–[4.64], and thus coincides with the unique algebra map S_p, $p = hf$, given by Theorem 4.8.

Furthermore, if $\alpha \in \mathcal{H}^\circ$ with

$$\Delta(\alpha) = \sum_{j=1}^{m} \beta_j \otimes \gamma_j,$$

then, for each $(u, v) \in \mathcal{F} \times \mathcal{F}$,

$$\alpha_{uv} = \langle \alpha, uv \rangle = \langle \Delta\alpha, u \otimes v \rangle = \sum_{j=1}^{m} (\beta^{[j]})_u \otimes (\gamma^{[j]})_v,$$

and thus, by virtue of Proposition 4.6, for all $g_1, g_2 \in C^\infty(\mathbb{R}^d)$, we have

$$S_{hf}(\alpha) g_1 g_2 = \sum_{j=1}^{m} \left(S_{hf}(\beta^{[j]}) g_1 \right) \left(S_{hf}(\gamma^{[j]}) g_2 \right).$$

4.4.4. *A universal property of the commutative Hopf algebra of rooted trees*

Consider the vector space $\mathcal{V} = \mathbb{R}\,\mathcal{T}$. The restriction to $\mathcal{V}$ of $\Delta - (\mathrm{id}_{\mathcal{H}} \otimes e)$ is a linear map $\rho : \mathcal{V} \to \mathcal{H} \otimes \mathcal{V}$ (the coassociativity of Δ implies that ρ is a left $\mathcal{H}$-comodule map).

DEFINITION 4.13.– *Following Loday and Ronco (2010), we say that a commutative Hopf algebra* $\widetilde{\mathcal{H}}$ *is a* combinatorial right-sided Hopf algebra *if the following conditions hold:*

– *as a coalgebra,* $\widetilde{\mathcal{H}}$ *is conilpotent;*

– *as an algebra,* $\widetilde{\mathcal{H}}$ *is freely generated by a subspace* $\widetilde{\mathcal{V}}$ *having a left* $\widetilde{\mathcal{H}}$*-comodule structure with comodule map* $\tilde{\rho} : \widetilde{\mathcal{V}} \to \widetilde{\mathcal{H}} \otimes \widetilde{\mathcal{V}}$ *satisfying* $\tilde{\Delta}(v) = v \otimes \tilde{e} + \tilde{\rho}(v)$.

The following result is a dual version of Theorem 4.7

PROPOSITION 4.13.– *Let* $\widetilde{\mathcal{H}}$ *be a commutative combinatorial right-sided Hopf algebra, and let* $\widetilde{\mathcal{V}}$ *be the corresponding vector space* $\widetilde{\mathcal{V}}$. *Given a linear form* $\tilde{\eta} : \widetilde{\mathcal{V}} \to \mathbb{R}$, *there exists a unique Hopf algebra morphism* $\phi : \mathcal{H} \to \widetilde{\mathcal{H}}$, *such that*

$$\phi(\widetilde{\mathcal{V}}) \subset \mathcal{V} \quad and \quad \tilde{\eta} = \eta \circ \phi,$$

where $\eta : \mathcal{V} \to \mathbb{R}$ *is given by* $\eta(\bullet) = 1$, $\eta(u) = 0$ *if* $|u| > 1$.

The following result allows constructing the Hopf algebra morphism ϕ in Proposition 4.13 in a recursive way.

PROPOSITION 4.14.– *Under the assumptions of Proposition 4.13, the following diagram commutes*

$$\begin{array}{ccc} \widetilde{\mathcal{V}} & \xrightarrow{\tilde{\rho}} & \widetilde{\mathcal{H}} \otimes \widetilde{\mathcal{V}} \\ \phi \downarrow & & \downarrow \phi \otimes \tilde{\eta} \\ \mathcal{V} & \xleftarrow{[\cdot]} & \mathcal{H} \end{array}$$

4.4.5. *The substitution law*

Given a B-series $B_{hf}(\alpha, y)$, it is of interest to study how this B-series is transformed if hf is substituted by $h\tilde{f}(y) = B_{hf}(\beta, y)$, where $\beta \in \mathfrak{g}$. As shown in Chartier *et al.* (2005, 2010),

$$B_{h\tilde{f}}(\alpha, y) = B_{hf}(\beta \star \alpha, y),$$

where the coefficients $(\beta \star \alpha)_u$ for $u \in \mathcal{T}$ are given as polynomials in β_v and α_w.

More generally, we may be interested in studying the series of differential operators $S_{h\tilde{f}}(\alpha)$ ($\alpha \in \mathcal{H}^*$) under the substitution $h\tilde{f}(y) = B_{hf}(\beta, y)$.

PROPOSITION 4.15.– *Given* $\beta \in \mathfrak{g}$ *and* $h\tilde{f} = B_{hf}(\beta, y)$, *for each* $\alpha \in \mathcal{H}^*$,

$$B_{h\tilde{f}}(\alpha, y) = B_{hf}(\beta \star \alpha, y), \quad S_{h\tilde{f}}(\alpha) = S_{hf}(\beta \star \alpha),$$

where $\beta \star \alpha \in \mathcal{H}^*$ is determined as follows: $\forall u \in \mathcal{H}$,

$$\langle \beta \star \alpha, u \rangle = \langle \alpha, \phi(u) \rangle,$$

where $\phi : \mathcal{H} \to \mathcal{H}$ is the algebra map given by Proposition 4.13 with $\widetilde{\mathcal{H}} = \mathcal{H}$, $\widetilde{\mathcal{V}} = \mathcal{U}$ and $\tilde{\eta}(u) = \beta_u$ for each $u \in \mathcal{T}$.

Hence, $\phi : \mathcal{H} \to \mathcal{H}$ is recursively determined with the help of Proposition 4.14 as follows: For each $u \in \mathcal{T}$,

$$\phi(u) = \beta_u\,[e] + \sum_{\substack{(V \succ W) \subset U \\ V, W \neq \emptyset}} \beta_w\,[\phi(v)].$$

4.4.6. *Bibliographical comments*

The commutative Hopf algebra of rooted trees $\mathcal{H}$ was first identified by Dür (1986), who realized that Butcher's group (Butcher 1972) was an affine group scheme, and as such, was equivalent to a commutative Hopf algebra. Grossman and Larson (1989) constructed a commutative Hopf algebra $\mathcal{H}_{GL}$ of rooted trees related to the linear differential operators in section 4.3.3, which was eventually shown (Foissy 2002; Hoffman 2003) to be the graded dual of $\mathcal{H}$ (i.e. the universal enveloping algebra of $\mathfrak{g}_0$, hence a Hopf subalgebra of the dual Hopf algebra $\mathcal{H}^\circ$). Later on, Connes and Kreimer (1998) rediscovered the commutative Hopf algebra $\mathcal{H}$ in the context of renormalization in quantum field theory (Connes and Kreimer 1998; Kreimer 1998). Brouder (2000) seems to be the first author to note the relationship of Kreimer's work with Butcher's theory. The substitution law in section 4.4.5 was first considered (in the context of numerical integrators expanded as B-series) in Chartier *et al.* (2005) (see also Chartier *et al.* 2010). A bialgebra and a Hopf algebra associated with the substitution law $\star$ (with coproduct obtained by dualizating $\star$) is studied in Calaque *et al.* (2011). Proposition 4.13 can be obtained by combining Chapoton's result (Chapoton and Livernet 2001) (that says that the pre-Lie algebra on the linear span of rooted trees is the free pre-algebra on one generator) and Loday and Ronco's Theorem 5.8 in Loday and Ronco (2010). Proposition 4.14 is to our knowledge, new. Proposition 4.15 can be seen as a dualization of a recursion for the dual of $\star$ found in Manchon and Saidi (2011) and Calaque *et al.* (2011).

4.5. References

Brouder, C. (2000). Runge–Kutta methods and renormalization. *Eur. Phys. J. C*, 12, 512–534.

Butcher, J.C. (1972). An algebraic theory of integration methods. *Math. Comp.*, 26, 79–106 [Online]. Available at: https://doi.org/10.2307/2004720.

Butcher, J.C. (2008). *Numerical Methods for Ordinary Differential Equations*, 2nd edition. John Wiley & Sons, Chichester [Online]. Available at: https://doi.org/10.1002/9780470753767.

Butcher, J.C. and Chan, T.M.H. (2002). A new approach to the algebraic structures for integration methods. *BIT*, 42(3), 477–489.

Calaque, D., Ebrahimi-Fard, K., Manchon, D. (2011). Two interacting Hopf algebras of trees: A Hopf-algebraic approach to composition and substitution of B-series. *Adv. Appl. Math.*, 47(2), 282–308 [Online]. Avaiable at: https://doi.org/10.1016/j.aam.2009.08.003.

Calvo, M.P., Murua, A., Sanz-Serna, J.M. (1994). Modified equations for ODEs. *Contemp. Math.*, 172, 63–74 [Online]. Available at: https://doi.org/10.1090/conm/172/01798.

Chan, T.M.H. (1998). Algebraic structures for the analysis of numerical methods. PhD Thesis, University of Auckland, Auckland.

Chapoton, F. and Livernet, M. (2001). Pre-Lie algebras and the rooted trees operad. *Int. Math. Res. Not.*, 2001, 395–408.

Chartier, P. and Murua, A. (2007). Preserving first integrals and volume forms of additively split systems. *IMA J. Numer. Anal.*, 27(2), 381–405 [Online]. Available at: https://doi.org/10.1093/imanum/drl039.

Chartier, P., Hairer, E., Vilmart, G. (2005). A substitution law for *b*-series vector fields. *INRIA*, 5498, February.

Chartier, P., Faou, E., Murua, A. (2006). An algebraic approach to invariant preserving integrators: The case of quadratic and Hamiltonian invariants. *Numer. Math.*, 103(4), 575–590 [Online]. Available at: https://doi.org/10.1007/s00211-006-0003-8.

Chartier, P., Hairer, E., Vilmart, G. (2010). Algebraic structures of B-series. *Found. Comput. Math.*, 10(4), 407–427 [Online]. Available at: https://doi.org/10.1007/s10208-010-9065-1.

Connes, A. and Kreimer, D. (1998). Hopf algebras, renormalization and noncommutative geometry. *Comm. Math. Phys.*, 199, 203–242.

Dür, A. (1986). *Möbius Functions, Incidence Algebras and Power Series Representations*. Springer-Verlag, Berlin, Heidelberg.

Foissy, L. (2002). Les algèbres de Hopf des arbres enracinés décorés II. *Bull. Sci. Math.*, 126(4), 249–288.

Grossman, R. and Larson, R.G. (1989). Hopf-algebraic structure of families of trees. *J. Algebra*, 126, 184–210.

Grossman, R. and Larson, R.G. (1990). Hopf-algebraic structure of combinatorial objects and differential operators. *Hopf Algebras*, 72, 109–117 [Online]. Available at: https://doi.org/10.1007/BF02764614.

Hairer, E. (1994). Backward analysis of numerical integrators and symplectic methods. *Annals of Numerical Mathematics*, 1, 107–132.

Hairer, E. and Wanner, G. (1974). On the Butcher group and general multi-value methods. *Computing (Arch. Elektron. Rechnen)*, 13(1), 1–15 [Online]. Available at: https://doi.org/10.1007/bf02268387.

Hochschild, G.P. (1981). *Basic Theory of Algebraic Groups and Lie Algebras*. Springer-Verlag, New York, Berlin.

Hoffman, M.E. (2003). Combinatorics of rooted trees and Hopf algebras. *Trans. Amer. Math. Soc.*, 355, 3795–3811.

Kreimer, D. (1998). On the Hopf algebra structure of perturbative quantum field theories. *Adv. Theor. Math. Phys.*, 2(2), 303–334 [Online]. Available at: https://doi.org/10.4310/ATMP.1998.v2.n2.a4.

Loday, J.-L. and Ronco, M. (2010). Combinatorial Hopf algebras. *Clay Math. Proc.*, 11.

Manchon, D. and Saidi, A. (2011). Lois Pr-Lie en interaction. *Communications in Algebra*, 39(10), 3662–3680 [Online]. Available at: https://doi.org/10.1080/00927872.2010.510813.

Merson, R.H. (2001). An operational method for the study of integration processes. *Proc. Symp. Data Processing, Weapons Res. Establ. Salisbury*, Salisbury, 110–125.

Murua, A. (1999). Formal series and numerical integrators I. Systems of ODEs and symplectic integrators. *Appl. Numer. Math.*, 29(2), 221–251 [Online]. Available at: https://doi.org/10.1016/S0168-9274(98)00064-6.

Murua, A. (2006). The Hopf algebra of rooted trees, free Lie algebras, and Lie series. *Found. Comput. Math.*, 6(4), 387–426.

Sweedler, M.E. (1969). *Hopf Algebras*. W. A. Benjamin Inc., New York.

5

Combinatorial Algebra in Controllability and Optimal Control

Matthias KAWSKI

Arizona State University, Tempe, USA

Control theory models, analyzes and designs purposeful interactions with dynamical systems so that they behave in a desired way. Fundamental to the ability to synthesize complex responses by strategically combining simple control building blocks is a desired lack of commutativity of the flows associated with control actions. In its most simple terms, this means that the dynamics are not constrained to evolve on lower-dimensional integral manifolds. Mathematically, this lack of integrability is manifested in terms of (nonabelian) Lie algebras. In recent years, it has been recognized that Zinbiel algebras and combinatorial Hopf algebras play an important unifying role for a deeper understanding of the underlying structures.

This chapter presents fundamental concepts in control theory that are inherently linked to combinatorial and algebraic structures. It demonstrates how modern combinatorial algebraic tools provide both deeper insight and facilitate analyses, computations and designs. The emphasis is on exhibiting the algebraic structures that map combinatorial structures to geometric and dynamic objects.

5.1. Introduction

Control theory models, analyzes and designs purposeful interactions with dynamical systems, with the objective of making them behave in a desired way. In the most simple systems, this interaction may be as simple as selecting the value of a

Algebra and Applications 2,
coordinated by Abdenacer MAKHLOUF.

constant control parameter. In applications, this may be a flow rate or temperature in a chemical plant, a vaccination rate in a model monitoring the spread of infectious diseases, an interest rate in a financial system or a torque or force in a simple mechanical setting. In the classical setting, such constant control parameters may parameterize time-varying control inputs via their frequency or decay rate. More generally, modern control theory considers classes of time-varying control inputs (usually measurable) that need not necessarily be parameterized by a finite number of parameters.

Very simple systems may have a corresponding input channel for each degree of freedom. The first complication arises when these inputs are not decoupled, but instead interact with others. But the main challenge is to control systems that are characterized by a high-dimensional state space (many degrees of freedom), together with a very small number of control input channels. In this case, the interactions between the controls play a critical role and they may allow us to independently *control* a large number of states with very few controls, possibly even a single scalar control. The analysis and exploitation of such interactions of the controls, with each other and with themselves, is at the heart of modern control theory.

A decisive role is played by the lack of commutativity: in general, control actions taken in different orders result in different outcomes. This demands that we develop and employ corresponding mathematical structures that both give insight into the underlying structures, and allow us to efficiently perform computations for analysis and design. The classical linear theory developed in the 1960s was largely based on linear algebra and complex analysis. Since the 1970s, Lie algebras have been routinely employed for the study of nonlinear systems. More recently, it has become clear that more can be gained by using further refinements. Indeed, the roots of using combinatorial and Hopf-algebraic tools in control theory can be traced back to the late 1980s; see, for example, Grossman and Larson (1989, 1992, 2004).

The subsequent sections give a high-level survey of fundamental concepts and problems of control theory, with special focus on items that are intrinsically linked to deeper algebraic and combinatorial structures. Starting from the very conceptual descriptions and problems, the sections quickly progress to more abstract structures. The development is intended to be understandable to an audience that has no prior technical background in control theory. The main contribution consists of unifying the many findings from the last 30 years and framing them in a form that highlights their common algebraic and combinatorial foundations, with special focus on the mappings from combinatorial and algebraic objects to geometric, analytic and dynamic objects. Throughout, whenever feasible, the notation and viewpoint of the chronological calculus of Agračev and Gamkrelidze (1978, 1980) are employed, and we use sub- and superscripts to denote contra- and covariant objects, respectively.

5.1.1. *Motivation: idealized examples*

The following examples will be revisited. Control theory is just as applicable in other frameworks, from electrical, biological to financial systems. This article uses simple, idealized mechanical systems, as their nontrivial dynamics are intuitive and familiar from everyday life and more easily accessible than other applications. They illustrate the usefulness of a systematic geometric description using the advanced combinatorial and algebraic tools developed in this article.

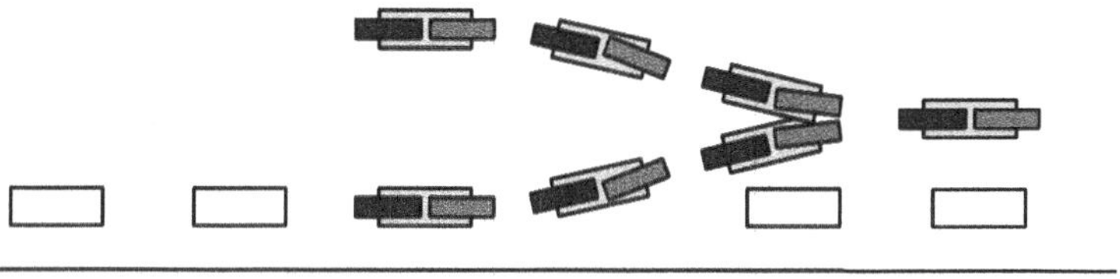

Figure 5.1. *Parallel parking a car (bicycle). For a color version of this figure, see www.iste.co.uk/makhlouf/algebra2.zip*

The task of *parallel parking a car* is familiar, but not trivial as any first-time driver will attest (see Figure 5.1). The natural choices for the controls are the angular velocity of the steering wheel and the position of the gas and brake pedals (or we may even consider the torque on the steering wheel as control input).

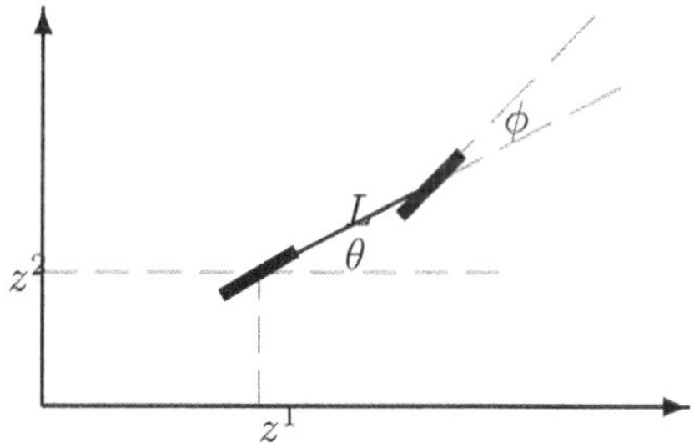

Figure 5.2. *The states of the bicycle*

To model the system by a set of differential equations, first define the states (see Figure 5.2). To avoid complications due to the different angular velocities *inside* and *outside* the wheels in real cars, consider the simpler case of a bicycle. More specifically, denote the position of the center of the rear wheel by $(z^1, z^2) \in \mathbb{R}^2$, and identify the angle of the bicycle with the z^1-axis and the angle of the front wheel with respect to the bicycle with points θ and φ on the circle S^1. Given the length

L between the axes of the front and back wheels, the position of the front wheel is $(z^1 + L\cos\theta, z^2 + L\sin\theta)$.

The dynamics are determined by a non-slip condition [5.1] that imposes that the wheels can only roll in the natural direction, and cannot slip sideways

$$\begin{cases} 0 = \sin\theta \, dz^1 - \cos\theta \, dz^2, \quad \text{and} \\ 0 = \sin(\theta+\varphi) \, d(z^1 + L\cos\theta) - \cos(\theta+\varphi) \, d(z^2 + L\sin\theta). \end{cases} \qquad [5.1]$$

A more challenging second example is an inverted pendulum. Its control was dramatically illustrated in the Apollo program in the 1960s, with the 105 m tall Saturn V rocket being kept vertical by the sophisticated control of the directions of the thrust generated by powerful nozzles at its base. Among the many variations of inverted pendulums and double pendulums, consider the case recently studied by Hauser (2010). The reference state that is to be stabilized, if possible, is a vertical inverted pendulum attached to the end of a horizontal arm. The control is the angular acceleration about the shoulder.

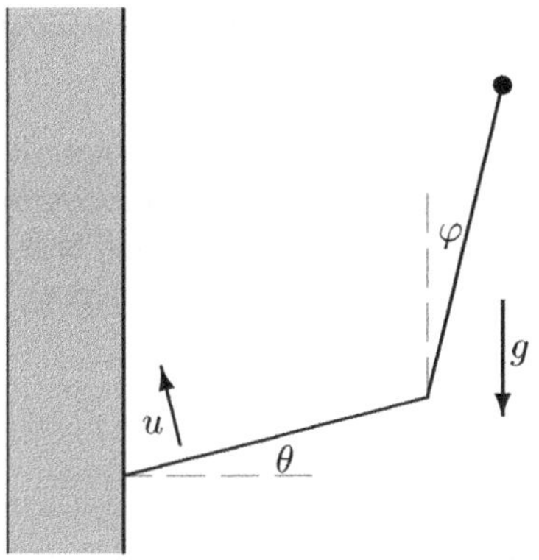

Figure 5.3. *An inverted pendulum*

The equations of motion are readily derived from the first principles. Denoting the ratio of the lengths of the arms by β and the control (angular acceleration) by u, we obtain:

$$\ddot{\theta} = u, \quad \text{and}$$

$$\ddot{\varphi} = \sin\varphi - \beta\cos(\varphi-\theta)\dot{\theta}^2 + \beta u \sin(\varphi-\theta). \qquad [5.2]$$

These examples will be revisited in sections 5.2.1 and 5.3.1 through 5.3.3 where their controllability properties will be analyzed, and suitable approximating systems will be constructed.

Many similar introductory examples including rolling pennies, sliding skates and balls rolling on each other may be found in the literature, particularly in popular textbooks such as those by Jurdjevic (1997), Bloch (2003), Bullo and Lewis (2005), and Bressan and Piccoli (2007).

5.1.2. *Controlled dynamical systems*

Control theory is a broad discipline that studies a broad range of dynamical systems, ranging from discrete systems to systems governed by stochastic partial differential equations. What they have in common and what formally distinguishes them from generic dynamical systems are that they admit inputs (controls) and generally, specified outputs (observed, or measured quantities). Even more distinctive are the questions asked and the problems studied: instead of analyzing, describing and predicting behavior, control theory fundamentally asks inverse questions such as determining, if feasible, an input that will result in a desired future behavior.

The primary concern of this study is the analysis of systems that are governed by deterministic ordinary differential equations, generally evolving on smooth manifolds. In various places, it will be apparent how this work may extend, generalize or translate into systems that are infinitely dimensional (governed by partial differential equations), or that evolve in discrete time steps, or with stochastic signals. The objective is not utmost generality, i.e. the weakest analytic hypotheses, but instead to illuminate the underlying geometric, algebraic and eventually combinatorial structures. The simple mechanical systems presented in the preceding section serve as prototypes for applications. These can be quickly made more demanding, such as the attitude control of a satellite in space, which may itself consist of multiple linked rigid bodies.

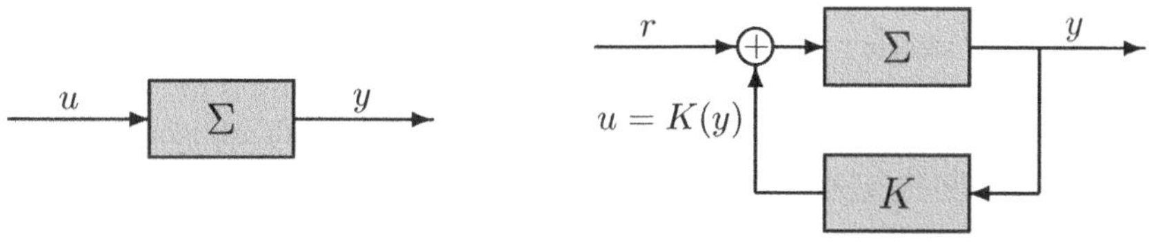

Figure 5.4. *Open-loop and closed-loop controls with a feedback controller K*

Classically, control theory distinguishes two complementary ways of modeling systems. On one hand, we consider input signals $u(t)$ that are fed into a box (possibly a black box) which produces corresponding output signals $y(t)$. Mathematically, this point of view corresponds to maps between the spaces of functions of a single continuous (or discrete) variable (time). Careful considerations of the domain, range and regularity of the operators are critical. This point of view is particularly suitable for considering system interconnections, where the output of one system is the input for another system. Of particular interest are feedback interconnections in which the output y of the plant, the physical system, is the input for the (automatic) controller, which computes and generates an output $u = K(y)$, which is fed back as the input to the plant (usually added to an external reference input r, see Figure 5.4).

On the other hand, systems may be modeled as (ordinary) differential equations which generally depend on time-varying parameters (inputs), and which evolve on a manifold, called the state-space. Outputs, or observed measurements, are modeled by smooth functions on the state space. Generally, these are not injective, i.e. at any instant of time, they only provide partial information about the actual state of the system.

5.1.3. *Fundamental questions in control*

Arguably, three of the most central objectives and concepts are

– controllability;

– optimal control;

– feedback stabilization.

In (uncontrolled) dynamical systems, the uniqueness of the solutions of initial value problems leaves the primary remaining question: where do the solutions go? One studies the limit sets. In contrast, in controlled dynamical systems, the central objective is to *make* a system (trajectory) go to a desired state. The first question is whether the system is controllable, i.e. decide whether it is possible to transfer the state from any initial state to any desired terminal state using some admissible control input $u(t)$.

Rather than steering the system with *open-loop* controls, i.e. inputting controls that are functions of time, the main objective in automatic control is to *close the loop* by feeding back control signals $u(x)$ or $u(y)$ that are automatically generated from information about the state x or the output y, respectively.

Algebraic and combinatorial tools play an important role for deciding controllability and studying related properties. The algebraic and analytic background of feedback stabilization of nonlinear systems is comparatively less well understood. This chapter primarily addresses issues related to controllability.

In order to study controllability, optimality and stabilizability, we need to develop other fundamental notions such as

– observability;

– system identification;

– state-space realization;

– equivalence and normal forms;

– system approximation.

Given a collection of input–output pairs, system identification is concerned with identifying (the parameters of) the mapping from inputs to outputs. Going one step further, state-space realizations typically represent the system by a collection of first-order differential equations (vector fields) on a smooth manifold, together with functions on the state space that provide the outputs. Roughly speaking, a system is called observable if the outputs (over some time intervals) provide sufficient information to determine the state of the system. From an algebraic operator point of view, observability (output-to-state) is, in a natural way, a notion dual to controllability (input-to-state). Two systems in state-space form that generate the same input–output behavior may be considered equivalent. Formally, we study the equivalence of state-space systems under the action of groups of diffeomorphisms (local coordinate changes) and feedback transformations, identifying invariants under such actions, and singling out normal forms for each orbit.

From the modeling perspective (*"no model is perfect"*) and for analytical work, it becomes important to develop suitable notions when one system is a good approximation of another.

Much of the algebraic and combinatorial work in this chapter deals with the interplay of input–output operators with state-space realizations, and with approximations and normal forms. The emphasis is on matching the control inputs, outputs and observations to corresponding algebraic and combinatorial structures. System identification will not be addressed explicitly.

Once we have generated a solution for a control problem, it is natural to refine it, and ask, for example, for

– optimization;

– robustness.

From a geometric point of view, optimality is in a well-defined sense dual to controllability, and as such, will appear in the following. Going beyond the mathematical solution of the idealized problems (model, system), control engineering is as much an art as it is a science, and beyond optimizing solutions, much effort is devoted to addressing modeling errors, rejecting disturbances, practical implementations of digital controllers and so on. Such aspects go much beyond the framework addressed in this chapter.

5.2. Analytic foundations

For the input–output operator point of view, the domain and range of the operators play key technical roles. In the case of linear systems, the systems represented by bounded linear operators that map L^2 input signals (Lebesgue square integrable) to L^2 output signals are the best studied. In the case of nonlinear systems, the situation is considerably less clear-cut. This work will concentrate on systems that can be represented by formal power series which have desirable convergence properties. The focus here are the algebraic and combinatorial aspects. For in-depth treatments of analytic issues, especially convergence, the interested reader may refer to (Li and Gray 2006; Gray and Wang 2002; Gray and Li 2005). A detailed description of the Faa di Bruno approach is given in Gray and Duffaut Espinosa (2011), whereas the convergence issues of feedback interconnections (Thitsa and Gray 2012) were only resolved in 2012. For more recent Hopf-algebraic constructions for output feedback, see, for example, Gray *et al.* (2014).

For systems represented in state-space form, i.e. by differential equations, or time-varying vector fields on manifolds, the first questions are about the existence of solutions, and regularity properties of the *reachable sets*. The existence proofs that basically refine the Picard iteration already foreshadow the Hopf-algebraic structures, and will be developed in detail in the next section. Similarly, the variation of parameters is a key tool for arguments about controllability and optimality, and is intimately related to combinatorial structures.

Throughout this chapter, notations and viewpoints of chronological calculus are employed. This takes a functional analytic perspective: instead of considering the solutions of differential equations (i.e. the flows) as families of diffeomorphisms on a manifold, they are considered as families of automorphisms on the algebra of smooth functions on the manifold.

Before addressing central control issues, the next sections first review and develop some mathematical foundations.

5.2.1. *State-space models and vector fields on manifolds*

This work focuses on the rich class of smooth systems on manifolds that are affine in the control. These commonly arise in applications. The way that the dynamics split into time-varying controls that may be manipulated, and static vector fields that determine the geometry of the system is very attractive from a mathematical point of view, as well as for engineering applications. For practical implementations, we may be able to compute objects related to the geometry in advance and off-line, while only updating the less demanding operations on the time-varying signals in real time. In traditional notation[1], we write

$$\dot{q} = f_0(q) + \sum_{i=1}^{m} u^i f_i(q), \quad \text{and} \tag{5.3}$$

$$y = \varphi(q). \tag{5.4}$$

The state $q \in M^n$ takes values in a smooth manifold M^n. The space $\mathcal{U}$ of controls consists of measurable functions $u\colon [0,T] \mapsto U \subseteq \mathbb{R}^m$ that are typically assumed to take values in a compact and convex subset $U \subseteq \mathbb{R}^m$ that contains $0 \in \operatorname{int} U$ in its interior. The output function $\varphi\colon M \mapsto \mathbb{R}^p$ is smooth, and the dynamics are defined by smooth vector fields $f_i \in \Gamma^\infty(M)$ on the manifold M^n.

For the example of parallel parking a car [5.1], a natural state space $M = R^3 \times S^1 \times S^1$ may be locally identified with $\mathbb{R}^5$. The state may be chosen in local coordinates as $q = (v, \varphi, z^1, \theta, z^2) \in \mathbb{R} \times S^1 \times \mathbb{R} \times S^1 \times \mathbb{R}$, where the forward speed v is defined by $dz^1 = v\cos\theta\, dt$ and $dz^2 = v\sin\theta\, dt$. Manipulating the second constraint in [5.1], we quickly obtain, using elementary trigonometry

$$d\theta = \frac{v\,dt}{L} \cdot \frac{\cos\theta \cdot \tan(\theta+\varphi) - \sin\theta}{\sin\theta \cdot \tan(\theta+\varphi) + \cos\theta} = \frac{v}{L}\tan\varphi\, dt. \tag{5.5}$$

From practical considerations, the natural choices for the control inputs are the forward/backward acceleration $\dot{v} = u^1$ and the angular velocity of the steering angle $\dot{\varphi} = u^2$. Combining these differential equations, the controlled dynamics may be written in the standard form of [5.3] with $m = 2$, by introducing the vector fields f_0, f_1, f_2 written as column vectors in terms of the coordinates q

$$f_0(q) = \begin{pmatrix} 0 \\ 0 \\ q^1\cos q^4 \\ q^1 \tan q^2 \\ q^1 \sin q^4 \end{pmatrix}, \quad f_1(q) = \begin{pmatrix} 1 \\ 0 \\ 0 \\ 0 \\ 0 \end{pmatrix}, \quad \text{and} \quad f_2(q) = \begin{pmatrix} 0 \\ 1 \\ 0 \\ 0 \\ 0 \end{pmatrix}. \tag{5.6}$$

For the example of the pendubot [5.2], a natural state space is the tangent bundle $M = T\mathbb{T}^2$ of the torus may be locally identified with R^4. The state may be chosen in local coordinates as $q = (\dot{\theta}, \theta, \dot{\varphi}, \varphi)$. The controls take values in a set $U = [-u_0, u_0] \subset \mathbb{R}$, and the dynamics are determined by a drift vector field f_0

1. While most of the subsequent theoretical discussion will use the chronological notation, in this introduction and in concrete examples, we will use the traditional notation, including column vectors.

and a controlled vector field f_1, which written as column vectors in local coordinates are

$$f_0(q) = \begin{pmatrix} 0 \\ q^1 \\ \sin q^4 - \beta\cos(q^4 - q^2)(q^1)^2 \\ q^3 \end{pmatrix}, \quad \text{and} \quad f_1(q) = \begin{pmatrix} 1 \\ 0 \\ \beta\sin(q^4 - q^2) \\ 0 \end{pmatrix}. \qquad [5.7]$$

5.2.2. *Chronological calculus*

In the geometric control theory, it is standard to *define* (or identify) vector fields as (with) first-order partial differential operators. For example, the vector fields, written as column vectors in the preceding example, are more precisely understood as the linear maps $f_0, f_1 \colon C^\infty(T\mathbb{T}^2) \mapsto C^\infty(T\mathbb{T}^2)$

$$\begin{aligned} f_0 &= q^1\tfrac{\partial}{\partial q^2} + (\sin q^4 - \beta\cos(q^4 - q^2)(q^1)^2)\tfrac{\partial}{\partial q^3} + q^3\tfrac{\partial}{\partial q^4}, \quad \text{and} \\ f_1 &= \tfrac{\partial}{\partial q^2} + \beta\sin(q^4 - q^2)\tfrac{\partial}{\partial q^4}. \end{aligned} \qquad [5.8]$$

It is convenient to go further and also consider points in the state-space and flows of vector fields (solutions of differential equations) as operators on the commutative algebra of smooth functions on the state space. This functional analytic approach was pioneered in the nonlinear control theory by Agračev and Gamkrelidze (1978, 1980). For a detailed discussion, see the recent book by Agračhev and Sachkov (2004), or the technical details provided in Kawski and Sussmann (1997) and Sarychev (2001).

In particular, for a point $p \in M$, denote by $\hat{p}$ the multiplicative linear functional $\hat{p} \colon C^\infty(M) \mapsto \mathbb{R}$ defined by $\hat{p}(\varphi) = \varphi(p)$. For $\varphi, \psi \in C^\infty(M)$ and $c \in \mathbb{R}$, this satisfies

$$\hat{p}(\varphi + c\psi) = \hat{p}(\varphi) + c\hat{p}(\psi) \;\text{ and }\; \hat{p}(\varphi \cdot \psi) = \hat{p}(\varphi) \cdot \hat{p}(\psi). \qquad [5.9]$$

This chapter will suppress the notational distinction between p and $\hat{p}$. The most immediate effect is that what is traditionally written as $\varphi(p)$, $f(p)$ and $(f\varphi)(p)$ for $p \in M$, $\varphi \in C^\infty(M)$ and $f \in \Gamma^\infty(M)$, a smooth vector field on M is simplified by the consistent notations $p\varphi$, pf and $pf\varphi$. Moreover, for a point $q \in M$ and a set S of vector fields, differential operators or diffeomorphisms on M, we also use the notation $qS = \{qf \colon f \in S\}$. This is most commonly used when S is the Lie algebra $L(f, g)$ generated by vector fields f and g (or a subspace thereof).

For the (local) flow of a vector field $f \in \Gamma^\infty(M)$, introduce the notation $(t, q_0) \mapsto q_0e^{tf}$. This notation will be further justified in the next section. In the chronological framework, rather than considering the local flow of a local diffeomorphism of M, it is considered locally as an automorphism on the algebra of smooth functions on M. It is (locally) defined for $\varphi \in C^\infty(M)$ by

$$e^{0f}\varphi = \varphi \quad \text{and} \quad \tfrac{d}{dt}e^{tf}\varphi = fe^{tf}\varphi. \tag{5.10}$$

Evaluating the latter at any point $q \in M$ yields the formula $\frac{d}{dt}qe^{tf}\varphi = qe^{tf}f\varphi$. Note that $e^{tf}f\varphi = fe^{tf}\varphi$. This notation becomes particular handy when differentiating the compositions of flows. For example, for $t \in \mathbb{R}$, $q \in M$, $\varphi \in C^\infty(M)$ and $f, g \in \Gamma^\infty(M)$ in the following calculation

$$\tfrac{d}{dt}qe^{tf}e^{tg}\varphi = qe^{tf}fe^{tg}\varphi + qe^{tf}e^{tg}g\varphi. \tag{5.11}$$

the e^{tg} in the first term has to be interpreted as the tangent map (or differential) of the flow e^{tg}. Indeed, in traditional notation we may write

$$\tfrac{d}{dt}\varphi(\Phi_t^g \circ \Phi_t^f(q)) = (\Phi_{t*}^g(f))(\varphi)(\Phi_t^f(q)) + (g\varphi)(\Phi_t^g \circ \Phi_t^f(q)). \tag{5.12}$$

In the chronological notation, it is completely clear how to interpret any exponential, depending on whether we read it as acting on smooth functions on the right, or on differential operators or points (which are considered zeroth-order differential operators) on the left. For example, it makes perfect sense to calculate

$$\begin{aligned}
&\tfrac{d}{dt}\Big|_{t=0} qe^{tf}e^{tg}e^{-tf}e^{-tg}\varphi = \\
&= q\left(e^{tf}fe^{tg}e^{-tf}e^{-tg} + e^{tf}e^{tg}ge^{-tf}e^{-tg}\right. \qquad [5.13]\\
&\quad \left.-e^{tf}e^{tg}e^{-tf}fe^{-tg} - e^{tf}e^{tg}e^{-tf}e^{-tg}g\right)\varphi\Big|_{t=0}\\
&= qf\varphi + qg\varphi - qf\varphi - qg\varphi = 0. \qquad [5.14]
\end{aligned}$$

It is straightforward (Kawski and Sussmann 1997) to calculate the next derivative and formally derive

$$\tfrac{1}{2}\tfrac{d^2}{dt^2}\Big|_{t=0} qe^{tf}e^{tg}e^{-tf}e^{-tg}\varphi = q(fg - gf)\varphi = q[f, g]\varphi \tag{5.15}$$

which defines the Lie bracket of two smooth vector fields. Recall that every smooth vector field $f \in \Gamma^\infty(M)$ is a derivation on the algebra $C^\infty(M)$, i.e. for $\varphi, \psi \in C^\infty(M)$ it satisfies $f(\varphi \cdot \psi) = (f\varphi) \cdot \psi + \varphi \cdot (f\psi)$. At a point $q \in M$, this is written as

$$qf(\varphi \cdot \psi) = qf\varphi \cdot q\psi + q\varphi \cdot qf\psi. \tag{5.16}$$

As we readily verify, in general, the composition $(g \circ f) \colon \varphi \mapsto gf\varphi$ is not a derivation. However, the Lie bracket $[f, g] \colon \varphi \mapsto fg\varphi - gf\varphi$ is always a derivation, i.e. a first-order partial differential operator, or a smooth vector field.

Another example that demonstrates how this formalism simplifies appearance and calculations is a version of the *ad* formula. This formula is used frequently in control to calculate the effect of moving control variations along the flow of another vector field

$$\tfrac{d}{dt} q e^{tf} g e^{-tf} = q e^{tf} f g e^{-tf} - q e^{tf} g f e^{-tf} = q e^{tf} [f, g] e^{-tf}. \qquad [5.17]$$

In particular, evaluating this identity at $t = 0$ yields

$$\tfrac{d}{dt}\Big|_{t=0} q e^{tf} g e^{-tf} = q[f, g]. \qquad [5.18]$$

In traditional notation, using Φ^f for the local flow of f, this formula appears in terms of the differential of the pullback and gives rise to the Taylor expansion (assuming analyticity)

$$\Phi^f_{-t*}(g(\Phi^f_t(q))) = \sum_{k=0}^{\infty} \tfrac{t^k}{k!} (\mathrm{ad}^k f, g)(q) \qquad [5.19]$$

where $(\mathrm{ad}^k f, g)$ is defined recursively by $(\mathrm{ad}^0 f, g) = g$ and $(\mathrm{ad}^{k+1} f, g) = [f, (\mathrm{ad}^k f, g)]$ for $k \in \mathbb{Z}_0^+$.

This work is mainly concerned with the algebraic and combinatorial aspects. But we note that the chronological calculus is also an effective tool for the topological and analytic aspects. In particular, the commutative algebra $C^\infty(M)$ of smooth functions has a natural topology: a sequence $\{\varphi_k\}_{k=0}^\infty \in C^\infty(M)$ converges to $\psi \in C^\infty(M)$ if and only if for every compact set $K \subseteq M$ and every finite set of smooth vector fields $f_1 \dots f_s \in \Gamma^\infty M$, the sequence $\{f_s \dots f_2 f_1 \varphi_k\}_{k=0}^\infty$ converges uniformly to $f_s \dots f_2 f_1 \psi$ on K. Thus, every tangent vector qf at any point $q \in M$ is clearly a continuous linear functional, and is also an element of the dual space of $C^\infty(M)$, the space of compactly supported Schwartz distributions.

For a more detailed technical description of the topology and structure of $C^\infty(M)$ as a Fréchet space in terms of seminorms, see, for example, the introduction in Agračhev and Sachkov (2004). Here, we will routinely make use of the fact that smooth vector fields on M and diffeomorphisms of M may be regarded as continuous linear operators on $C^\infty(M)$ as a topological vector space.

Finally, we remark that the chronological calculus of Agračhev and Gamkrelidze achieves its full power when working with time-varying vector fields. However, for the purposes of this work, with its focus on elements of combinatorial Hopf algebra in control, the split into time-varying controls and static vector fields that define the geometry is critically important. It is this splitting that allows us to develop powerful algebraic and combinatorial tools that apply in this important special case.

REMARK 5.1.– *In the following, we will write* $q_0 f$ *when we want to emphasize an initial value* $q(0) = q_0$, *but otherwise, in theoretical work, we will appropriately use lower and upper indices like* u^i, x^i, f_i *wherever feasible and meaningful. But when dealing with specific examples, we keep some in the traditional notation, such as* $f(0)$, $[f_0, f_1](0)$, x_1^3 *when readability would suffer too much from a change.*

5.2.3. *Piecewise constant controls and the Baker–Campbell–Hausdorff formula*

Briefly returning to the parking example, we illustrate how the exponential products provide insight into the controlled dynamics. Observe that in the coordinates chosen, the Jacobian matrices of partial derivatives of the vector fields [5.6] are all strictly lower triangular. Consequently, due to the special *cascade form* of the associated differential equation, it is possible to explicitly write out the solution for any control using simple quadratures only (i.e. unlike in the case of general nonlinear differential equations, here, the solution can be expressed in terms of a finite number of iterated integrals of scalar functions only). In particular, for any admissible control $\eta\colon [0,T] \mapsto U \subseteq \mathbb{R}^2$, the corresponding iterated integrals are immediately calculated as: first $q^1(t) = \int_0^t \eta_s^1\, ds$ and $q^2(t) = \int_0^t \eta_s^2\, ds$. Then, iteratively

$$
\begin{aligned}
q^4(t) &= \int_0^t \left(\int_0^s \eta_\sigma^1 d\sigma\right) \tan\left(\int_0^s \eta_\sigma^2 d\sigma\right) ds, \\
q^3(t) &= \int_0^t \left(\int_0^s \eta_\sigma^1 d\sigma\right) \cos\left(\int_0^s \left(\int_0^s \eta_\tau^1 d\tau\right) \tan\left(\int_0^\sigma \eta_\tau^2 d\tau\right) d\sigma\right) ds, \\
q^5(t) &= \int_0^t \left(\int_0^s \eta_\sigma^1 d\sigma\right) \cdot \sin\left(\ldots q_s^3 \ldots\right) ds.
\end{aligned} \qquad [5.20]
$$

To parallel park a car, we need to choose η to achieve a purely sideways motion, i.e. so that $q(T) = (0,0,0,0,C)$ for some $C \neq 0$. As everyday experience shows, this is possible, but not completely trivial. The complexity of the required choices corresponds to the number of integrations needed. One possible solution using piecewise constant controls is given by the following choice (for some $u^1, u^2 > 0$) and results in the pictured solution curves. The first three show graphs as a function of time, and the fourth is a phase-plot showing the location of the car in the $(z^1, z^2) = (q^3, q^5)$-plane

$$
\eta(t) = \begin{cases} (u^1, u^2) & \text{if } 0 \leq t < 2 \\ (0, -u^2) & \text{if } 2 \leq t < 6 \\ (-u^1, u^2) & \text{if } 6 \leq t < 10 \\ (0, -u^2) & \text{if } 10 \leq t < 14 \\ (u^1, u^2) & \text{if } 14 \leq t \leq 16. \end{cases} \qquad [5.21]
$$

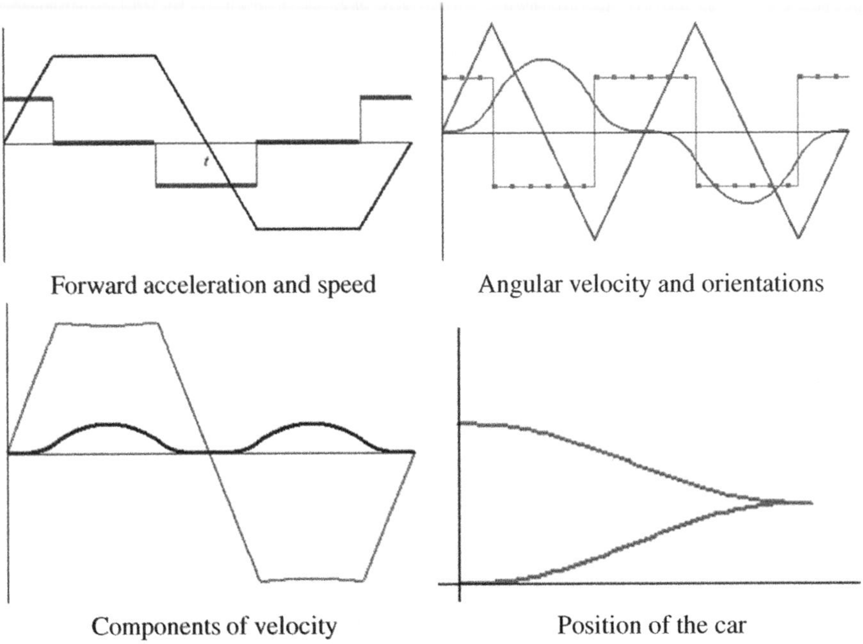

Figure 5.5. *Parallel parking a car (bicycle). For a color version of this figure, see www.iste.co.uk/makhlouf/algebra2.zip*

The solution of the corresponding sequence of differential equations

$$\begin{cases} \dot{q} = f_0 + u^1 f_1 + u^2 f_2 & \text{if } 0 \leq t < 2 \\ \dot{q} = f_0 \qquad\qquad\; - u^2 f_2 & \text{if } 2 \leq t < 6 \\ \dot{q} = f_0 - u^1 f_1 + u^2 f_2 & \text{if } 6 \leq t < 10 \\ \dot{q} = f_0 \qquad\qquad\; - u^2 f_2 & \text{if } 10 \leq t < 14 \\ \dot{q} = f_0 + u^1 f_1 + u^2 f_2 & \text{if } 14 \leq t \leq 16 \end{cases} \qquad [5.22]$$

may be written in exponential notation using $\tau_i = t_1 - t_{i-1}$ for the time between the switching times t_i in the example

$$q(t) = q_0 e^{\tau_1(f_0+u^1 f_1+u^2 f_2)} e^{\tau_2(f_0-u^2 f_2)} e^{\tau_3(f_0-u^1 f_1+u^2 f_2)}$$
$$e^{\tau_4(f_0-u^2 f_2)} e^{\tau_5(f_0+u^1 f_1+u^2 f_2)}. \qquad [5.23]$$

This product of exponentials may be formally rewritten using the classical Baker–Campbell–Hausdorff formula (see Campbell (1897)),

$$e^X e^Y = e^{X+Y+\frac{1}{2}[X,Y]+\frac{1}{12}[X,[X,Y]-\frac{1}{12}[Y,[X,Y]]+\frac{1}{24}[X,[Y,[X,Y]]]\ldots} \qquad [5.24]$$

where each X and Y takes the form $\tau_1(f_0 \pm u^1 f_1 \pm u^2 f_2)$.

The resulting formula as a single exponential is extremely unwieldy and not useful for practical computations, even in this very simple example of parking a car. Nonetheless, it was a popular tool in the 1970s and early 1980s when the first higher-order conditions for nonlinear controllability and optimality were derived. However, it clearly shows that each summand in the exponent is an iterated Lie bracket of the vector fields f_i, whose coefficient is a homogeneous polynomial in u^i and τ_j. Moreover, the degrees of each of these polynomials in u and τ are equal to the numbers of controlled vector fields $f_1, \dots f_m$ and equal to the total number of all vector fields $f_0, \dots f_m$ involved in the iterated bracket, respectively. This will be made precise in the next sections, which aim to provide more efficient alternatives to this classical expansion. Specifically, in realistic control applications, the number of switchings may be very large, and more generally, the controls do not need to be piecewise constant. Moreover, the exponent in the classical Baker–Campbell–Hausdorff formula is written as a linear combination of all right-to-left iterated Lie brackets. But these are always linearly dependent because of relations resulting from the Jacobi identity, and hence, the coefficients in this linear combinations are not well defined. For recent alternatives, see, for example, Strichartz (1987), Gehrig (2007), Gehrig and Kawski (2008) and the later sections of this chapter.

5.2.4. *Picard iteration and formal series solutions*

For vector fields that are locally Lipschitz continuous, the classical method of the Picard iteration, together with a fixed-point argument, yields the existence and uniqueness of local flows. In the context of affine control systems, it is motivated by first rewriting the system [5.3] as an equivalent integral equation. More specifically, suppose that u is a measurable function taking values in a compact set U, $f_i \in \Gamma^\infty(M)$ are smooth vector fields, $\varphi \in C^\infty(M)$ is a smooth function on a smooth manifold M and $q_0 \in M$ is an initial point. For notational convenience, introduce the additional control $u^0 \equiv 1$. Then, the output φ evaluated along a solution curve $t \mapsto q_t \in M$ of [5.3] for sufficiently small t satisfies the equivalent integral equation

$$q_t\varphi = q_0\varphi + \sum_{i=0}^{m} \int_0^t u_s^i \cdot q_s f_i \varphi \, ds. \qquad [5.25]$$

This motivates us to define and study iterates of the mapping P from absolutely continuous curves q_t in M, to absolutely continuous curves $(P[q])_t$, defined for $\varphi \in C^\infty(M)$ by

$$(P[q])_t\varphi = q_0\varphi + \sum_{i=0}^{m} \int_0^t u_s^i \cdot q_s f_i \varphi \, ds. \qquad [5.26]$$

The fixed points of this mapping P correspond to the solution curves of [5.25], and thus of [5.3]. Using the chronological formalism, it can again be shown that on sufficiently small-time intervals, the iterates of this map converge to such a fixed point, which is indeed again an absolutely continuous curve. This curve is the unique solution (on that interval) for the initial value problem [5.3]. For a meticulous exposition of all technical details in terms of seminomas on $C^\infty(M)$, see Agračhev and Sachkov (2004).

A small modification of the Picard iteration process yields a formal series expansion of the local flow. Under the same hypotheses as above, formally iterate the integral equation [5.25], by substituting the right-hand side of the corresponding equation for $q_t f_i \varphi$ (instead of $q_t \varphi$) into [5.25]

$$
\begin{aligned}
q_t\varphi &= q_0\varphi + \sum_{i=0}^{m} \int_0^t u_s^i \cdot \left(q_0\varphi + \sum_{j=0}^{m} \int_0^s u_\sigma^j \cdot q_\sigma f_j f_i \varphi \, d\sigma \right) ds \\
&= q_0\varphi + \sum_{i=0}^{m} \left(\int_0^t u_s^i \, ds \right) \left(q_0 f_i \varphi \right) + \sum_{i,j=0}^{m} \int_0^t \int_0^s u_s^i u_\sigma^j \cdot q_\sigma f_j f_i \varphi \, d\sigma \, ds \\
&= q_0\varphi + \sum_{i=0}^{m} \left(\int_0^t u_s^i \, ds \right) \left(q_0 f_i \varphi \right) + \sum_{i,j=0}^{m} \left(\int_0^t \int_0^s u_s^i u_\sigma^j \, d\sigma \, ds \right) \left(q_0 f_j f_i \varphi \right) \\
&\quad + \sum_{i,j,\ell=0}^{m} \int_0^t \int_0^s \int_0^\sigma u_s^i u_\sigma^j u_\tau^\ell \cdot q_\tau f_\ell f_j f_i \varphi \, d\tau \, d\sigma \, ds \qquad [5.27] \\
&= \dots
\end{aligned}
$$

Note that at each stage, every term except the last one splits into an iterated integral that depends on the controls u only, and a higher-order partial derivative of the output functions evaluated at the initial point.

Continuing this iteration process yields a formal infinite series

$$
\widetilde{CF}(t, u, f, \varphi, q_0) = \sum_I \left(\int_0^t u^I \right) \cdot q_0 f_I \varphi \qquad [5.28]
$$

where the sum is taken over all multi-indices $I = (i_1, i_2, \dots, i_s) \in \{0, 1, \dots m\}^s$ with $s \geq 0$, $\int_0^t u^\emptyset = 1$, $f_\emptyset \varphi = \varphi$, and inductively for $(I, j) \in \{0, 1, \dots m\}^{s+1}$

$$
\int_0^t u^{(I,j)} = \int_0^t u_s^j \int_0^s u_\sigma^I \, d\sigma \, ds \;\text{ and }\; f_{(I,j)}\varphi = f_I(f_j\varphi). \qquad [5.29]
$$

Note that the convenient indexing is the reverse order from the iteration illustrated above. In particular, when f_I acts on the left of a point or on the right of a smooth output function, this corresponds to reverse orders of composition. For the iterated integrals, the notation is more ambiguous as in our setting the controls commute (they are scalars), and the products $u_{t_1}^{i_1} u_{t_2}^{i_2} = u_{t_2}^{i_2} u_{t_1}^{i_1}$ can be written in either order. However, the order of integration (indexing of integration variables) must be consistent with the order of application of the partial differential operators

$$\int_0^t u^{(i_1,i_2,i_3)} = \int_0^t u_{t_3}^{i_3} \int_0^{t_3} u_{t_2}^{i_2} \int_0^{t_2} u_{t_1}^{i_1} \, dt_1 \, dt_2 \, dt_3 \quad \text{and} \tag{5.30}$$

$$q f_{(i_1,i_2,i_3)} \varphi = q f_{i_1} f_{i_2} f_{i_3} \varphi. \tag{5.31}$$

This apparent reversal of the indexing of these terms is not purely a choice of notation. Indeed, it may be interpreted as the appearance of the antipode in the formal Hopf-algebraic treatment.

This expansion for q_t as a curve in the dual of $C^\infty(M)$ considered as a formal series of partial differential operators $\widetilde{CF}(t, u, f, \cdot, q_0)$ *never* converges, even weakly (if f_i are nontrivial) to a well-defined functional defined on all of $C^\infty(M)$. Instead, the expression [5.28] must be considered as an asymptotic series for the output φ along the solution q_t of [5.3]; see, for example, Sussmann (1983). Technical details may be found in Sussmann (1983) and section 2.4.4 of Agračhev and Sachkov (2004).

5.2.5. *The Chen–Fliess series and abstractions*

A central theme of this survey is that commonly laborious manipulations performed on analytic or dynamic objects are fundamentally of a combinatorial and algebraic nature. The structure becomes more clearly discernible and the work much easier when the manipulations are done at the level of the corresponding combinatorial objects. The main objective here is to abstract the asymptotic series [5.28] for the evolution of an output function φ along the solution curves of a control system to a combinatorial abstract form of the Chen–Fliess series, which may be considered an unevaluated form $\widetilde{CF}(t, u, f, \varphi, q_0)$ in [5.28]. This section introduces the basic terminology and notation used, and discusses some analytic properties of the evaluation maps. Their algebraic properties will be elaborated in section 5.4.3.

To facilitate readability and to conform to standard verbiage in the combinatorics literature, it is convenient to allow indexing sets (for vector fields, coordinates, etc.) that are not necessarily sets of integers $\{0, 1, \ldots m\}$. Start with a set Z, which will generally be finite in this chapter. Call the set Z an alphabet and its elements letters. The set of all finite nonempty sequences taking values in Z is denoted by Z^+. The empty sequence $\emptyset$ is sometimes also written as e. Define $Z^* = Z^+ \cup \{e\}$. The concatenation of sequences endows Z^* with a natural noncommutative monoid structure. For notational simplicity, abbreviate sequences and their products by juxtaposition, for example, if $w = (a_{i_1}, a_{i_2}, \ldots a_{i_s})$ and $z = (b_{j_1}, b_{j_2}, \ldots b_{j_t})$, then simply write, for example, $w = a_{i_1} a_{i_2} \cdots a_{i_s}$ and $wz = a_{i_1} a_{i_2} \cdots a_{i_s} b_{j_1} b_{j_2} \ldots b_{j_t}$. Next, introduce the associative algebra $A(Z)$ of all finite linear combinations of elements of Z^* with coefficients in a base field, which, in this chapter, is always the field of real numbers $\mathbb{R}$. In other words, $A(Z)$ is the algebra of noncommutative polynomials over Z with coefficients in $\mathbb{R}$. Write $A^+(Z)$ for the subalgebra of all finite linear combinations of elements of Z^+, the subalgebra of noncommutative polynomials with zero constant term. The Lie algebra $L(Z)$ is the smallest subspace of $A(Z)$ that contains Z and is closed under the Lie bracket $[\cdot\,,\,\cdot]\colon A(Z)\times A(Z)\mapsto A(Z)$, $[\cdot\,,\,\cdot]\colon (w,z)\mapsto [w,z]=wz-zw$.

The algebras $A(Z)$, and $L(Z)$ have natural graded structures. For each $a\in Z$, the monoid homomorphism $\delta_a\colon Z^*\mapsto \mathbb{Z}_0^+$ defined on the generators $b\in Z$ by $\delta_a(b)=\delta_{a,b}$ (Kronecker delta) *counts* the number of occurrences of the letter a in a word in Z^*. If $Z=\{a_0,a_1,\ldots a_m\}$, it is convenient to introduce for $k\in\mathbb{Z}_0^+$ and $\ell\in(\mathbb{Z}_0^+)^m$, the subsets $Z^{(k,\ell)}\subseteq Z$, defined by

$$Z^{(k,\ell)}=\{w\in Z^*\colon k=\delta_{a_0}(w) \text{ and } \ell_j=\delta_{a_j}(w) \text{ for } 1\le j\le m\}. \qquad [5.32]$$

The distinguished letter a_0 plays an important role for controllability, and corresponds to an uncontrolled drift vector field. The corresponding homogeneous subspaces $A^{(k,\ell)}\subseteq A(Z)$ and $L^{(k,\ell)}\subseteq L(Z)$ are spanned by (Lie) polynomials that are linear combinations of words in $Z^{(k,\ell)}$.

Following Kawski and Sussmann (1997), the algebra $A(Z)$ has a natural topology as a uniform space, where the uniform structure is the collection of the sets (called *entourages*) $E_S=\{(\sum_{w\in Z^*}c_w w,\sum_{w\in Z^*}d_w w)\;:\;c_w,d_w\in\mathbb{R},$ and $c_w=d_w$ if $w\in S\}$, indexed by all finite subsets $S\subseteq Z^*$ of words. The completion of $A(Z)$ with respect to this uniform structure is the algebra $\hat{A}(Z)$ of formal power series over Z. The multiplication on $A(Z)$ is uniformly continuous and thus extends to a continuous map from $\hat{A}(Z)\times\hat{A}(Z)$ to $\hat{A}(Z)$. The basis Z^* of $A(Z)$ induces a natural pairing $\langle\cdot,\cdot\rangle\colon A(Z)\times\hat{A}(Z)\mapsto\mathbb{R}$, defined by

$$\langle\sum_{w\in Z^*}c_w w\,,\,\sum_{w\in Z^*}d_w w\rangle=\sum_{w\in Z^*}c_w d_w \qquad [5.33]$$

where all but a finite number of the coefficients c_w vanish. With this pairing and upon identifying $p\in A(Z)$ with the linear functional $q\mapsto\langle p,q\rangle$ on $\hat{A}(Z)$, $\hat{A}(Z)$ is the algebraic dual of $A(Z)$, and $A(Z)$ is the topological dual of $\hat{A}(Z)$.

Chen (1957) (see also Chen (1954, 1958)) introduced the Chen series as a geometric invariant of curves in Euclidean space $\mathbb{R}^m$. He associated a formal power series $\sum_{w\in Z^*}c_w w$ (on an alphabet Z with m letters) with every curve, where the coefficients $c_w\in\mathbb{R}$ were defined as iterated line integrals along the curve. Fliess (1980, 1981) reinterpreted the Chen series in the context of control systems. The curves are the solution curves of the nonlinear system [5.3] that is affine in the controls, and the coefficients c_w are iterated integrals of the controls. For every fixed curve $\gamma\colon[0,T]\mapsto\mathbb{R}^m$ or every fixed control $\eta\colon[0,T]\mapsto U$, either series is one element in $\hat{A}(Z)$, or a curve in $\hat{A}(Z)$, when considering variable end-times $t \in [0, T]$. Recently, in the context of rough paths and stochastic signals, a large body of closely related theory has been developed in which these series are commonly referred to as *signatures* of signals (see, for example, Lyons and Qian (2002); Hambly and Lyons (2010)).

To make best use of the underlying Hopf-algebra structures, it is beneficial to further abstract the Chen–Fliess series, and to start with the formal representation of the identity map on $\hat{A}(Z)$, by identifying the representation of $\mathrm{Id}_{\hat{A}(Z)}(\cdot) = \sum_{w \in Z^*} \langle w\,,\,\cdot\rangle w$. with the formal power series

$$CF = \sum_{w \in Z^*} w \otimes w \in A(Z) \otimes \hat{A}(Z). \qquad [5.34]$$

There are numerous ways to map this formal object to a partially evaluated series [5.28] by specifying some or all of: a set of vector fields, an output function, an initial point, a set of control inputs and a terminal time. For the most important mappings, introduce the following notation.

For a sequence of smooth vector fields $F = \{f_{a_0}, \ldots, f_{a_m}\} \subseteq \Gamma^\infty(M)$ indexed by a set $Z = \{a_0, \ldots a_m\}$, define the map

$$\mathcal{F}\colon Z \mapsto \Gamma^\infty(M) \quad \text{by} \quad \mathcal{F}(a) = f_a \qquad [5.35]$$

where $f_a \in \Gamma^\infty(M)$ is considered as a linear operator (indeed, a derivation), mapping $C^\infty(M)$ to itself. Next, for a word $w \in Z^*$ and a letter $a \in Z$, and any function $\varphi \in C^\infty(M)$, define $\mathcal{F}(aw)(\varphi) = f_a\,(\mathcal{F}(w)\varphi)$. Reusing the same symbol, extend this map linearly to a Lie algebra homomorphism $\mathcal{F}\colon L(Z) \mapsto L(F) \subseteq \Gamma^\infty(M)$ and to an associative algebra homomorphism that maps $A(Z)$ to linear combinations of partial differential operators on $C^\infty(M)$. Under suitable growth conditions on the coefficients and conditions on the vector fields (e.g. compactly supported), this map further extends (still using the same symbol) to a map on formal power series $\hat{A}(Z)$ to, for example, compactly supported Schwartz distributions on M. For technical details, see, for example, Agračev and Gamkrelidze (1978, 1980); Sussmann (1983); Kawski and Sussmann (1997); Agrachev and Sachkov (2004).

On the other hand, rather than directly evaluating on control inputs, it is advantageous to map $A(Z)$ to an algebra of iterated integral functionals (see Kawski and Sussmann (1997)). From an algebraic point of view, it is most convenient to work with the primitives $t \mapsto U^a(t) = \int_0^t u^a(s)\,ds$ of the control inputs rather than the controls themselves, as done in Kawski and Sussmann (1997). However, in this context, to enhance readability and facilitate connections with control applications, this survey will stay with the traditional notions of the inputs $u \in \mathcal{U}$. Note that these are usually presumed to be essentially bounded integrable functions of time, whereas their primitives are absolutely continuous functions. The main interest lies in a Zinbiel subalgebra (to be introduced later) of the set $\mathcal{M}_{\mathcal{U}}$ of all mappings from a space of controls $\mathcal{U}$ (typically to be taken as $L^1([0,T],U)$, where $U \subseteq \mathbb{R}^m$ is a compact subset containing the origin in its interior; compare the assumptions for system [5.3] to the set of absolutely continuous functions $AC([0,T],\mathbb{R})$.

Thus, without further specifying topological or algebraic structures on the range at this time, introduce the map $\Upsilon\colon A(Z) \mapsto \mathcal{M}_{\mathcal{U}}$ first for a letter $a \in Z$, any word $w \in Z^*$, for any control $u \in \mathcal{U}$ and $t \in [0,T]$ by

$$\Upsilon^a(u)(t) = \int_0^t u^a(s)\,ds \text{ and } \Upsilon^{wa}(u)(t) = \int_0^t u^a(s) \cdot \Upsilon^w(u)(s)\,ds \qquad [5.36]$$

Extend this map linearly to an associative algebra homomorphism that maps $A(Z)$ to a subset of iterated integral functionals in $\mathcal{M}_{\mathcal{U}}$.

Recognize that with this notation, the infinite series solution [5.28] obtained from the Picard iteration is the image of the identity on $A(Z)$ under distinguished evaluation maps

$$\widetilde{CF}(t,u,f,\varphi,q_0) = \sum_{w \in Z^*} \Upsilon^w(u)(t) \cdot q_0 \mathcal{F}(w)\varphi. \qquad [5.37]$$

Rather than thinking of I or w indexing some iterated integrals or partial differential operators, this chapter takes interest in the algebraic properties of the mappings Υ and $\mathcal{F}$ when the domain and codomain are equipped with suitable algebraic structures (the analytic and topological properties of these maps are largely beyond the scope of this survey).

The duality between the *partially evaluated* images of the abstract series CF is of particular interest. In particular, rather than considering CF itself, or the fully evaluated expansion $(\Upsilon \otimes \mathcal{F})(CF)$, the two expansions $(\mathrm{Id} \otimes \mathcal{F})(CF)$ and $(\Upsilon \otimes \mathrm{Id})(CF)$ are of special interest. The first associates, with every control system defined by a sequence of smooth vector fields, a formal power series whose coefficients are partial differential operators on the state space. Upon further evaluation on an output function, and possibly an initial state q_0, this view associates a formal power series with every system. The second is a formal power series whose coefficients are iterated integral functionals. Upon further evaluation of control inputs, this view associates a formal power series with every control. These two different kinds of power series may be naturally paired to yield the output response for a specific control when fed into a specific system. While analytically, these appear as very different objects, algebraically, the series representations of controls and of systems exhibit a natural duality.

Further refinements are possible by considering fixed terminal times T, or curves parameterized by $t \in [0,T]$, or by considering local flows (local diffeomorphisms) by leaving q_0 open, or considering specific solution curves of the Cauchy problem for a specified initial condition q_0. The special case where a vector-valued φ is a local diffeomorphism $\varphi\colon M \mapsto \mathbb{R}^n$ for some nonempty open set $V \subseteq M$ yields the familiar form of the system [5.3] represented in local coordinates.

5.3. Controllability and optimality

A control system is (globally) controllable if, for every pair of states $q_0,\ q_1$, there exists an admissible control that steers the system from the initial state q_0 to the terminal state q_1. Many variations of this concept have been proposed, investigated and proven useful. Here, we focus on the well-studied property of *small-time local controllability* (STLC), which is dual to optimality in a geometric way.

In separate sections, the following items are addressed. The first step is to relate controllability to a lack of integrability and Frobenius' theorem. The following section accounts for the presence of a distinguished drift vector field that breaks the symmetry and thus necessitates more refined statements to accurately describe the reachable sets. For most controllable systems of practical relevance, a simple algorithm allows us to construct a nilpotent approximating system that preserves important geometric properties, but is much easier to use. Theorems for controllability can be rewritten as statements about optimality, generalizing the Pontryagin maximum principle. Their proofs rely on approximating cones to reachable sets and families of control variations, and use the graded structures and the correspondence of Lie brackets and iterated integrals.

5.3.1. *Reachable sets and accessibility*

Consider a control system $\Sigma_{F,\mathfrak{U}}$ defined by a finite sequence $F = (f_0, f_1 \dots f_m)$ of smooth vector fields on a manifold M of dimension n, a compact convex set $U \subset \mathbb{R}^m$ that contains $0 \in \operatorname{int} U$ in its interior, and the set $\mathfrak{U}$ of all measurable functions u taking values in U, and each defined for some $T_u > 0$ on the interval $[0, T_u]$. For each such control $u \in \mathfrak{U}$, the dynamics are governed by the differential equation

$$\dot{q}_t = q_t f_0 + \sum_{i=1}^{m} u_t^i \, q_t f_i. \qquad [5.38]$$

For $q_0 \in M$, $\eta \in \mathfrak{U}$ and $T_\eta > 0$ sufficiently small, there exists a unique curve $\xi(\cdot; q_0, \eta)\colon [0, T] \mapsto M$, such that $\xi(0; q_0, \eta) = q_0$ and for almost all $t \in [0, T]$ (in traditional notation)

$$\dot{\xi}(t; q_0, \eta) = f_0(\xi(t; q_0, \eta)) + \sum_{i=1}^{m} \eta^i(t) f_i(\xi(t; q_0, \eta)). \qquad [5.39]$$

We may abbreviate $\xi(t; q_0, \eta)$ and write $\xi(t)$ when there is no danger of confusion.

DEFINITION 5.1.– *For a control system $\Sigma_{F,\mathfrak{U}}$ as above, $q_0 \in M$ and $T \geq 0$ sufficiently small, the reachable sets from q_0 at time T, and in time at most T, are*

$$q_0\mathfrak{R}_T = \{\xi(T; q_0, \eta), \eta \in \mathfrak{U}\}, \text{ and } q_0\mathfrak{R}_{\leq T} = \bigcup_{0 \leq t \leq T} q_0\mathfrak{R}_t. \qquad [5.40]$$

DEFINITION 5.2.– *For $q_0 \in M$, the control system $\Sigma_{F,\mathfrak{U}}$ is accessible from q_0 if for every $T > 0$,* int $q_0\mathfrak{R}_T \neq \emptyset$.

By concatenating motions along the integral curves of the control vector fields f_i, we may generate infinitesimal motion in the direction of the Lie brackets of the vector fields. In particular, to interpret the equation [5.15] in terms of control systems, consider a system without drift (i.e. $f_0 \equiv 0$) and $m = 2$ controls. For the piecewise constant control, $\eta\colon [0,T] \mapsto U = [-1,1]^2$ is defined by

$$\eta(t) = \begin{cases} (1,0) & \text{if } \quad 0 \leq t < \ T \\ (0,1) & \text{if } \quad T \leq t < 2T \\ (-1,0) & \text{if } \ 2T \leq t < 3T \\ (0,-1) & \text{if } \ 3T \leq t \leq 4T. \end{cases} \qquad [5.41]$$

We calculate $\xi(4T; q_0, \eta) = q_0 + T^2\, q_0[f_1, f_2] + O(T^3)$. By increasing the number of switchings, it is possible to generate infinitesimal movements in the directions of higher-order iterated Lie brackets.

The Frobenius integrability theorem guarantees that for every sufficiently small time $T > 0$, the reachable set from q_0 at time T is contained in the integral manifold of the involutive closure, the Lie algebra $L(f_0, f_1, \ldots f_m)$ generated by the vector fields f_i. The Hermann–Nagano theorem (Nagano 1966; Sussmann 1983) guarantees that in the case of analytic vector fields, the reachable set has a relatively open interior in this manifold, too. In particular, this implies the following algorithmic criterion to decide whether an analytic system is accessible.

THEOREM 5.1.– *For $q_0 \in M$, a real analytic control system $\Sigma_{F,\mathfrak{U}}$ is accessible from q_0 if and only if the vector fields $f_0, f_1, \ldots f_m$ satisfy the Lie algebra rank condition [LARC]*

$$\dim q_0 L(f_0, f_1, \ldots f_m) = n. \qquad [5.42]$$

For an easily accessible proof, see section 4.3 of Sontag (1998).

Returning to the parallel parking example, we readily compute the iterated Lie brackets $f_{\pi_3} = [f_1, f_0]]$, $f_{\pi_4} = [f_0, [f_1, f_2]]$ and $f_{\pi_5} = [[f_1, f_0]], [f_0, [f_1, f_2]]]$ in local coordinates (written in traditional notation as column vectors)

$$qf_{\pi_3} = \begin{pmatrix} 0 \\ 0 \\ \cos q^4 \\ \tan q^2 \\ \sin q^4 \end{pmatrix}, \; qf_{\pi_4} = \begin{pmatrix} 0 \\ 0 \\ 0 \\ \sec^2 q^2 \\ 0 \end{pmatrix}, \; \text{and } qf_{\pi_5} = \begin{pmatrix} 0 \\ 0 \\ -\sin q^4 \sec^2 q^2 \\ 0 \\ \cos q^4 \sec^2 q^2 \end{pmatrix}.$$

Evaluating $f_{\pi_1} = f_1$, $f_{\pi_2} = f_2$ together with these three iterated Lie brackets at $q_0 = 0$ yields five independent vectors $q_0 f_{\pi_k} = \frac{\partial}{\partial q^k}\Big|_{q_0}$, which are a basis for the tangent space at q_0. Thus, the system satisfies the LARC and it is accessible from q_0.

It is noteworthy that the length of the iterated Lie brackets computed above matches the order of the first nonvanishing term in the Taylor expansions (with respect to time) of the iterated integrals in [5.20]. For example, the five factors in the iterated Lie bracket f_{π_5} correspond to the formula for $q^5(t)$ in [5.20] being fifth-order in time: the lowest order for which $\frac{d^k}{dt^k} q^5(t)|_{t=0}$ does not vanish is $k = 5$.

5.3.2. *Small-time local controllability*

Accessibility means that the reachable set from a fixed starting point is not constrained to lie in a lower-dimensional integral submanifold, this does not guarantee controllability. Consider the example in the plane with a scalar control

$$\dot{q}^1 = u, \quad \text{and} \tag{5.43}$$

$$\dot{q}^2 = (q^1)^k. \tag{5.44}$$

Here, for k even and $q_0 = (q_0^1, q_0^2) \in \mathbb{R}^2$, the reachable set $q_0 \mathcal{R}_T$ is contained in the half-plane $\{q \in \mathbb{R}^2 : q^2 \geq q_0^2\}$. In terms of iterated integrals, it is the even power k in $\int_0^T \left(\int_0^t u(s)\, ds \right)^k dt$ that is responsible. In terms of the vector fields $f_0 = \frac{\partial}{\partial q^1}$, $f_1 = (q^1)^k \frac{\partial}{\partial q^2}$, we can easily compute that $(\mathrm{ad}^k f_1, f_0) = k!\, \frac{\partial}{\partial q^2}$. Hence, this system satisfies the Lie algebra rank condition [5.42] and hence by theorem 5.1, the reachable sets for every positive time, from any initial condition have a nonempty interior in the plane. However, corresponding to the iterated integral $\int_0^T \left(\int_0^t u(s)\, ds \right)^k dt$, it is the even number of *factors* f_1 in the (*first* and only) iterated Lie bracket $(\mathrm{ad}^k f_1, f_0)$ that *gives* the $\frac{\partial}{\partial q^2}$ direction.

This is the first appearance where the invariance of an iterated integral under an input symmetry, here $u \mapsto -u$, results in an obstruction to controllability. The detailed study of such obstructions to controllability goes back to the 1970s, and was very active in the 1980s (see, for example, Hermes (1982), Sussmann (1983, 1987) and Stefani (1985, 1986)). There is still no complete solution to this problem, and it requires careful analysis of the combinatorial structure of iterated formal Lie brackets. There have been numerous different notions of local controllability of nonlinear systems, from Haynes' and Hermes' notion of locally–locally controllable (Haynes and Hermes 1970), to the most recent notion of small-time local controllability with small controls by Beauchard and Marbach (2018), which emphasizes the choice of the norm, for example, Sobolev norms with negative exponents. Because of both its duality for standard problems in optimal control (especially minimum time problems) and the interest in engineering applications to reach nearby states in short time (without long excursions), one of the best studied concepts is small-time local controllability. We now state its definition and a few major theorems that suffice for most practical applications.

DEFINITION 5.3.– *For $p \in M$, the control system $\Sigma_{F,\mathfrak{U}}$ is small-time locally controllable (STLC) from p, if for every $T > 0$, $p \in \operatorname{int} p\mathfrak{R}_T$.*

THEOREM 5.2.– *(Hermes (1977) and Sussmann (1983)) Suppose $m = 1$, $U = [-1, 1]$, and the system $\Sigma_{F,\mathfrak{U}}$ is accessible from p. If for all $k \geq 0$*

$$\sum_{\ell \geq 0} pL^{(\ell,2k)}(f_0, f_1) \subseteq \sum_{\ell \geq 0} pL^{(\ell,2k-1)}(f_0, f_1) \qquad [5.45]$$

then the system $\Sigma_{F,\mathfrak{U}}$ is STLC from p.

In practical terms, this means that when successively calculating iterated Lie brackets containing an increasing number of factors f_1, and evaluating these at the initial point p, new directions are only obtained from brackets with an odd number of factors of the controlled vector field. The rationale is that the corresponding iterated integrals contain the control $u = u^1$ an odd number of times, and hence are mapped to their negatives by the action of the input symmetry $u \mapsto -u$. We consider the terms which are of the lowest order in the control u, which for $\|u\|$ sufficiently small are expected to dominate all terms that are of the higher order in u. For more details, see the discussion on control variations and approximating cones in section 5.3.5. A corresponding necessary condition for STLC was also proven in the mid-1980s.

THEOREM 5.3.– *(Stefani 1986) If $m = 1$ and $U = [-1, 1]$, and the system $\Sigma_{F,\mathfrak{U}}$ is STLC about q_0, then for all $k \geq 1$*

$$q_0(ad^{2k} f_0, f_1) \in \sum_{\ell \geq 0} q_0 L^{(\ell,2k-1)}(f_0, f_1). \qquad [5.46]$$

Allowing different weights on the time and the magnitude of the control leads to different ways of bounding some iterated integrals by others. This is formalized in the following theorem, yielding sufficient conditions for STLC that cover most applications of interest.

THEOREM 5.4.– *(Adapted from Sussmann (1987)) Suppose* $U = [-1,1]^m$*, and the system* $\Sigma_{F,\mathcal{U}}$ *is accessible from* q_0*. If* $\theta \in (0,1]$ *and for all* k *odd, and all* $\ell_1, \ldots \ell_m$ *even,*

$$q_0 L^{(k,\ell_1,\ldots\ell_m)}(f_0,\ldots f_m) \subseteq \sum_{(k^i,\ell^i)} q_0 L^{(k^i,\ell^i_1,\ldots\ell^i_m)}(f_0,\ldots f_m) \qquad [5.47]$$

where the sum extends over all (k^i, ℓ^i)*, such that*

$$\theta k^i + \ell^i_1 + \ldots + \ell^i_m < \theta k + \ell_1 + \ldots + \ell_m \qquad [5.48]$$

then the system $\Sigma_{F,\mathcal{U}}$ *is STLC from* q_0*.*

More general statements allow different weights θ_i for different vector fields, but the version above suffices for most applications. The first proofs of theorems of this form explicitly constructed families of control variations that generated tangent vectors to the reachable sets from the initial point. The earliest constructions depended heavily on analyzing the effects of piecewise constant control variations using tools, such as for the pullback of a vector field [5.18] and the Baker–Campbell–Hausdorff formula, as illustrated in section 5.2.2. The resulting coefficients of each iterated Lie bracket are polynomial in the parameters of the piecewise constant control variations (amplitude, switching times, see section 5.3.5). The tangent vectors arise as the leading terms in these Taylor series expansions by these parameters.

The next improvement uses the Chen–Fliess series directly and associates iterated Lie brackets with iterated integrals of the controls. Highly technical estimates of these iterated integrals are key to the proof of the necessary condition theorem 5.3. However, these calculations can be much simplified once we take into account the underlying structure as an exponential Lie series and use combinatorial bases for free Lie algebra (see section 5.4).

The proof of theorem 5.4 further introduces and exploits the concept of *input symmetries*, such as mapping a control u to its negative $-u$, or its time reversal $u^-(t) = u(T-t)$, or possibly permutations such as $(u^1,u^2) \mapsto (u^2,u^1)$. Every such input symmetry induces actions on both the iterated brackets of the vector fields and on the associated iterated integrals. *Possible obstructions to controllability* are identified with fixed points of these symmetries. However, if they do not vanish at the starting point, these possible obstructions may be *neutralized* by lower-order brackets, where the order of a (formal) bracket is determined, as in theorem 5.4, by one of the various possible graded and filtered structures on the Lie algebra of vector fields, and correspondingly on the iterated integrals. To make these informal descriptions precise, we need to abstract to free Lie algebras and their enveloping algebras, and work with formal combinatorial objects (see section 5.4).

For an authoritative discussion of controllability that goes from the finite dimensional systems considered here all the way to systems governed by partial differential equations, see the recent monograph by Coron (2007).

Revisiting the parallel parking example, the everyday experience suggests that the system is STLC. Maybe the only question that does not have an evidently affirmative answer is whether, for example, sideways motions can be achieved in arbitrarily small time without large excursions of the state from the reference point. Indeed, this is confirmed analytically. The inspection of the iterated Lie brackets evaluated above, which yield new directions, confirms that none are of the type (odd, totally even) that could be possible obstructions to controllability. Thus, by theorem 5.4, this system is STLC from the initial point $q_0 = 0$. However, from a very large collection of combinatorially possible Lie brackets of the fields f_i, the above calculation managed to *get by* by only considering a few iterated Lie brackets. For practical purposes, we would prefer to have an algorithm that minimizes the number of iterated Lie brackets that need to be calculated. Thus, in general, we should start with formal brackets that form a basis for a free Lie algebra (see definition [5.6]).

The iterated Lie brackets that need to be considered for the pendubot system are an example of the delicate internal structure of iterated Lie brackets (and of the associated iterated integral functionals) that may decide controllability. Indeed, in this case, the established sufficient conditions that are found in the literature cannot establish STLC using only a count of the factors. Instead, finer combinatorial studies and analytic estimates are needed.

In the local coordinates in the traditional column notation, the system vector fields for the pendubot were given in equation [5.7]. It is immediate that $q_0 f_1 = \left.\frac{\partial}{\partial q^1}\right|_{q_0}$ and $q_0[f_0, f_1] = -\left.\frac{\partial}{\partial q^2}\right|_{q_0}$. Calculating increasingly longer iterated brackets, and evaluating them at q_0 does not yield any new linearly independent tangent vectors until we get to iterated Lie brackets involving five factors. While in general, for lower-order brackets, we only need to count the number of times each vector field appears as a factor, at this length the internal combinatorial structure of the Lie brackets begins to play a role. Indeed, in the free Lie algebra $L(a, b)$ over a two-element set $\{a, b\}$, each of the homogeneous subspaces $L^{(2,3)}(a, b)$ and $L^{(3,2)}(a, b)$ has a dimension two. Among the many combinatorial choices for the internal structure of the iterated Lie brackets, the following four arise from commonly used Hall sets (see definition 5.6). Two brackets computed from different numbers of copies of the vector fields f_0 and f_1 both vanish at the reference point,

$$q_0[f_0, [f_0, [[f_0, f_1], f_1]]] = q_0[f_0, [[f_0, f_1], f_1], f_1]] = 0 \qquad [5.49]$$

while complementary iterated brackets with the same numbers of copies of f_0 and f_1 both yield the same new direction at the reference point,

$$q_0[[f_0,[f_0,f_1]],[f_0,f_1]] = -\beta \frac{\partial}{\partial q^3}\Big|_{q_0}, \quad \text{and}$$

$$q_0[[f_0,f_1],[[f_0,f_1],f_1]] = -2\beta^2 \frac{\partial}{\partial q^3}\Big|_{q_0}. \qquad [5.50]$$

Many other choices for the bases of the homogeneous subspaces are possible, but they are related to each other by the Jacobi identity and anticommutativity.

One additional bracket with the drift vector field, corresponding to one additional integration with respect to time, yields the final direction

$$q_0[f_0,[[f_0,[f_0,f_1]],[f_0,f_1]]] = -\beta \frac{\partial}{\partial q^4}\Big|_{q_0}, \quad \text{and}$$

$$q_0[f_0,[[f_0,f_1],[[f_0,f_1],f_1]]] = -2\beta^2 \frac{\partial}{\partial q^4}\Big|_{q_0}. \qquad [5.51]$$

These two are chosen for simplicity, but they do not arise from evaluating Hall trees. Indeed, the dimensions of the corresponding homogeneous subspaces $L^{(3,3)}(a,b)$ and $L^{(4,2)}(a,b)$ in the free Lie algebra are three and two, respectively. The numbers of combinatorially not equivalent iterated Lie brackets involving six factors demand that we make strategic choices that are based on the underlying combinatorial structure.

The calculations above prove that the pendubot is accessible from the given initial condition, but theorem 5.4 alone does not allow us to conclude that the system is STLC. By using intricate constructions of controls, Hauser (2010) proved that the pendubot is indeed STLC, provided that the upper bounds on the magnitude of the control are not too small. This appears to be the first nonacademic, at least reasonably realistic, example in which the general theorem 5.4 does not allow us to decide controllability. Further academic examples were analyzed in Kawski (1987) and Kawski (1988b), including a notion of two possible obstructions to controllability *balancing* each other. But this is the first system of practical interest that lies in the still not completely known territory between sufficient and necessary conditions for STLC.

5.3.3. *Nilpotent approximating systems*

For many control applications such as feedback stabilization and path planning, it is desirable to approximate a system by a simpler model, but one that still captures the critical geometric properties, such as controllability. Such approximating systems may, in turn, also facilitate the analysis of the original system – and the proofs of theorems such as theorem 5.4 intrinsically use nilpotent approximations inside the technical work. For highly nonlinear control systems, nilpotent approximating systems play a role similar to that of (Jacobian) linearizations of nonlinear dynamical systems about their singularities; see, for example, Hermes (1986, 1991), Stefani (1986), Bellaïche (1996), Jean (2001) and Sklyar and Ignatovich (2002, 2007, 2008).

The latter qualitatively determines the local behavior about singularities, such as stability, except in more degenerate cases. While (Jacobian) linearizations have an important role in control systems too, as they cannot capture the characteristic features of highly nonlinear systems, most notably the angular momentum equations of a rigid body with two controls. Nilpotent approximating systems are the next most simple systems, and this class is sufficiently rich so that approximating systems chosen from this class can preserve desirable local properties.

DEFINITION 5.4.– *A control system of form [5.3] is called nilpotent if the Lie algebra* $L(f_0, f_1, \ldots f_m)$ *generated by the system vector fields is nilpotent, i.e. if there exists an integer* N*, such that every iterated Lie bracket* $[f_{i_1}, [f_{i_2}, [f_{i_3}, \ldots [f_{i_{s-1}}, f_{i_s}] \ldots]]])$ *with* $i_j \in \{0, 1, \ldots m\}$ *and* $s \geq N$ *vanishes identically.*

Nilpotent systems are particularly attractive as solution curves corresponding to controls that are piecewise elementary functions that may be computed via simple quadratures only. Compare this to the complexity of the solution curves of even simple quadratic polynomial systems, such as the chaotic Lorenz system.

THEOREM 5.5.– *(Kawski 1988c) Suppose that* $f_0, f_1 \ldots f_m$ *are analytic vector fields on a manifold* M^n *that generate a nilpotent Lie algebra, which at a point* $q_0 \in M^n$ *spans the tangent space* T_pM*. Then, there exists a local coordinate chart about* p*, such that in these coordinates, each of the vector fields has polynomial components and the Jacobian matrices of partial derivatives are strictly lower triangular.*

For uncontrolled dynamical systems, whose linearization about an equilibrium does not have full rank, many efforts have been devoted to classifying normal forms of the singularities. However, no general procedure is available that identifies the *leading part* in cases of higher orders of degeneracy. However, for affine control systems of form [5.3] that satisfy the sufficient conditions for STLC in theorem 5.4, a simple algorithmic procedure allows us to construct a nilpotent approximating system (on the same space) that preserves controllability. Normal forms are addressed in detail in section 5.4.4.

Moreover, feedback laws that are homogeneous in a well-defined sense, and which asymptotically stabilize the approximating system, also locally asymptotically stabilize the original system (see for example, Hermes (1991)). This generalizes Lyapunov's indirect method and supports the claim that the nilpotent approximating system is the natural generalization of (Jacobian) linearizations.

The following algorithm follows the presentation given in Hermes (1986). For more details about adapted coordinate charts, see Stefani (1986) and Hermes (1991). For a more recent abstract algebraic description, see Sklyar and Ignatovich (2007, 2008).

Since vector fields f_i on a manifold may satisfy relationships such as $[f_0, f_1] = f_0$, it is generally meaningless to talk about the length of a Lie bracket of vector fields, or its degree of homogeneity. Thus, technically, the following algorithm needs to be defined in terms of the free magma $\mathcal{M}(Z)$, that is the set of formal iterated brackets, or equivalently, rooted binary trees labeled by the alphabet Z. See definition [5.79] and section 5.4.1 for the technical details.

ALGORITHM FOR CONSTRUCTING NILPOTENT APPROXIMATIONS.– Suppose that the vector fields $f_0, f_1, \ldots f_m$ are analytic and define a control system of form [5.3] on $M^n = \mathbb{R}^n$ that is accessible from $q_0 = 0$, i.e. it satisfies the LARC (theorem 5.1), and $q_0 f_0 = 0$. For simplicity of exposition here, only consider the case when all controlled vector fields f_i, $i \geq 1$ are assigned the same weights $\theta_i = 1$.

1) Choose a weight $\theta = \theta_0 \in (0, 1]$ for the drift vector field f_0 (in general, the weight $\theta = 0$ yields solvable, not necessarily nilpotent approximating systems). Let $\delta^\theta \colon \mathcal{M}(\{0, 1, \ldots m\}) \mapsto \mathbb{R}^+$ be the homomorphism defined on generators by $\delta^\theta(0) = \theta$ and $\delta^\theta(i) = 1$ if $i \geq 1$, i.e. for formal brackets $\ell, \ell' \in \mathcal{M}(\{0, 1, \ldots m\})$, $\delta^\theta([\ell, \ell']) = \delta^\theta(\ell) + \delta^\theta(\ell')$.

2) Choose formal Lie brackets $\ell_{\pi_1}, \ell_{\pi_2}, \ldots \ell_{\pi_n} \in \mathcal{M}(0, 1, \ldots m)$, such that:

- the corresponding Lie brackets of vector fields $f_{\pi_i} = \mathcal{F}(\ell_{\pi_i})$, $i = 1, \ldots n$ span the tangent space $T_{q_0}\mathbb{R}^n$ at 0, i.e. $\{q_0 f_{\pi_1}, \ldots q_0 f_{\pi_n}, \}$ are linearly independent; and

- the weights $r = \left(\delta^\theta(\ell_{\pi_1}), \ldots, \delta^\theta(\ell_{\pi_n})\right)$ are minimal among all such n-tuples of formal Lie brackets (in the lexicographical order).

3) Perform a constant linear coordinate change, such that in the new coordinates, $q_0 f_{\pi_i} = \frac{\partial}{\partial x^i}\Big|_{q_0}$ for $1 \leq i \leq n$.

4) Define the group of dilations $\{\Delta_s \colon s > 0\}$ on $\mathbb{R}^n$ by $\Delta_s(x) = (s^{r_1} x_1, \ldots, s^{r_n} x_n)$.

5) Expand each component $\langle dx^j, f_i \rangle$ for $0 \leq i \leq m$ in a power series $f_i^j = \sum\limits_\alpha c_{i\alpha}^j x^\alpha$ in the new x-coordinates and truncate it, only keeping terms of order at most $r_i - r_j$, i.e. define $\tilde{c}_{i\alpha}^j = c_{i\alpha}^j$ if $\alpha \cdot r \leq r_i - r_j$, and $\tilde{c}_{i\alpha}^j = 0$ otherwise. Define the approximating vector fields $\tilde{f}_i = \sum\limits_j \sum\limits_\alpha \tilde{c}_{i\alpha}^j x^\alpha \frac{\partial}{\partial x^j}$.

6) If desired, perform a finite sequence of strictly triangular polynomial coordinate changes to an adapted chart that *homogenizes* the approximating vector fields. For more details, see, for example, Stefani (1986) and Hermes (1991) and the example below.

THEOREM 5.6.– *(Hermes 1986, 1991) The vector fields $\tilde{f}_0, \tilde{f}_1, \dots \tilde{f}_m$ constructed above generate a nilpotent Lie algebra. They define a control system of form [5.3] that is STLC if it follows from theorem 5.4 (with the same weight θ as in the construction above) that the original system is STLC.*

Moreover, if the fields have been homogenized by choosing an adapted chart as in the last step, then there is a natural (but coordinate-dependent) way to measure the distance between the trajectories of the original and approximating systems.

Regarding the last, optional step of further adapting the coordinates, consider the simple example of $f_0(x) = x_1 \frac{\partial}{\partial x^2} + (x_2 x_1 + x_1^{99}) \frac{\partial}{\partial x^3}$ and $f_1(x) = \frac{\partial}{\partial x^1}$. The term $x_2 x_1$ is *not seen* by any Lie brackets, and it has no impact on the controllability properties, i.e. the *leading term* is x_1^{99}. The simple coordinate change $(y_1, y_2, y_3) = (x_1, x_2, x_3 - \frac{1}{2}x_2^2)$ *eliminates* this *ghost term* and renders the same vector fields homogeneous in the new coordinates $f_0(y) = y_1 \frac{\partial}{\partial y^2} + y_1^{99} \frac{\partial}{\partial y^3}$ and $f_1(y) = \frac{\partial}{\partial y^1}$.

Such *ghost terms* can be completely avoided by initially defining adapted coordinates via the flows of the selected vector fields f_{π_i}. However, in principle, that would require the exact solution of initial value problems for nonlinear differential equations. The above procedure avoids any such use of the implicit function theorem and only uses constant and triangular polynomial coordinate changes.

We illustrate the algorithm for the example of the pendubot.

1) Given the iterated Lie brackets evaluated previously, choose the weight $\theta = 1$, which maximizes the importance of the number of integration, i.e. the role of the time variable, and minimizes the importance of the control amplitude.

2) Corresponding to the iterated Lie brackets calculated, compare [5.50] and [5.51], which are linearly independent at $q_0 = 0$, and select the formal iterated Lie brackets (or *binary trees* labeled by the alphabet by $Z = \{0, 1\}$)

$$\begin{aligned} \ell_1 &= 1, \\ \ell_2 &= (0,1), \\ \ell_3 &= ((0,(0,1)),(0,1)), \\ \ell_3' &= ((0,1),((0,1),1)), \\ \ell_4 &= (0,((0,(0,1)),(0,1))), \text{ and} \\ \ell_4' &= (0,((0,1),((0,1),1))). \end{aligned} \qquad [5.52]$$

With the choice of the weight $\theta = 1$, either choice of binary tree ℓ_i or ℓ_i' for $i = 3, 4$ yields the same exponents $r = (1, 2, 5, 6)$ for the group of dilations.

3) In this case, no constant linear coordinate change is needed, as the selected iterated brackets of vector fields are already aligned with the coordinate axes at the origin.

For further computations (with a computer algebra system), it is convenient to straighten out the field f_1 so that in the new coordinates $f_1 \equiv \frac{\partial}{\partial x_1}$ is constant. This is achieved by the triangular coordinate change $x_1 = q_1$, $x_2 = q_3$, $x_3 = q_3 - \beta q_1 \sin(q_4 - q_2)$, $x_4 = q_4$).

4) Using the x coordinates, define the dilation group $\Delta_s(x) = (sx_1, s^2 x_2, s^5 x_3, s^6 x_4)$.

5) Expand each component of the vector fields f_0 in the x-coordinates in a Taylor series, and for each $1 \leq i \leq 4$, truncate the ith components at terms of orders 0, 1, 4 and 5 relative to the dilation.

6) This yields the approximating vector fields

$$\tilde{f}_0 = x_1 \frac{\partial}{\partial x_2} + (\beta^2 x_1^2 x_2 - \tfrac{1}{2}\beta x_2^2)\frac{\partial}{\partial x_3} + x_3 \frac{\partial}{\partial x_4} \quad \text{and} \quad \tilde{f}_1 = \frac{\partial}{\partial x_1}. \qquad [5.53]$$

This nilpotent approximating system and the original system will be small-time-controllable, depending on the bound on the control values relative to the physical scale parameter β (ratio of the length of the arms). Since both the sign-definite and the sign-indefinite terms in the third component of $\tilde{f}_0$ correspond to the same number of integrations, controllability is determined by comparing the sizes of the two iterated integral functionals

$$\int_0^T \left(\int_0^{t_1} \int_0^{t_2} u(t_3)\, dt_3\, dt_2 \right)^2 dt_1 \ \text{and}\ \int_0^T \left(\int_0^{t_1} \int_0^{t_2} u(t_3) dt_3\, dt_2 \right)\left(\int_0^{t_1} u(t_2)\, dt_2 \right) dt_1. \qquad [5.54]$$

Such a comparison is similar to that carried out in Kawski (1987, 1990), using integration by parts while imposing the constraints such as

$$\int_0^T \int_0^{t_1} u(t_2)\, dt_2\, dt_1 = \int_0^T u(t) dt = 0. \qquad [5.55]$$

The small-time local controllability of this model of a pendubot was also analyzed by Hauser (2010) via direct construction of control variations.

5.3.4. *Optimality and the maximum principle*

Unlike classical dynamical systems that are characterized by unique solutions to initial value problems, in controlled dynamical systems, if there is one admissible trajectory that goes from the initial point q_0 to the final point q_T, then generally, there are many others. Throughout mathematics, picking one solution from many may be a difficult task. A commonly employed strategy is to add additional constraints or performance criteria, so that the new problem only has one solution, or one best solution, which may then be computed algorithmically. There are many possible choices of cost functionals. Some arise naturally from physical considerations or variational principles, or are dictated by specific applications. Others may be chosen simply because they do the job (singling out a unique trajectory that steers the system to a desired equilibrium state) and are mathematically tractable. For an introductory discussion, it is convenient and customary to consider more general dynamic constraints of the form

$$\dot{q}_t = f(t, q_t, u_t). \tag{5.56}$$

These include the special class of control-affine systems [5.3] considered in most of this chapter. On the space $\mathcal{U}$ of admissible controls, we add to [5.56], now considered as a dynamic constraint, a cost functional $J\colon \mathcal{U} \mapsto \mathbb{R}$, which is to be minimized over the space $\mathcal{U}$. The general form of the cost functional in the Bolza problem in classical calculus variations is

$$\text{minimize } J(u) = \psi(q_T) + \int_{t_0}^{T} L(t, q_t, u_t)\, dt. \tag{5.57}$$

It is straightforward to rewrite this as an equivalent problem, for example, by adding additional states, in a form that incorporates the running cost $L(q_t, u_t)$ into the terminal cost $\psi(q_t))$, or the other way around. These problems are known as the Mayer form ($L \equiv 0$) and the Lagrange form ($\psi \equiv 0$). For the purposes of this introduction, it is convenient to add the time and running cost as new states and consider $\tilde{q} = (q^0, q, q^{n+1}) \in \mathbb{R} \times M^n \times \mathbb{R}$. Add the additional dynamic constraints $\dot{q}^0 = 1$ and $\dot{q}^{n+1} = -L(q^0, q, u)$, and the initial conditions $(q^0, q^{n+1})(t_0) = (t_0, 0)$. For simplicity, we relabel the new state $\tilde{q}$ again as q. Numerous technical difficulties that largely deal with the topology, the analytic and regularity properties of the various functionals and the existence of solutions, are treated in the large body of literature on the calculus of variations and optimal control theory.

The main basic tool to characterize optimal control-trajectory pairs is the Pontryagin maximum principle. It is commonly formulated in terms of a Hamiltonian function $H\colon T^*M \times U \mapsto \mathbb{R}$ that adjoins the dynamic constraints [5.56] to the Lagrangian in [5.57], by means of a time-varying Lagrange multiplier $p\colon [t_0, T] \mapsto T^*M$

$$H(q, p, u) = \langle p, f(q, u)\rangle - L(q, u). \tag{5.58}$$

Note that in the Mayer form with the enlarged state, $L \equiv 0$. The following statement of the maximum principle dispenses with many technical hypotheses, for which we refer the reader to the literature. Bressan and Piccoli (2007) give a concise statement at an introductory level; Clarke (1990) is a standard reference in the setting of nonsmooth analysis, whereas Sussmann (2001) illustrates the state-of-the-art at the interface of differential geometric tools with those from nonsmooth analysis.

THEOREM 5.7.– *If* (ξ^*, η^*) *is an admissible trajectory-control pair that minimizes the cost* J*, then there exists an absolutely continuous curve* $\lambda^* \colon [t_0, T] \mapsto T^*M \setminus \{0\}$*, such that it is* λ^* *never zero, and satisfies the following conditions:*

(i) For almost every $t \in [t_0, T]$ *and every* $u \in U$ *(pointwise maximization)*

$$H^*(\xi_t^*, \lambda_t^*) \stackrel{\text{def}}{=} H(\xi_t^*, \lambda_t^*, \eta_t^*) \geq H(\xi_t^*, \lambda_t^*, u). \qquad [5.59]$$

(ii) $\psi(\xi_T^*) + H(\xi_T^*, \lambda_T^*, \eta_T^*) = 0$ *(terminal constraint).*

(iii) For almost every $t \in [t_0, T]$*, the co-state satisfies the adjoint equation*

$$\dot{\lambda}_t^* = -\frac{\partial H^*}{\partial q}(\xi_t^*, \lambda_t^*). \qquad [5.60]$$

(iv) If the final state q_T *is free in the optimal control problem, then* $\dot{\lambda}_T^* = \frac{\partial \psi}{\partial q}(\xi_T^*)$.

In the following, we illustrate the connection of the maximum principle with families of control variations and with controllability. The analysis of both relies on tools from combinatorial algebra. In that framework, the adjoint equation [5.60], too, is of interest and deserves more attention, but this is beyond the scope of this work.

For a brief pictorial description of the items in, and the ideas behind the maximum principle, consider Figure 5.6. Suppose that (ξ^*, η^*) is an optimal trajectory-control pair. When assuming that for $t \geq 0$, the reachable sets $q_0\mathcal{R}_t$ are compact, it is easy to see that the minimum must be attained at a boundary point of the reachable set $q_0\mathcal{R}_T$ at time T from q_0. Using continuous dependence of the flow (on initial conditions) associated with the fixed control η^* and the *invariance of domain principle*, it is clear that if a trajectory lies in the interior of the reachable set $q_0\mathcal{R}_t$ at some time $t \in (t_0, T]$, then it must also lie in the interior $q_0\mathcal{R}_\tau$ at all later times $\tau \geq t$. Hence, the contrapositive implies that any trajectory that is optimal on the interval $[t_0, T]$ must lie at each time $t \in [t_0, T]$ on the boundary of $q_0\mathcal{R}_t$.

Now, consider a one-parameter family of admissible controls $\eta_s \colon [t_0, T] \mapsto U$, such that $\eta_0 = \eta^*$. Assuming that these *control variations* are sufficiently smoothly parameterized so that the resulting end points $\xi(T; \eta_s)$ define a (not necessarily continuous) curve in $q_0\mathcal{R}_T$ that is differentiable at $s = 0$, the derivative

$$\frac{d}{ds}\Big|_{s=0} \xi(T; \eta_s)$$

defines a tangent vector ζ to the reachable set $q_0\mathcal{R}_T$ at the point ξ_T^*. Under suitable technical hypotheses, the set of all such *variational vectors* can be shown to be a convex cone that approximates the reachable set in a natural way, and which is contained in a half-space. Consequently, there exists a *separating hyperplane*, labeled S in Figure 5.6, and a nonzero co-state vector λ_T^*, such that for all of the variational vectors ζ above, the inequality $\langle \lambda_T^*, \zeta \rangle \leq 0$ holds. The adjoint equation [5.60] *transports* this co-state vector back along the reference trajectory $\xi^*(\cdot)$ to the initial point q_0. The maximization condition encapsulates that at every time $t \in (t_0, T]$ any variations of the reference trajectory $\xi^*(\cdot)$ must point inwards to the interior of the reachable set at time t.

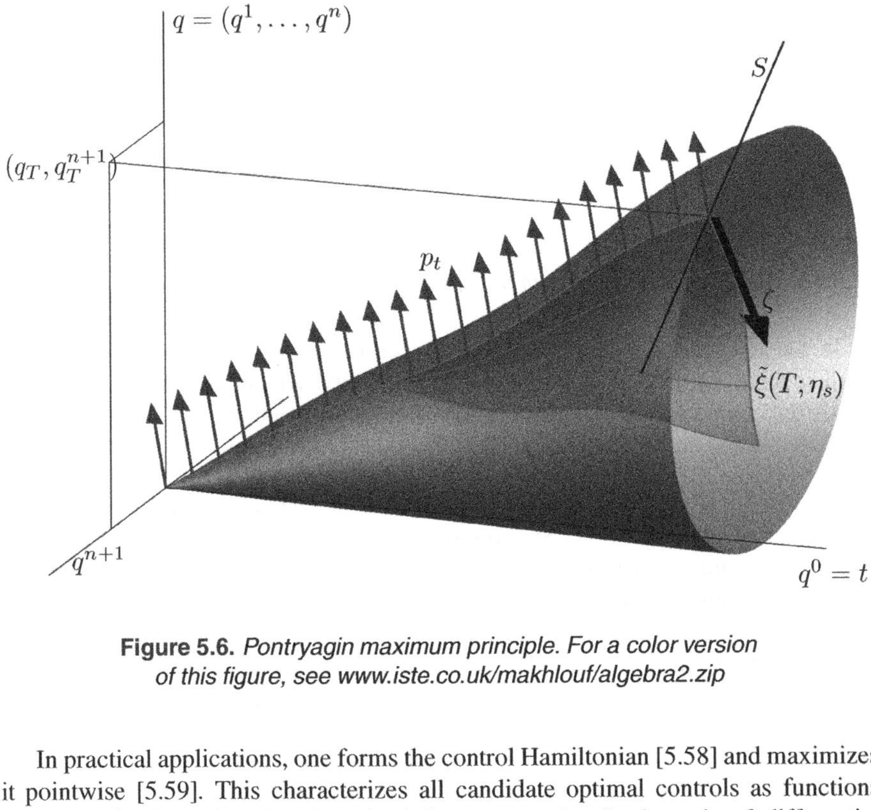

Figure 5.6. *Pontryagin maximum principle. For a color version of this figure, see www.iste.co.uk/makhlouf/algebra2.zip*

In practical applications, one forms the control Hamiltonian [5.58] and maximizes it pointwise [5.59]. This characterizes all candidate optimal controls as functions $u^* = u^*(q, p)$. Using the thus singled out controls u^*, the pair of differential equations [5.56] and [5.60], the adjoint equation (for all possible initial states q_0 and terminal co-states λ_T) characterize all candidate optimal trajectories. These include, in particular, all trajectories that lie on the boundary of the *funnel* $\bigcup_{0 \leq t \leq T} \{t\} \times q_0 \mathcal{R}_t$ (for each choice of the initial state q_0) at all times.

In contrast, the system [5.56] is controllable about a not necessarily constant reference trajectory ξ_t if this trajectory lies in the interior of the reachable sets $\xi_0 \mathcal{R}_t$ for all $t > 0$. In this case, we expect that suitably defined approximating cones of variational vectors, as above, do not lie in any half-space, but their positive linear combinations span the whole tangent space. Much of the literature on this topic investigates the possible notions of families of control variations and associated approximating cones, together with high-order open-mapping theorems that guarantee that if the approximating cones are not contained in a half-space, then the reference trajectory lies in the interior, and hence, cannot be optimal. All of these are generalized to systems whose data are only required to satisfy ever weaker regularity hypotheses.

The next section elaborates the connection between control variations and iterated Lie brackets and iterated integrals, again in the special case of affine systems defined by smooth vector fields, and a constant reference trajectory $\xi_t^* \equiv q_0$. Whereas the original Pontryagin maximum principle is a first order necessary condition for optimality, the emphasis here is on higher-order conditions for optimality and small-time local controllability.

5.3.5. *Control variations and approximating cones*

To best illustrate the connections between approximating cones of variational vectors and combinatorial iterated Lie brackets, consider the special case of controllability about (or optimality of) the stationary reference $\xi_t^* \equiv q_0$ for the system [5.3], with smooth vector fields f_i and the control set $U = [-1,1]^m$.

In this case, a very simple notion of variational vectors to the family of reachable sets from q_0 suffices to obtain strong theorems on controllability and optimality. Since this discussion is of a local nature, we may assume that $M^n = \mathbb{R}^n$ without loss of generality, and identify the tangent spaces $T_q\mathbb{R}^n$ with the state-space $\mathbb{R}^n$.

DEFINITION 5.5.– *For $\alpha > 0$, a vector $\zeta \in \mathbb{R}^n$ is called a tangent vector of order α at q_0 to the family of reachable sets from q_0, if there exists a family of admissible controls $\eta_s \colon [0,s] \mapsto U$ and the end points satisfy*

$$\xi(s;\eta_s) = q_0 + s^\alpha \zeta + o(s^\alpha). \qquad [5.61]$$

Let $\mathcal{C}^\alpha$ be the set of all such tangent vectors and define $\overline{\mathcal{C}}^\alpha = \{s\zeta \colon s > 0, \zeta \in \mathcal{C}^\alpha\}$.

Here, we use the notation $o(s^\alpha)$ for functions $r \colon [0,s_0] \mapsto \mathbb{R}^n$, for which there exist C, s_1, $\beta > 0$, such that for all $0 < s < s_1$ the upper bound $\|r(s)\| \leq Cs^{\alpha+\beta}$ holds.

This definition is a minor adaption from Kawski (1988a), and it is a very special case of a much more general notion of variational vectors investigated in, for example, Frankowska (1986) and Clarke (1990). It is easy to prove that sets $\mathcal{C}^\alpha$ form an increasing sequence of truncated cones (Kawski 1988a). We present the short proof as it gives insight into the scaling properties of control variations, and their concatenations.

LEMMA 5.1.– *Using the hypothesis on the system and definitions as above,*

(i) if $\tau \in [0,1]$ *and* $\zeta \in \mathcal{C}^\alpha$, *then* $\tau\zeta \in \mathcal{C}^\alpha$;

(ii) if $\alpha \leq \beta$, *then* $\mathcal{C}^\alpha \subseteq \mathcal{C}^\beta$;

(iii) if $\zeta_1, \zeta_2 \in \mathcal{C}^\alpha$ *and* $\tau \in [0,1]$, *then* $\tau^\alpha\zeta_1 + (1-\tau)^\alpha\zeta_2 \in \mathcal{C}^\alpha$.

PROOF.– Without loss of generality, assume $q_0 = 0$. Suppose $s \in [0,1]$, $\zeta \in \mathcal{C}^\alpha$ and choose a one-parameter family $\{\eta_s\}_{s\geq 0}$ of control variations $\eta_s : [0,s] \mapsto \mathcal{U}$, such that the end points satisfy $\xi(s;\eta_s) = s^\alpha\zeta + o(s^\alpha)$. Define the scaled controls $\tilde{\eta}_s : [0,s] \mapsto \mathcal{U}$ by

$$\tilde{\eta}_s(t) = \begin{cases} 0 & \text{if } 0 \leq t \leq (1-\tau)s, \quad \text{and} \\ \eta_s(t-(1-\tau)s) & \text{if } (1-\tau)s < t \leq s. \end{cases} \qquad [5.62]$$

Then clearly $\xi(s;\tilde{\eta}_s) = \xi(\tau s;\eta_{\tau s}) = s^\alpha(\tau^\alpha\zeta) + o(s^\alpha)$, proving (i).

Suppose τ, ζ and $\{\eta_s\}_{s\geq 0}$ are as above and $\alpha \geq \beta$. Without loss of generality, $s \leq 1$ and, using $\sigma = s^{\frac{\alpha}{\beta}}$, define

$$\overline{\eta}_s(t) = \begin{cases} 0 & \text{if } 0 \leq t \leq s-\sigma, \quad \text{and} \\ \eta_s(t-\sigma) & \text{if } \sigma - s < t \leq s. \end{cases} \qquad [5.63]$$

Then, $\xi(s,\overline{\eta}_s) = \xi(\sigma,\eta_\sigma) = \sigma^\beta\zeta + o(\sigma^\beta) = s^\alpha\zeta + o(s^\alpha)$, proving (ii).

Suppose $\{\eta_s^{(1)}\}_{s\geq 0}$ and $\{\eta_s^{(2)}\}_{s\geq 0}$ are families of control variations that generate the α-th order tangent vectors ζ_1 and ζ_2, respectively. For $\tau \in [0,1]$, define the family $\{\eta_s^\tau\}_{s\geq 0} : [0,s] \mapsto \mathcal{U}$ by

$$\eta_s^\tau(t) = \begin{cases} \eta_{\tau s}^{(1)}(t) & \text{if } 0 \leq t \leq \tau s, \quad \text{and} \\ \eta_{1-\tau s}^{(2)}(t-\tau s) & \text{if } \tau s < t \leq s. \end{cases} \qquad [5.64]$$

The conclusion $\xi(s,\eta_s^\tau) = s^\alpha(\tau^\alpha\zeta_1 + (1-\tau)^\alpha\zeta_2) + o(s^\alpha)$ is an immediate consequence of Gronwall's lemma. □

Note that with this very simple definition, even if $s \mapsto \xi(s,\eta_s^{(1)})$ and $s \mapsto \xi(s,\eta_s^{(1)})$ are smooth curves, in general, the map $s \mapsto \xi(s,\eta_s^\tau)$ as in (iii) need not be continuous.

The properties (i) and (ii) of the truncated cones $\mathcal{C}^\alpha$ strongly depend on $\xi_t \equiv 0$ being an equilibrium solution. For more general settings, especially nonstationary reference trajectories, it is a very delicate process to rescale or combine families of control variations. This applies, in particular, to *higher-order* conditions for optimality and controllability. Classical work detailing when such combinations are feasible dates back to Krener's high-order maximal principle (Krener 1977), and the careful conditions provided in Knobloch (1981); see also Bianchini and Stefani (1984). As shown in Bianchini and Kawski (2003), in general, meaningful higher-order approximating *cones* along nonstationary reference trajectories need not be convex.

The value of an approximating cone of variational vectors stems from an associated open-mapping theorem. In general, we distinguish topological arguments that apply in finite dimensions from tools used in infinite dimensional settings (see the discussion about Type L and Type T arguments in Sussmann (2007)). But in this simple setting, it is even possible to give simple constructive proofs of the following (Kawski 1988a) (which is again a special case of Frankowska (1986)).

THEOREM 5.8.– *Using the hypothesis on the system and definitions as above, if* $\overline{\mathcal{C}'}$ *is a closed convex cone (with vertex* $0 \in \mathbb{R}^n$*), such that* $\overline{\mathcal{C}'} \setminus \{0\} \subseteq \operatorname{int} \overline{\mathcal{C}^\alpha}$ *for some* $\alpha < \infty$*, then there are constants* $C > 0,\ T > 0$*, such that* $\overline{\mathcal{C}'} \cap B(0, Ct^\alpha) \subseteq q_0\mathcal{R}_t$ *for all* $0 \leq t \leq T$.

COROLLARY 5.1.– *Using the hypothesis on the system and definitions as above, if* $\overline{\mathcal{C}^m} = \mathbb{R}^n$*, then there are constants* $C > 0,\ T > 0$*, such that* $B(0, Ct^m) \subseteq q_0\mathcal{R}_t$ *for all* $0 \leq t \leq T$.

The corollary gives a lower bound on infinitesimal growth rates of the reachable sets and a sufficient condition for small-time local controllability. The theorem provides high-order necessary conditions for optimality of a stationary reference trajectory. In general, statements of this form become stronger if they are combined with constructions that yield more variational vectors, and thus larger approximating cones. The remainder of this section is devoted to linking the variational vectors to iterated Lie brackets and gradings of the associated Hopf algebra.

At an elementary level, the classical calculus of the variations derives the first-order necessary conditions for optimality by considering the directional derivatives of the functional [5.57]. In very general terms, to derive the necessary conditions for optimality of an admissible curve u^*, we fix a variation v and then manipulate the condition

$$\frac{d}{\varepsilon}\Big|_{\varepsilon=0} J(u^* + \varepsilon v) = 0 \qquad [5.65]$$

to derive the familiar Euler–Lagrange equations. In many settings, such as classical mechanics or when finding length minimizing curves (geodesics) locally in classical differential geometry, these may suffice to single out unique candidates u^* that satisfy the desired boundary conditions. However, in nonlinear control theory, as in many nonholonomic mechanic systems, such first-order conditions may be too weak: they have too many solutions, most of which are not optimal. In this case, we look for stronger necessary conditions that may combine higher derivatives and more general variations than $u_\varepsilon = u^* + \varepsilon v$.

One key innovation of optimal control theory introduces *needle variations* that are scaled by time, rather than by amplitude. In their most simple form, for a candidate reference control $\eta^* \colon [t_0, T] \mapsto U$, and a Lebesgue point $t_1 \in (t_0, T)$ of u^* and $v \in U$, define a family of control variations η_δ for some $\delta_0 > 0$ and $0 \leq \delta \leq \delta_0$ by

$$\eta_\delta(t) = \begin{cases} \eta^*(t) & \text{if } t_0 \leq t < t_1, \\ v & \text{if } t_1 \leq t < t_1 + \delta, \text{ and} \\ \eta^*(t) & \text{if } t_1 + \delta \leq t < T. \end{cases} \qquad [5.66]$$

Differentiating the end-point map $\delta \mapsto \xi(T; \eta_\delta)$ and evaluating at $\delta = 0$ again yields a tangent vector to the reachable set $q_0\mathfrak{R}_T$ at $\xi^*(T, \eta^*)$. In some cases, we may think of these needle variations as being associated with the L^1 norm in the space of controls, as opposed to the L^∞ norm associated with the variations of the form $\eta^* + \varepsilon v$, familiar from the classical calculus of variations.

In order to get usable approximating cones and associated open-mapping principles, we need to be able to take *convex combinations* of such control variations. These may be defined by finite sequences of times $(t_1, t_2, \ldots t_r) \in (0, T)^r$ and control values $(v_1, v_2, \ldots v_r) \in U^r$. As long as the times t_i at which the variations are taken are pairwise distinct, for sufficiently small $\delta_{0,i}$, the intervals $[t_i, t_i + \delta_{0,i}]$ on which different control values are taken will not overlap. Many articles in optimal control theory try to prove that under increasingly weaker regularity conditions, such combinations of families of control variations indeed yield the expected convex combinations of the associated tangent vectors to the reachable set $q_0\mathfrak{R}_T$, and prove the associated open-mapping theorems.

To obtain ever stronger necessary conditions for optimality, or, equivalently, stronger sufficient conditions for small-time local controllability, we may devise more complex families of control variations; consider higher derivatives, and try to combine any of the above.

As previously mentioned, the technical details very quickly become intricate, and we refer the interested reader to Krener (1977), Knobloch (1981) and Bianchini and Stefani (1984). One simple issue deals with generating convex combinations of the tangent vectors generated by two different families of needle variations taken at the same time $t_i = t_j$. Our intuition may suggest that we might be able to move the support of one of the variations ever so slightly, for example, to construct a new family of needle variations that takes the values v_i and v_j on the intervals $[t_i, t_i + \delta_i]$ and $[t_i + \delta_i, t_i + \delta_i + \delta_j]$, respectively. When using continuous dependence (or even smooth dependence for smooth controls), we may be tempted to expect that this vanishing shifting of the support only produces higher-order effects in the end-point map. However, a counterexample constructed in Bianchini and Kawski

(2003) demonstrates that for higher-order variations (higher-order derivatives), this need not be the case, even in systems with polynomial vector fields. Indeed, even for arbitrarily short times, the counterexample has reachable sets, one of whose natural cross sections is best approximated by the *union* of two half-spaces akin to the set $\{x \in \mathbb{R}^2 \colon x^1 \geq 0 \text{ or } x^2 \geq 0\}$. In some cases, this counterexample justifies the highly technical conditions on admissible control variations that are proposed in, for example, Krener (1977), Knobloch (1981) and Bianchini and Stefani (1984).

The remainder of the section considers the case of a stationary reference trajectory $\xi^*_t \equiv q_0$ corresponding to an equilibrium point $f_0(q_0) = 0$ of the drift vector field and trivial control $\eta^* \equiv 0$, in which none of these problems arise. The simple notion of tangent vectors of definition 5.5, together with the open-mapping theorem 5.8, rely very much on the bounds on the growth rates of analytic vector fields near equilibria, together with the ability to shift a control variation in time without changing the terminal points (by prepending an interval with zero control): $\xi(T;\eta) = \xi(T+\tau,\eta_\tau)$, where $\eta_\tau(t) = 0$ for $0 \leq t < \tau$ and $\eta_\tau(t+\tau) = \eta(t)$ for $0 \leq t \leq T$.

To obtain algorithmically computable conditions for controllability and optimality, we relate the control variations, as above, to relations between the iterated Lie brackets of the vector fields f_i defining the system. We may construct and combine the increasing numbers of families of needle variations, as above, so that their lower effects cancel and then relate the higher-order effects to iterated Lie brackets. The classical Baker–Campbell–Hausdorff formula [5.24] in section 5.2.3 is clearly not suited for such analysis. Instead, we may directly use the iterated integrals in improved versions of the Chen–Fliess series [5.28] to design control variations that generate specific directions.

The first step is to consider the homogeneity and scaling properties of the iterated functionals. For a fixed control $\eta \colon [0,T] \mapsto U$, considered as a variation of the zero reference control $\eta^*_t \equiv 0$, introduce an $(m+1)$-parameter family of scaled controls as follows (see Figure 5.7). For $(\delta,\varepsilon) \in [0,1]^{1+m}$ and $0 \leq t \leq (1-\delta)T$, define $\eta^i_{\varepsilon,\delta}(t) = 0$ and for $1 \leq i \leq m$ and $t \in [0,T)$ define

$$\eta^i_{\varepsilon,\delta}(T-\delta t) = \varepsilon_i \eta^i(T-t). \qquad [5.67]$$

PROPOSITION 5.1.– *Suppose* $(k,\ell) = (k,\ell_1,\dots\ell_m) \in (\mathbb{Z}^+_0)^{m+1}$ *and* $w \in A^{(k,\ell)}(Z)$, *and* $\eta_{\varepsilon,\delta}$ *is a* $(m+1)$*-parameter family of control variations, as above, then*

$$\Upsilon^w(\eta_{\varepsilon,\delta})(T) = \varepsilon^\ell \delta^{k+|\ell|} \Upsilon^w(\eta_{1,1})(T) \;, \textit{where } \; \varepsilon^\ell = \varepsilon_1^{\ell_1} \cdot \ldots \cdot \varepsilon_m^{\ell_m}. \qquad [5.68]$$

Recall that $w \in A^{(k,\ell)}(Z)$ means that w is a word (linear combination of words each of) which contains k times the letter a_0, and ℓ_i times the letter a_i for $1 \leq i \leq m$. The proof is immediate by induction and using the chain rule (substituting for the time variables of integration).

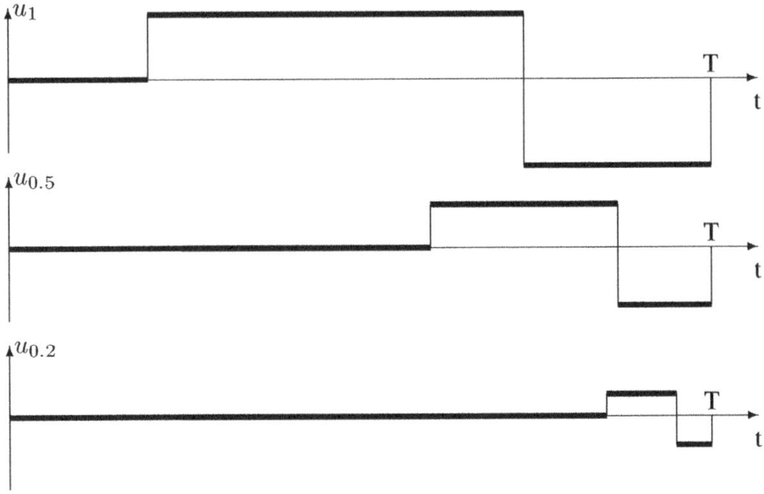

Figure 5.7. *Needle variations also scaled by amplitude*

To generate a single tangent vector ζ to the reachable set $q_0\mathcal{R}(T)$, define a sufficiently smooth curve $s \mapsto (\varepsilon(s), \delta(s))$ and evaluate the first nonvanishing derivative

$$\frac{d^r}{ds^r}\Big|_{s=0} \xi(T; \eta_{\varepsilon(s),\delta(s)}). \qquad [5.69]$$

Using the Chen–Fliess series [5.28] to expand the end-point map, the iterated integrals are polynomial in $(\varepsilon(s), \delta(s))$. Commonly, we employ weightings $s \mapsto (\varepsilon(s), \delta(s))$ in which each component is a nonnegative power of s. This induces a partial order on the words in Z^*, which descends to a partial order of the iterated integral functionals. A higher-order Lie bracket is a tangent vector to the reachable sets if all of the *lower-order terms* (in the induced partial order) vanish, either because the iterated integrals vanish by construction of the control variation, or the corresponding iterated Lie brackets vanish at the initial point q_0, or because certain linear combinations of iterated Lie brackets and their associated iterated integrals vanish. The rewriting of the higher-order partial differential operators in the Chen–Fliess series in terms of first-order differential operators, i.e. as iterated Lie brackets, will be presented in the next section.

Some commonly used weightings include the following:

– If $s \mapsto (\varepsilon(s), \delta(s)) = (s, \ldots, s, 1)$, then the support of the control variations is fixed, and only their amplitude $\|\eta_\varepsilon\|_\infty$ goes to zero as $s \searrow 0$. This corresponds to the above-described variations familiar from the classical calculus of variations, and this is instrumental in the proof of, for example, the Hermes condition in theorem 5.2.

– The choice $s \mapsto (\varepsilon(s), \delta(s)) = (1, \ldots, 1, s)$ corresponds to the original needle variations, whose *amplitude* is fixed. As $s \searrow 0$, only the measure of the support of the control variations goes to zero, and hence $\|\eta_\varepsilon\|_1 \searrow 0$ as $s \searrow 0$. All control vector fields and the drift vector field are weighted equally. This scheme was recently employed for a necessary and sufficient condition for systems that are homogeneous in a very narrow sense (Aguilar and Lewis 2012).

– If $s \mapsto (\varepsilon(s), \delta(s)) = (s^{1-\theta}, \ldots, s^{1-\theta}, s^\theta)$, then for $w \in A^{(k,\ell)}(Z)$ the iterated integral satisfies

$$\Upsilon^w(\eta_{\varepsilon(s),\delta(s)}) = s^{(1-\theta)\sum_{i=1}^m \ell_i} \cdot s^{\theta(k+\sum_{i=1}^m \ell_i)} \Upsilon^w(\eta_{1,1}) = s^{\theta k + \sum_{i=1}^m \ell_i} \Upsilon^w(\eta_{1,1}). \quad [5.70]$$

This balancing of the rates at which the amplitude and the support of the control variations vanish as $s \searrow 0$ corresponds to the scaling in the version of Sussmann's general theorem stated in theorem 5.4.

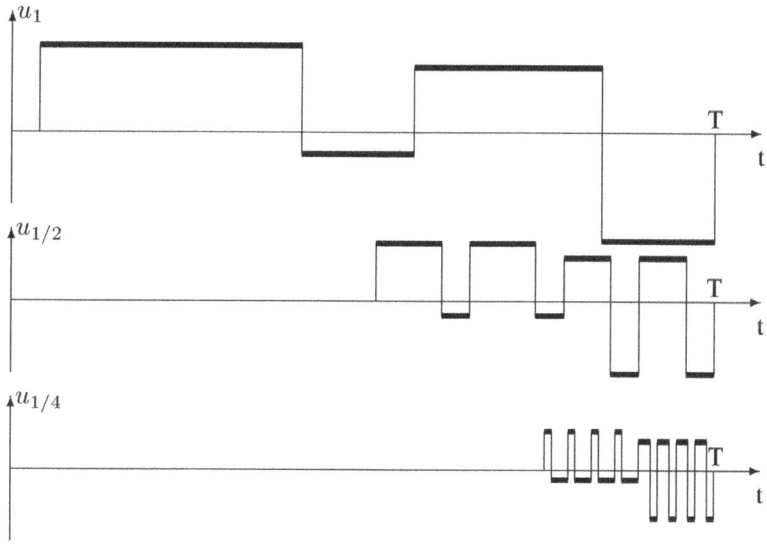

Figure 5.8. *More complex family of control variations*

Common in all of the above constructions and conditions is that they only depend on the finely homogeneous degree of the iterated Lie bracket or iterated integral functional. The next section demonstrates how the Chen series may be rewritten in terms of first-order differential operators only. It will become apparent that the combinatorial structure of iterated integral functionals and formal iterated Lie brackets is much more delicate than what can be captured by their finely homogeneous degrees (the number of times each letter appears in the corresponding word). Indeed, in

Kawski (1988b), an example of a simple polynomial single-input system on $\mathbb{R}^4$ was presented and shown to be small-time locally controllable, but none of the above weightings and scalings suffice to establish controllability. This example uses families of control variations that effectively switch faster and faster as the parameter $s \searrow 0$ goes to zero. It was shown in Kawski (1988b) that such a strategy is necessary: if there is a bound on the number of changes of sign of the antiderivative $t \mapsto \int_{t_0}^{T} u(s)\,ds$ of the control, then for sufficiently small time and bound on the magnitude $\|u\| \leq u_0$ of the control, the system is not controllable with such controls. The example is built around the two functionals

$$u \mapsto \int_0^T \Big(\int_0^t \Big(\int_0^s u(\sigma)d\sigma \Big)^3 ds \Big)^2 dt \quad \text{and} \quad u \mapsto \int_0^T \Big(\int_0^t \int_0^s u(\sigma)d\sigma\, ds \Big)^7 dt \qquad [5.71]$$

which correspond to the formal iterated Lie brackets (labeled binary trees)

$$\begin{array}{l} (((b,(b,(b,a)))),(((b,(b,(b,a)))),a)) \quad \text{and} \\ ((a,b),((a,b),((a,b),((a,b),((a,b),((a,b),((a,b),a))))))) \end{array} \qquad [5.72]$$

whose multi-degrees in (a, b) are $(6, 3)$ and $(8, 7)$, respectively. Where the first iterated integral functional is clearly sign definite, it was shown that by the judicious choice of the fast switching control variations that its *order* could be made larger than that of the second, sign-indefinite integral functional – even though the first corresponds to fewer integrations (lower order in time) and is of a lower order in the control. However, as $s \searrow 0$ and both the support and the control amplitude go to zero, the latter can be made to dominate the first (Kawski 1988b).

In the context of this chapter, this example demonstrates the need to use advanced combinatorial tools to investigate more sophisticated partial orders on Z^*, or rather, on suitably chosen bases for $L(Z)$ that correspond to meaningful constructions of families of control variations, and thereby yield stronger sufficient conditions for controllability or equivalently, stronger necessary conditions for optimality.

5.4. Product expansions and realizations

Solving a differential equation by iteration of the equivalent integral equation, as in sections 5.2.4 and 5.2.5, yields an infinite series expansion of the solution. An alternative is to iterate the method of variation of parameters and obtain an infinite (directed) product expansion for the solution. Using bases for free Lie algebras, such exponential product expansions are better suited as tools for analysis and design in control. A further alternative is to write the solutions as the exponential of a Lie series. In the setting of matrix differential equations, the corresponding expansions are known as classical (Magnus 1954; Fer 1958) expansions. In a formal algebraic framework, their close relationship was demonstrated in Ebrahimi-Fard and Manchon (2009), which used dendriform and pre-Lie algebras to provide a uniform approach to both expansions. This section aims to provide a bridge between the classical dynamical system setting and the formal framework of Ebrahimi-Fard and Manchon (2009), with an emphasis on the mappings and correspondences between formal combinatorial objects and geometric and dynamic objects in control.

5.4.1. *Variation of parameters and exponential products*

First, recall the method of variation of parameters for solving differential equations. A discussion about the possible meanings of the objects and symbols will be postponed until after the proposition and the outline of a formal proof.

PROPOSITION 5.2 (Formal variation of parameters).– *If* q_t *and* p_t *formally satisfy* $q_0 = p_0 = \mathbf{1}$, $\dot{q}_t = q_t(f_t + g_t)$ *and* $\dot{p}_t = p_t f_t$, *and* $q_t = r_t p_t$, *then* r_t *satisfies*

$$\dot{r}_t = r_t \sum_{k=0}^{\infty} \tfrac{1}{k!}(\mathrm{ad}^k f_t, g_t). \qquad [5.73]$$

PROOF.– In a completely formal way, using only the Leibniz rule and distributivity and assuming that p is invertible, the equations $\dot{q} = q(f+g)$, $\dot{p} = pf$ and $q = rp$ yield $rp(f+g) = \dot{r}p + rpf$, and hence $\dot{r} = rpgp^{-1}$.

Using the chronological exponential notation (Agračev and Gamkrelidze 1978, 1980), we may write $p_t = e^{\overrightarrow{\int_0^t} f_s\, ds}$. When differentiated, the *Ansatz* $q_t = r_t e^{\overrightarrow{\int_0^t} f_s\, ds}$ yields

$$r_t e^{\overrightarrow{\int_0^t} f_s\, ds}(f_t + g_t) = \dot{r}_t e^{\overrightarrow{\int_0^t} f_s\, ds} + r_t e^{\overrightarrow{\int_0^t} f_s\, ds} f_t. \qquad [5.74]$$

After canceling two terms, the result is easily solved to yield a differential equation for the *correction term* r_t. Using the *ad-formula* [5.19] (and using the fact that the exponential (the flow of f_t) is invertible), this simplifies to

$$\dot{r}_t = r_t e^{\overrightarrow{\int_0^t} f_s\, ds} g_t e^{-\overrightarrow{\int_0^t} f_s\, ds} = r_t e^{(\mathrm{ad}\overrightarrow{\int_0^t} f_s\, ds,\cdot)} g_t = r_t \sum_{\nu=0}^{\infty} \tfrac{1}{\nu!}(\mathrm{ad}^{\nu} \overrightarrow{\textstyle\int_0^t} f_s\, ds, g_t). \qquad [5.75]$$

□

Solving differential equations by variation of parameters is most familiar in the context of time-invariant linear differential equations $\dot{x} = Ax + b$, where $A \in \mathbb{R}^{n \times n}$ and $b \in \mathbb{R}^n$. The solution above is recovered by interpreting $f_t \longleftrightarrow Ax$ and $g_t \longleftrightarrow b$ (which may be time varying), yielding $x(t) = e^{tA}x_0 + \int_0^t e^{(t-s)A} b(s)\, ds$. The case of time-varying linear systems $\dot{x} = A(t)x + B(t)x$ is analogous, where $B(t)$ is frequently interpreted as a *perturbation* of $A(t)$.

In the setting of affine control systems [5.3], of particular interest is the case when the right-hand side splits into time-varying controls and fixed geometric vector fields $f_t \leftrightarrow u_t^1 f_1$ and $g_t \leftrightarrow u_t^2 f_2$. The previous linear case is included by combining the components in the form

$$A(t)x \longleftrightarrow \sum_{i,j=1}^{n} a_j^i(t) x^j \frac{\partial}{\partial x^i} \quad \text{and} \quad b(t) \longleftrightarrow \sum_{j=1}^{n} b^j(t) \frac{\partial}{\partial x^j}. \qquad [5.76]$$

The main difference between the Magnus and Fer expansions and the formulas preferred in control is that in the former, there are n^2 many unrelated terms $a_j^i(t)$ playing the role of the inputs, whereas in control, there are typically much fewer controls and states, i.e. the main interest is in matrices $A(t) = \sum_{i=1}^{m} u^i(t) A_i$ that have a very rigid structure with $m << n^2$.

Further abstracting the setting to the algebra $\hat{A}(Z)$ of formal power series over the alphabet Z, the solution q_t is a curve of group-like elements $q_t \in \hat{A}(Z)$, characterized in terms of the coproduct by $\Delta q_t = q_t \otimes q_t$. The vector fields f_t and g_t correspond to generators $a \in Z$, or more generally, Lie elements $\ell_t \in L(Z)$ characterized by $\Delta \ell_t = \ell_t \otimes 1 + 1 \otimes \ell_t$. In the case of general time-varying vector fields f_t, as in Agračhev's chronological calculus, the exponentials also become directed as in general $f_s f_t \neq f_t f_s$, i.e. the fields do not commute with themselves at different times.

In the case that $f_t = u_t^1 f$ and $f_t = u_t^2 g$ split into time-dependent scalar controls and constant geometric vector fields, the differential equation [5.75] for the correction term also splits

$$\dot{r}_t = r_t \sum_{\nu=0}^{\infty} \frac{1}{\nu!} \left(\int_0^t u_s^1 \, ds \right)^{\nu} u_t^2 \, (\mathrm{ad}^{\nu} f, \, g). \qquad [5.77]$$

The corresponding formal analogue identifies $q_t, \ f_t, \ p_t$ with elements in $\hat{A}(Z)$ or in $A(Z) \otimes \hat{A}(Z)$. For example, for $a \in Z$, the vector $a \otimes a$ is mapped to the product $u_t^a f_a$, whereas q_t corresponds to the series $(\Upsilon \otimes \mathrm{Id})(CF)(u_t) = \sum_{w \in Z^*} \Upsilon^w(u_t) \otimes w$ of the control u_t.

The variation of parameters has resulted in the new differential equation [5.77], which is of the same form as the original equation $\dot{q}_t = q_t(f_t + g_t)$. Its right-hand side is again a linear combination of constant vector fields with time-varying coefficients. Hence, we may iterate this procedure using the *Ansatz* $r_t^k = r_t^{k+1} p_t^k$ with $r_t^0 = q_t$. This yields the expansion $q_t = q_0 r_t^k p_t^{k-1} \cdots p_t^1$, where each p_t^i is of the form of a *known* simple exponential

$$q_t = r_t^s \cdot e^{\int_0^t u^{h_k}(s) ds \, f_{h_k}} \cdots \cdot e^{\int_0^t u^{h_2}(s) ds \, f_{h_2}} \cdot e^{\int_0^t u^{h_1}(s) ds \, f_{h_1}}. \qquad [5.78]$$

The *vector fields* f_{h_j} are selected from the differential equations for the j-th correction term r_t^j, and the *new controls* u^{h_i} are calculated from the controls at the previous stage, according to the formula [5.77]. At first sight, it may appear that at every stage, we replace one differential equation by one whose right-hand side has infinitely many more terms, and each of these is of a higher complexity. But a clever construction assures that this process converges when continued *ad infinitum*. The objective is to obtain simple recursive formulas for the new vector fields and new controls (iterated integrals), and to prove convergence. The resulting formulas were first presented by Schützenberger (1958) and were then rediscovered in various different forms. This presentation follows the proof given by Sussmann (1986), but suppresses most of the differential equation notations and emphasizes a combinatorial symbolism. For the construction to work, it is critical to work with formal iterated brackets or, equivalently, rooted binary trees labeled by the alphabet Z. Technically, they are elements of the free magma $\mathcal{M}(Z)$ (Bourbaki 1998), which is recursively defined by $\mathcal{M}^0(Z) = \{e\}$ (the set containing the empty word or empty tree), and

$$\mathcal{M}^1(Z) = Z,\ \mathcal{M}^k(Z) = \bigcup_{j=0}^{k} \mathcal{M}^{k-j}(Z) \times \mathcal{M}^j(Z), \text{ and } \mathcal{M}(Z) = \bigcup_{k=0}^{\infty} \mathcal{M}^k(Z). \qquad [5.79]$$

For two trees $\tilde{\tau}, \tilde{\tau}' \in \mathcal{M}(Z)$, denote their noncommutative, nonassociative product by $(\tilde{\tau}, \tilde{\tau}')$, and write $(\tilde{\tau}, e) = \tilde{\tau} = (e, \tilde{\tau})$. It is convenient to also recursively introduce notation for iterated left products $\lambda_{\tilde{\tau}}^0(\tilde{\tau}') = \tilde{\tau}'$ and $\lambda_{\tilde{\tau}}^{k+1}(\tilde{\tau}') = (\tilde{\tau}, \lambda_{\tilde{\tau}}^k(\tilde{\tau}'))$.

Denote the natural deparenthesization map that sends each tree to its *foliage* by *dropping the tilde*: for example, for a tree $\tilde{\tau} = (\tilde{\tau}', \tilde{\tau}'') \in \mathcal{M}(Z)$ denote by $\tau = \tau'\tau'' \in Z^*$ its image under the foliage map (this notation allows that $\tilde{\tau} \neq \tilde{\tau}'$ while $\tau = \tau'$ – but this will not be a problem). Similarly, denote the natural map that sends each binary tree $\lambda_{\tilde{\tau}}(\tilde{\tau}') = (\tilde{\tau}, \tilde{\tau}')$ to the commutator $(\text{ad}\,[\tilde{\tau}], [\tilde{\tau}']) = [[\tilde{\tau}], [\tilde{\tau}']] \in L(Z)$ by using unary brackets. This includes the natural identification of a letter $a \in Z$ with the Lie polynomial $[a] = a \in L(Z) \subseteq A(Z)$ and the rooted labeled tree consisting of a single leaf (which is labeled by a). Extend both maps to algebra homomorphisms, from the $\mathbb{R}$–vector space spanned by $\mathcal{M}(Z)$, to the algebra $A(Z)$, and the Lie algebra $L(Z) \subseteq A(Z)$. Note that, for example, if $a \neq b \in Z$, then the two trees $\tilde{\tau} = (b, (a, (a, b)))$ and $\tilde{\tau}' = (a, (b, (a, b)))$ are distinct, but because of the Jacobi identity, their images in the Lie algebra $L(\{a, b\}) \subseteq A(\{a, b\})$ are identical

$$[\tilde{\tau}] = [b, [a, [a, b]]] = bbaa - 2baba + 2abab - aabb = [a, [b, [a, b]]] = [\tilde{\tau}']. \qquad [5.80]$$

Using this formalism, the following iteration yields a converging infinite analogue of [5.78], and it provides explicit formulas for the exponents. The construction successively generates two sequences $\{E_k\}_{k=1}^{\infty}, \{F_k\}_{k=1}^{\infty} \subseteq \mathcal{M}(Z)$ of ordered subsets of labeled binary trees (formal brackets) and recursive formulas for the iterated integrals and the iterated Lie brackets that appear in the exponents of the infinite product expansion.

Start with the empty sets $E_0 = \emptyset$ and $F_1 = Z$, both considered as subsets of $\mathcal{M}(Z)$. At each stage, choose a well-ordering of the set F_k, which must satisfy some mild conditions to be discussed later. Select the smallest element $\tilde{h}_k \in F_k$ and adjoin it to the set E_k to form the new set $E_{k+1} = E_k \cup \{\tilde{h}_k\}$. At each stage, use the variation of the parameters to formally solve the formal differential equation

$$\dot{r}_t^k = r_t^k \left(u_t^{\tilde{h}_k}[\tilde{h}_k] + \sum_{\tilde{B}\in F_k} u_s^{\tilde{B}}[\tilde{B}] \right) \tag{5.81}$$

by making the *Ansatz*

$$r_t^k = r_t^{k+1} \cdot e^{\int_0^t u_s^{\tilde{h}_k} ds \, [\tilde{h}_k]}. \tag{5.82}$$

Analogous to the derivation of [5.77], this determines a new formal differential equation for r_t^{k+1}

$$\dot{r}_t^{k+1} = r_t^{k+1} \sum_{\tilde{B}\in F_{k+1}} u_s^{\tilde{B}}[\tilde{B}] \tag{5.83}$$

where the set F_{k+1} and the new controls $u^{\tilde{B}}$ for $\tilde{B} \in F_{k+1}$ are of the form

$$F_{k+1} = \{\lambda_{\tilde{h}_k}^{\nu}(\tilde{B})\colon \nu \geq 0,\ \tilde{B} \in F_k \setminus \{\tilde{h}_k\}\ \}, \quad \text{and} \tag{5.84}$$

$$u_t^{\lambda_{\tilde{h}_k}^{\nu}(\tilde{B})} = \tfrac{1}{\nu!} \left(\int_0^t u_s^{\tilde{h}_k} \, ds \right)^{\nu} u_t^{\tilde{B}} . \tag{5.85}$$

Upon well-ordering the new set F_{k+1} and singling out its smallest element $\tilde{h}_{k+1}$ on the right-hand side of the differential equation [5.83], the iteration may continue.

While infinitely many new trees are added to F_k at each stage to form $F_{k+1} \supseteq F_k \setminus \{\tilde{h}_k\}$, all of the new trees are larger than the one removed from the previous set. Since the original set $F_0 = Z$ was finite, for every k, the set $F_{k+1} \setminus F_k$ of newly generated trees only contains trees with more leaves than $\tilde{h}_k$, and for each $r \in \mathbb{Z}^+$, each set F_{k+1} contains only a finite number of trees with r leaves. Given the uniform structure on $A(Z)$ introduced in section 5.2.5, any well-orderings of each of the sets F_k yield convergence as long as they guarantee that for every $k \geq 0$, after a finite number of steps, all trees with at most k leaves will have appeared as the smallest element of some set F_j. It is traditional to choose the ordering of each set F_k to be compatible with the length of the trees (number of leaves), i.e., if $|\tilde{\tau}| < |\tilde{\tau}'|$, then $\tilde{\tau} \prec \tilde{\tau}'$. It is common (but not required) to choose the ordering such that if $\tilde{\tau}, \tilde{\tau}' \in F_{k+1} \cap F_k$ and $\tilde{\tau} \prec_k \tilde{\tau}'$ in F_k, then also $\tilde{\tau} \prec_{k+1} \tilde{\tau}'$ in F_{k+1}. The formal considerations above and the convergence in $A(Z)$ directly translate to corresponding results in terms of asymptotic expansions on the analytic side for controlled dynamical systems on smooth manifolds. For a detailed discussion and a proof that works as much as feasible on the dynamic system side, see Sussmann (1986). For analytic details on the variations of the parameter procedure in the case of time-varying vector fields, see Agračhev and Sachkov (2004).

5.4.2. *Computations using Zinbiel products*

This section establishes basic algebraic properties of the map Υ, from combinatorial objects to dynamical systems. These allow us to simplify the calculations for the Picard iteration presented in section 5.2.4, and those for iterating the variation of the parameter procedure, as in section 5.4.1.

The calculations in the preceding sections demonstrated that the *product*

$$(U \star V)(t) = \int_0^t U(s) \cdot V'(s)\, ds \qquad [5.86]$$

in control, a ubiquitous operation on absolutely continuous functions $U,\ V\colon [0,T] \mapsto \mathbb{R}$ (think of these as primitives of controls, i.e. $U' = u$). Analogous, for essentially bounded integrable functions $u,\ v\colon [0,T] \mapsto \mathbb{R}$, define

$$(u * v)(t) = \int_0^t u(s)\, ds \cdot v(t). \qquad [5.87]$$

It is an elementary exercise (using integration by parts) to show that both of these products satisfy the *right Zinbiel identity*

$$r * (s * t) = (r * s) * t + (s * r) * t. \qquad [5.88]$$

The name *Zinbiel* is *Leibniz* spelled backwards, reflecting the Koszul duality of the corresponding algebras (see Ginzburg and Kapranov (1994); Loday (1993)). This suggests that we should consider an analogous product on the combinatorial level. Inductively, define a binary operation $*\colon Z^* \times Z^+$ on words by setting for $a \in Z$ and $w, z \in Z^*$

$$e * w = w, \quad w * a = wa, \quad \text{and} \quad w * (za) = (w * z)a + (z * w)a \qquad [5.89]$$

and bilinearly extend the product to $A(Z) \times A^{+}(Z)$. The empty word is a left identity, but right multiplication by e remains undefined (for some applications, it may be convenient to extend the product by defining $w * e = 0$ for $w \in Z^{+}$). It is immediate that this product on its domain again satisfies the *right Zinbiel identity* [5.88]. With the above defined products, the spaces of absolutely continuous functions, integrable functions, iterated integral functionals and the algebra $A(Z)$ are right Zinbiel algebras. Using [5.29] and [5.36], it is immediately verified that the map Υ is a right Zinbiel algebra homomorphism, for example, when evaluated on an admissible control u, for words $w \in Z^{*}$ and $z \in Z^{+}$

$$\Upsilon^{w*z}(u) = \Upsilon^{w}(u) * \Upsilon^{z}(u). \qquad [5.90]$$

The symmetrization of the product $*$ on $A(Z)$ yields the familiar shuffle product, for example, for words $w \in A^{+}(Z)$

$$w \sqcup\!\sqcup z = w * z + z * w. \qquad [5.91]$$

This product is associative and indeed globally defined on all of $A(Z) \times A(Z)$, using the convention $w \sqcup\!\sqcup e = e \sqcup\!\sqcup w = w$. The corresponding product on integrable or absolutely continuous functions is pointwise multiplication, for example, for $w, z \in Z^{*}$

$$\Upsilon^{w \sqcup\!\sqcup z}(u)(t) = \Upsilon^{w}(u)(t) \cdot \Upsilon^{z}(u)(t). \qquad [5.92]$$

This may be restated as the fact that *the coefficients of the Chen series satisfy the shuffle relations*, and consequently, the Chen series is an exponential Lie series. This is known as Ree's theorem (Ree 1958; Reutenauer 1993).

To illustrate how much working on the combinatorial level simplifies the calculations and how it provides better insights, revisit the Picard iteration in section 5.2.4. For a set of admissible controls $\{u^{a} \colon a \in Z\}$ and their primitives $U^{a}(t) = \int_0^t u^{a}(s)\, ds$, consider the formal *universal* control system on $\hat{A}(Z)$ initialized at $S(0) = 1$

$$\frac{dS}{dt} = S(t) \cdot \sum_{a \in Z} u^{a}(t) a. \qquad [5.93]$$

Abbreviating $U = \sum_{a \in Z} U^{a}(t)\, a$, rewrite this as the equivalent integral equation

$$S = 1 + S * U. \qquad [5.94]$$

This may be solved by iteration using the Zinbiel product. The combinatorial analogue of the Picard iteration yields the Chen–Fliess series written in the form

$$S = 1 + U + (U * U) + ((U * U) * U) + \ldots = \sum_{k=1}^{\infty} U^{*k}. \qquad [5.95]$$

Here, the Zinbiel powers U^{*k} correspond to successive products taken on the right. Formally, for $w \in A^+(Z)$, define $w^{*1} = \lambda^{*1}(w) = w^{\sqcup\!\sqcup 1} = w$, and inductively for $n \geq 1$

$$\lambda^{*(n+1)}(w) = w * \lambda^{*n}(w), \quad w^{*(n+1)} = w^{*n} * w, \quad \text{and} \quad w^{\sqcup\!\sqcup (n+1)} = w \sqcup\!\sqcup w^{\sqcup\!\sqcup n}. \tag{5.96}$$

It is straightforward to verify the following identities for $w \in A^+(Z)$ and $n \in \mathbb{Z}^+$.

$$w * w^{*(n-1)} = (n-1) \cdot w^{*n}, \quad \lambda^{*n}(w) = (n-1)! \cdot w^{*n}, \quad w^{\sqcup\!\sqcup n} = n! \cdot w^{*n}. \tag{5.97}$$

Note that by rewriting a power using a different product, we may make factorials appear or disappear, and thus make, for example, the Taylor expansions of the exponential look like a geometric series, the Taylor expansion of the reciprocal $\frac{1}{1-x}$ (see Kawski (2000, 2002) and also Ebrahimi-Fard and Manchon (2009)). This appearance and disappearance of the factorials can also be explained in terms of the formal Laplace–Borel transform, as described in Li and Gray (2006). This is a familiar phenomenon in control engineering.

With these identities, the Chen series, as obtained by the formal Picard iteration at the combinatorial level, is immediately rewritten as

$$S = \sum_{k=1}^{\infty} U^{*k} = 1 + U + \frac{1}{2!} U^{\sqcup\!\sqcup 2} + \frac{1}{3!} U^{\sqcup\!\sqcup 3} \ldots = \exp^{\sqcup\!\sqcup}(U). \tag{5.98}$$

To further illustrate the simplifications at the combinatorial level, we may justify inverting the universal control system [5.94] using a product with the formal left Zinbiel inverse (noting that S is a group-like element, satisfying $\Delta S = S \otimes S$; see section 5.4.4, S is invertible)

$$U = \lambda^{*(-1)}(S) * (S - 1) \tag{5.99}$$

in terms of left Zinbiel powers, which is a preview of the logarithm of the Chen–Fliess series elaborated in section 5.4.4. Using the abbreviation $S^+ = (S-1)$,

$$\begin{aligned} U &= S^+ - S^+ * S^+ + S^+ * (S^+ * S^+) + \ldots = \sum_{k>0} (-)^{k+1} \lambda^{*k}(S^+) \\ &= \sum_{k>0} \tfrac{(-)^{k+1}}{k} (S-1)^{\sqcup\!\sqcup k} = \log^{\sqcup\!\sqcup}(S-1). \end{aligned} \tag{5.100}$$

5.4.3. *Exponential products and normal forms for nilpotent systems*

As noted in the previous section, since Υ is an associative homomorphism [5.92] of the shuffle algebra, the logarithm of the abstract Chen series [5.34] is a Lie series.

In other words, the Chen series may be expressed as the exponential of a Lie series, or as an infinite product of exponentials of Lie polynomials. This section exhibits the combinatorial analogue of the corresponding result obtained in section 5.4.1, which used the standard dynamical system technique of variation of parameters to obtain an exponential product expansion of the Chen–Fliess series. First, this requires a brief review of bases of free Lie algebra. The section concludes by rewriting the expansion in a form that immediately gives a coordinate representation of a universal free nilpotent system that *covers* every nilpotent control system.

As a consequence of Ree's theorem, for every basis $\mathcal{B} = \{[b]: b \in B\}$ (suitably indexed by a set $B \subseteq Z^*$) of the free Lie algebra $L(Z)$, there exist functions $\zeta, \xi: B \mapsto A(Z)$, such that

$$CF = \sum_{w \in Z^*} w \otimes w = \exp\left(\sum_{b \in B} \zeta^b \otimes [b]\right)$$

$$= \prod_{b \in B}^{\longrightarrow} \exp\left(\xi^b \otimes [b]\right) \in A(Z) \otimes \hat{A}(Z). \qquad [5.101]$$

In agreement with the established language from Lie groups and Lie algebras, refer to the functions ζ^h and ξ^h as *coordinates of the first and second kinds*, respectively. Of particular importance, for applications in the analysis and design of control systems, are the explicit formulas for the iterated integral functionals Υ^{ζ^h} and Υ^{ξ^h} that are the images of noncommutative polynomials under the Zinbiel homomorphism [5.90].

Numerous constructions of bases for free Lie algebras have been proposed in the literature (see, for example, Reutenauer (1993) for historic remarks; see also Hall (1950) and Širšov (1958, 1962)). The generalization of Hall bases as presented in the unifying work by Viennot (1978) has proven to be particularly useful for applications in control theory. It is distinguished by being intimately related to the dynamical system technique of iterating the method of variations of parameters, as presented in section 5.4.1. The common underlying algorithm is the Lazard elimination (Lazard 1954, 1960; Bourbaki 1998) (for recent related work, see Luque and Thibon (2007)), which is known to generate bases for the free Lie algebras (Bourbaki 1998). It is based on the fundamental observation that if $L(Z)$ is the Lie free algebra generated by the sets Z and $a_0 \in Z$, then $L(Z)$ is the direct sum of the one-dimensional subspace spanned by a_0 and the Lie algebra that is freely generated by the set

$$\{(\mathrm{ad}^\nu a_0, a): \nu \geq 0, \ \ a \in Z \setminus \{a_0\}\}. \qquad [5.102]$$

The construction of bases for free Lie algebras relies on being able to identify left and right factors, and thus uses the language of rooted labeled binary trees, making use of the *free magma* $\mathcal{M}(Z)$ introduced in [5.79].

DEFINITION 5.6.– *A* Hall–Viennot set over a finite set Z *is a strictly ordered subset* $\tilde{\mathcal{H}} \subseteq \mathcal{M}(Z)$ *that satisfies:*

(i) $Z \subseteq \tilde{\mathcal{H}}$.

(ii) Suppose $a \in Z$. *Then,* $(\tilde{t}, a) \in \tilde{\mathcal{H}}$ *if and only if* $\tilde{t} \in \tilde{\mathcal{H}}, \tilde{t} \prec a$. *and* $a \prec (\tilde{t}, a)$.

(iii) Suppose $\tilde{t}', \tilde{t}'', \tilde{t}'''.(\tilde{t}', \tilde{t}'') \in \tilde{\mathcal{H}}$.

Then, $(\tilde{t}', (\tilde{t}'', \tilde{t}''')) \in \tilde{\mathcal{H}}$ *if and only if* $\tilde{t}'' \leq \tilde{t}' \preceq (\tilde{t}'', \tilde{t}''')$ *and* $\tilde{t}' \prec (\tilde{t}', (\tilde{t}'', \tilde{t}'''))$.

The bases originally proposed by Hall (see Bourbaki (1998)) require that the ordering be compatible with the length, i.e. if $\tilde{t}'$ and $\tilde{t}''$ are Hall trees, and $\tilde{t}'$ has fewer leaves than $\tilde{t}''$, then $\tilde{t}' \prec \tilde{t}''$. Viennot replaced this condition (and other minor parts) by condition (ii) in definition 5.6.

The image of a Hall–Viennot set under the map $\tilde{h} \mapsto [h]$ is an ordered basis for $L(Z)$ (Bourbaki 1998; Hall 1950; Viennot 1973). Hall–Viennot sets have the unique factorization property that in some sense makes these bases *optimal* (Viennot 1978; Melançon and Reutenauer 1989; Melançon 1992; Reutenauer 1993).

THEOREM 5.9.– *Suppose* $\tilde{\mathcal{H}} \subseteq \mathcal{M}(Z)$ *is a Hall–Viennot set and* $\mathcal{H} = \{h \colon \tilde{h} \in \tilde{\mathcal{H}}\} \subseteq Z^+$ *is the corresponding set of Hall words with the induced ordering. Then, every word* $w \in Z^+$ *uniquely factors into a nonincreasing product of Hall words, i.e. there exist unique* $r \geq 0$, $h_j \in \mathcal{H}$, *such that*

$$w = h_1 h_2 \dots h_s \quad \text{and} \quad h_1 \succeq h_2 \succeq \dots \succeq h_r. \qquad [5.103]$$

In particular, the restriction of the parenthesization map $\mathcal{M}(Z) \mapsto Z^*$ to any Hall–Viennot set $\tilde{\mathcal{H}} \subseteq \mathcal{M}(Z)$ is injective, justifying the notation $[h]$ for Hall trees $\tilde{h}$.

This now allows us to obtain concise formulas for the coefficients ξ^b in the exponential product expansion [5.101], which map to the iterated integral functions Υ^{ξ^b}. Note that the injectivity of the foliage map (when restricted to a Hall–Viennot set) allows us to index these coefficients by Hall words rather than the more cumbersome Hall trees. Algebraically, these coefficients (up to scaling) form a dual basis to the Poincaré–Birkhoff–Witt basis for $A(Z)$, considered as the universal (associative) enveloping algebra of the free Lie algebra $L(Z)$.

THEOREM 5.10.– *(see Schützenberger (1958), Sussmann (1986), Melançon and Reutenauer (1989) and Reutenauer (1993)) Suppose* $\tilde{\mathcal{H}} \subseteq \mathcal{M}(Z)$ *is a Hall–Viennot set, and* $\mathcal{H} \subseteq A(Z)$ *is its canonical image under the parenthesization map that sends each binary labeled tree to its foliage. Then, the coefficients* ξ^h *for* $h \in \mathcal{H}$ *in the exponential product expansion [5.101] satisfy the following: if* $h^r h' \in \mathcal{H}$ *is such that* h *is not the left factor in the Hall–Viennot factorizations of* h', *then*

$$\xi^{h^r h'} = \frac{1}{r!}(\xi^h)^{\sqcup\!\sqcup\, r} * \xi^{h'}. \qquad [5.104]$$

For applications in control, it is convenient to rewrite the recursive equations [5.104] in the form of an infinite dimensional analogue of the system [5.3]. It is helpful to introduce for $a \in Z$, the adjoints of the left and right shift maps $w \mapsto wa$ and $w \mapsto aw$ on $A(Z)$, defined by

$$\text{for all } z \in Z^* \ \langle wa^{-1}, z\rangle = \langle w, za\rangle \quad \text{and} \quad \langle a^{-1}w, z\rangle = \langle w, az\rangle. \qquad [5.105]$$

A critical property is that on the Zinbiel algebra, the first is a linear map, whereas the second is a derivation (Kawski 2002): for all $a \in Z$ and $w, z \in Z^*$

$$(w * z)a^{-1} = w * (za^{-1}) \quad \text{and} \quad a^{-1}(w * z) = (a^{-1}w) * z + (a^{-1}z) * w. \qquad [5.106]$$

Related properties hold for Lie polynomials, and we may define w^{-1} for general $A(Z)$, but care needs to be taken to use the notation that identifies the map that sends a word $a_1a_2 \ldots a_r$ to the reversed word $a_n \ldots a_1$, which is intimately related to the antipode in the Hopf algebra. This section does not need these extensions from a^{-1} to w^{-1}.

THEOREM 5.11.– *Fix $r \geq 1$, $m \geq 2$ and a Hall–Viennot set $\mathcal{H} \subseteq Z^*$ over $Z = \{a_1, \ldots a_m\}$. Denote by $\mathcal{H}^{\leq r} \subseteq \mathcal{H}$ the set of all words in $\mathcal{H}$ that have a length strictly less than r. Assume that the ordering of $\mathcal{H}$ is compatible with the length of the words and let N be the cardinality of $\mathcal{H}^{<r}$. For $a \in Z$, define the vector fields*

$$f_a = \sum_{h \in \mathcal{H}^{<r}} q^{ha^{-1}} \frac{\partial}{\partial q^h} \qquad [5.107]$$

with polynomial components on $\mathbb{R}^N$ with its standard coordinates indexed by $\mathcal{H}^{\leq r}$. Then, the Lie algebra $L(\{f_{a_1}, \ldots f_{a_m}\})$ that is free nilpotent of step r.

This theorem effectively gives normal forms in coordinates for affine control systems whose fields generate free nilpotent Lie algebras Kawski (2002, see also Grayson and Grossman (1990)). Together with a theorem from Sussmann (1974) that states that the isotropy subalgebra at a point completely characterizes the control system locally, it follows as a corollary that this construction yields normal forms for all nilpotent systems (not necessarily free).

COROLLARY 5.2.– *If $g_1, \ldots g_m$ are analytic vector fields that generate a nilpotent Lie algebra then, there exists $r \in \mathbb{Z}^+$ and a linear map $\Phi \colon \mathbb{R}^N \mapsto \mathbb{R}^n$, such that the images of the fields f_i (both f_i and N as in the previous theorem) generate a Lie algebra that is isomorphic to $L(g_1, \ldots, g_n)$.*

The following example illustrates how straightforward it is to recursively generate an explicit version of the normal form of a free nilpotent system with two vector fields. The key is to index coordinate functions (or states) by Hall words rather than by natural numbers. The vector fields in the following system of the form $\dot{q} = q(u^0 f_0 + u^1 f_1)$ generate a maximally free nilpotent Lie algebra of rank 5

$$\begin{array}{llll}
\dot{q}^0 & = u^0, & & \\
\dot{q}^1 & = u^1, & & \\
\dot{q}^{01} & = q^0 \dot{q}^1 & = q^0\, u^1, & \\
\dot{q}^{001} & = q^0 \dot{q}^{01} & = (q^0)^2\, u^1 & \text{using the factorization } (0(01)), \\
\dot{q}^{101} & = q^1 \dot{q}^{01} & = q^1 q^0\, u^1 & \text{using the factorization } (1(01)), \\
\dot{q}^{0001} & = q^0 \dot{q}^{001} & = (q^0)^3\, u^1 & \text{using the factorization } (0(0(01))), \\
\dot{q}^{1001} & = q^1 \dot{q}^{001} & = q^1 (q^0)^2\, u^1 & \text{using the factorization } (1(0(01))), \\
\dot{q}^{1101} & = q^1 \dot{q}^{101} & = (q^1)^2 q^0\, u^1 & \text{using the factorization } (1(1(01))), \\
\dot{q}^{00001} & = q^0 \dot{q}^{001} & = (q^0)^4\, u^1 & \text{using the factorization } (0(0(0(01)))), \\
\dot{q}^{10001} & = q^1 \dot{q}^{0001} & = q^1 (q^0)^3\, u^1 & \text{using the factorization } (1(0(0(01)))), \\
\dot{q}^{11001} & = q^1 \dot{q}^{1001} & = (q^1)^2 (q^0)^2\, u^1 & \text{using the factorization } (1(1(0(01)))), \\
\dot{q}^{01001} & = q^{01} \dot{q}^{001} & = q^{01} (q^0)^3\, u^1 & \text{using the factorization } ((01)(0(01))), \\
\dot{q}^{01101} & = q^{01} \dot{q}^{101} & = q^{01} (q^1)^2 q^0\, u^1 & \text{using the factorization } ((01)(1(01))).
\end{array} \quad [5.108]$$

5.4.4. *Logarithm of the Chen series*

The exponential product expansion is useful for the theoretical analysis of controllability and optimality properties, and for proving that certain feedback schemes indeed stabilize a system. However, the computational cost of numerically computing each exponential suggests that other applications should combine all terms in a series and compute a single exponential. This applies, in particular, to path planning and tracking algorithms, as well as numerical calculations of system trajectories. This section deals with such an alternative way to simplify the Chen–Fliess series, by expressing it as single exponential of an infinite Lie series (see [5.101]).

Again, the existence of *coordinates of the first kind* ζ^b for any basis B of $L(Z)$ is *a priori* clear. Efficiently computable explicit formulas for these are of practical value. There is a long history of efforts to derive such *continuous Baker–Campbell–Hausdorff formulas* (see, for example, Bialynicki-Birula *et al.* (1969), Mielnik and Plebański (1970) and Strichartz (1987) for classical results, and Murua (2006) for recent work in the context of numerical integration). While the latter employs Hall sets, the older formulas generally do not use bases (only spanning sets). This section demonstrates how to use tools from combinatorial Hopf algebras to efficiently obtain explicit formulas for the ζ^h from those for the ξ^h for Hall words h. This presentation follows the approach taken in Gehrig (2007) and Gehrig and Kawski (2008). For a detailed discussion of elements of combinatorial Hopf algebras, see the corresponding chapters in this volume. This section briefly reviews some maps and properties that are directly used for the control-relevant results, and fixes the notation to match the presentation throughout this chapter, while staying close to established symbols (Reutenauer 1993).

The algebra $A(Z)$ is endowed with two Hopf-algebra structures, whose products are the concatenation and the shuffle product, respectively. They share the same antipode, denoted here by $\alpha\colon A \mapsto A$, which is the linear map that maps any word $w = a_1 a_2 \dots a_k \in Z^k$ to its signed reversal

$$\alpha(w) = (-1)^k \, a_k \dots a_2 a_1. \qquad [5.109]$$

The coproducts will be denoted by $\Delta, \Delta'\colon A(Z) \mapsto A(Z) \otimes A(Z)$ and are defined on generators $a \in Z$ by the same formula

$$\Delta(a) = \Delta'(a) = e \otimes a + a \otimes e. \qquad [5.110]$$

They are associative algebra homomorphisms with respect to the concatenation and the shuffle product, respectively. For example, for letters $a, b \in Z$, we have

$$\begin{aligned} \Delta(ab) &= e \otimes ab + a \otimes b + b \otimes a + ab \otimes e \quad \text{and} \\ \Delta'(ab) &= e \otimes ab + a \otimes b \qquad\qquad\;\; + ab \otimes e. \end{aligned} \qquad [5.111]$$

The coproduct Δ is the transpose of the shuffle product, i.e. for all $v, w, z \in \mathbb{Z}^*$,

$$\langle v \sqcup\!\sqcup w, z \rangle = \langle v \otimes w, \Delta(z) \rangle. \qquad [5.112]$$

A useful characterization of the second coproduct $\Delta'\colon A(Z) \mapsto A(Z) \otimes A(Z)$ is as the shuffle–algebra homomorphism that satisfies for every $w \in \mathbb{Z}^*$

$$\Delta'(w) = \sum_{u,v \in Z^*} \langle w, uv \rangle \; u \otimes v. \qquad [5.113]$$

Correspondingly, these two Hopf-algebra structures endow the algebra of linear endomorphism of the algebra $A(Z)$ with two different associative convolution products, which here are denoted $\star$ and $\star'$. Using conc to denote the concatenation map $\mathrm{conc}(w \otimes z) = wz$ and $\mathrm{shu}(w \otimes z) = w \sqcup\!\sqcup z$, for linear endomorphisms $f, g\colon A(Z) \mapsto A(Z)$, they are defined by

$$\begin{aligned} f \star g &= \mathrm{conc} \circ (f \otimes g) \circ \Delta, \quad \text{and} \\ f \star' g &= \mathrm{shu} \circ (f \otimes g) \circ \Delta', \end{aligned} \qquad [5.114]$$

i.e. for any word $w \in A(Z)$

$$(f \star g)(w) = \sum_{u,v \in Z^+} \langle w, u \sqcup\!\sqcup v \rangle \; f(u) g(v) \quad \text{and} \qquad [5.115]$$

$$(f \star' g)(w) = \sum_{u,v \in Z^+} \langle w, uv \rangle \; f(u) \sqcup\!\sqcup g(v) = \sum_{\substack{uv = w \\ u,v \in Z^+}} f(u) \sqcup\!\sqcup g(v). \qquad [5.116]$$

The formulas for the ζ^h are obtained from those for the ξ^h, by the adjoint of projections from $A(Z)$ onto the Lie algebra $L(Z)$, along subspaces spanned by the symmetric products of Lie polynomials. The key is to identify the kernel of the projection. For this, consider a composition of the algebra $A(Z)$ as a direct sum of *fundamental subspaces* U_k to be defined in the following, with associated projection maps

$$A(Z) = \bigoplus_{k=1}^{\infty} U_k \quad \text{and} \quad \pi_k \colon A(Z) \mapsto U_k. \tag{5.117}$$

Note that the projection maps π_k are *not* an orthogonal projection with respect to the standard inner product on $A(Z)$. The first fundamental subspace is the free Lie algebra $U_1 = L(Z) \subseteq A(Z)$. The higher fundamental subspaces U_k, $k > 1$ are spanned by *symmetric products* of k Lie polynomials. Here, the symmetric product of any finite sequence $p_1, \ldots p_k \in A(Z)$ of polynomials in $A(Z)$ is the multi-linear map defined by

$$\mathrm{Sym}(w_1, \ldots w_k) = \tfrac{1}{k!} \sum_{\sigma \in S_k} w_{\sigma(1)} w_{\sigma(2)} \ldots w_{\sigma(k)} \tag{5.118}$$

where the sum is taken over the symmetric group of order k.

As already observed in the formal calculation [5.100], the natural expansion of the logarithm as a formal power series involves the projection $I \colon A(Z) \mapsto A(Z)^+$ that maps the empty word to zero and that is the identity on Z^+. In terms of this map, we have a simple series expansion for the projection $\pi_1 \colon A(Z) \mapsto U_1$ in terms of the iterated convolution product

$$\pi_1 = \sum_{k \geq 1} (-1)^{k-1} \tfrac{1}{k} I^{\star k}. \tag{5.119}$$

For practical computations, using [5.115], we may expand these convolution products in terms of higher k-ary analogues of the coproduct and concatenation product

$$\pi_1 = \sum_{k \geq 1} (-1)^{k-1} \tfrac{1}{k} \,\mathrm{conc}^k \circ I^{\otimes k} \circ \Delta^k. \tag{5.120}$$

The projections $\pi_k \colon A(Z) \mapsto U_k$ onto the higher-order fundamental subspaces are normalized k-fold convolutions of π_1 with

$$\pi_k = \tfrac{1}{k!} \pi_1^{\star k}. \tag{5.121}$$

From the projections π_k on the side that maps to vector fields and higher-order partial differential operators, we obtain the corresponding projections for the dual space. These map to the iterated integral functionals ζ^h. Define the map $\pi_1' \colon A(Z) \mapsto A(Z)$ as the adjoint of π_1, i.e. the linear map that satisfies for all $w, z \in \mathbb{Z}$

$$\langle w, \pi_1(z) \rangle = \langle \pi_1'(w), z \rangle. \tag{5.122}$$

Using the fact that I is self-adjoint, and that conc and Δ' are adjoints of each other, as are ш and Δ, the definition [5.119] of π_1 in terms of the convolution product $\star$ gives the corresponding formula for π_1' in terms of the other convolution product $\star'$

$$\pi_1' = \sum_{k\geq 1}(-1)^{k-1}\tfrac{1}{k}I^{\star' k}. \qquad [5.123]$$

For practical computations, we may expand this formula for words $w \in Z$ in the form

$$\pi_1'(w) = \sum_{k\geq 1}\tfrac{1}{k}(-1)^{k-1}\sum_{u_1\ldots u_k = w} u_1 \,ш\, \ldots \,ш\, u_k.$$

The most simple nontrivial example is

$$\pi_1'(\xi^{ab}) = \pi_1'(ab) = ab - \tfrac{1}{2}\, a \,ш\, b = \tfrac{1}{2}\,(ab - ba). \qquad [5.124]$$

The main result of the thesis (Gehrig 2007) is that for Hall words $h \in \mathcal{H}$, the dual projection π_1' maps $\{\xi^h : h \in \mathcal{H}\}$ elementwise onto $\{\zeta^h : h \in \mathcal{H}\}$.

THEOREM 5.12 (Gehrig 2007).– *For every Hall set $\mathcal{H} \subseteq Z^*$, for all $h \in \mathcal{H}$*

$$\zeta^h = \pi_1'\,(\xi^h). \qquad [5.125]$$

Employing the function notation, this reads as $\zeta = \pi_1' \circ \xi$. Using the unique factorization property characteristic for Hall words (theorem 5.9), we may extend the domains of ξ and ζ to all of $A(Z)$ and obtain explicit formulas for the dual bases for Poincaré–Birkhoff–Witt bases of $A(Z)$, when regarded as the enveloping algebra of $L(Z)$. If $w = h_1^{r_1}h_2^{r_2}\cdots h_k^{r_k}$ is the factorization of the word $w \in A(Z)$, with $h_1 \succ h_2 \succ \ldots \succ h_k$ as a strictly decreasing sequence of Hall words h_j and $r_j \in Z^+$, then the dual basis elements are

$$\xi^w = \tfrac{1}{r_1!\cdots r_k!}(\xi^{h_1})^{ш\, r_1} \,ш\, (\xi^{h_2})^{ш\, r_2} \,ш\, \ldots \,ш\, (\xi^{h_k})^{ш\, r_k}. \qquad [5.126]$$

In general, the maps π_k' do not commute with the shuffle product, but they do on the products of this special form: as shown in Gehrig (2007), if $|r| = r_1 + \ldots + r_k$ with r_j as above, then

$$\zeta^w = \pi_{|r|}'(\xi^w) = \tfrac{1}{r_1!\cdots r_k!}(\zeta^{h_1})^{ш\, r_1} \,ш\, \ldots \,ш\, (\zeta^{h_k})^{ш\, r_k}. \qquad [5.127]$$

With the characterization [5.122] of π_1' and the formulas for it, it is now a straightforward calculation to obtain explicit formulas for the ζ^h (see Gehrig and Kawski (2008)). As an example, first consider the projection of an individual word

$$\begin{aligned}\pi_1'(aaabb) &= aaabb - \tfrac{1}{2}\,(aaab \,ш\, b + aaa \,ш\, bb + aa \,ш\, aab + a \,ш\, aabb)\\ &+\tfrac{1}{3}(aaa \,ш\, b \,ш\, b + aa \,ш\, ab \,ш\, b + 2aa \,ш\, a \,ш\, bb + a \,ш\, aab \,ш\, b + a \,ш\, a \,ш\, abb)\\ &-\tfrac{1}{4}(aa \,ш\, a \,ш\, b \,ш\, b + a \,ш\, aa \,ш\, b \,ш\, b + a \,ш\, a \,ш\, ab \,ш\, b + a \,ш\, a \,ш\, a \,ш\, bb)\\ &+\tfrac{1}{5}a \,ш\, a \,ш\, a \,ш\, b \,ш\, b.\end{aligned} \qquad [5.128]$$

For the projection of a polynomial ξ^{abaab}, from the previous section, use

$$\xi^{abaab} = \frac{1}{2}(ab \sqcup\!\sqcup a \sqcup\!\sqcup a)b = 3aaabb + 2aabab + abaab. \qquad [5.129]$$

Then, calculate ζ^{abaab}

$$\begin{aligned}\zeta^{abaab} &= \pi_1'(\xi^{abaab}) = \pi_1'(3aaabb + 2aabab + abaab)\\ &= \tfrac{1}{10}aaabb + \tfrac{1}{15}aabab + \tfrac{1}{15}aabba + \tfrac{1}{15}abaab - \tfrac{1}{10}ababa\\ &+ \tfrac{1}{15}abbaa - \tfrac{1}{10}baaab + \tfrac{1}{15}baaba + \tfrac{1}{15}babaa - \tfrac{1}{10}bbaaa.\end{aligned} \qquad [5.130]$$

The following system explicitly gives the polynomials ζ^h for the first 14 Hall words (see Kawski (2000) and Gehrig and Kawski (2008))

$$\begin{aligned}\zeta^{a} &= a,\\ \zeta^{b} &= b,\\ \zeta^{ab} &= \tfrac{1}{2}(ab - ba) = \tfrac{1}{2}[a,b],\\ \zeta^{aab} &= \tfrac{1}{6}\ (aab - 2aba + baa) = \tfrac{1}{6}\ [a,[a,b]],\\ &\tfrac{1}{6}\ [b,[a,b]],\\ \zeta^{aaab} &= \tfrac{1}{6}(abaa - aaba),\\ \zeta^{baab} &= \tfrac{1}{6}(abab - aabb + bbaa - baba),\\ \zeta^{bbab} &= \tfrac{1}{6}(bbab - babb),\end{aligned} \qquad [5.131]$$

$$\begin{aligned}\zeta^{aaaab} &= \tfrac{1}{30}(aaaab + aaaba - 4aabaa + abaaa + baaaa),\\ \zeta^{baaab} &= \tfrac{1}{30}(-2aaabb + 3aabab + 3aabba - 2abaab - 2ababa,\\ &\qquad + 3abbaa - 2baaab - 2baaba + 3babaa - 2bbaaa),\\ \zeta^{abaab} &= \tfrac{1}{30}(-3aaabb + 2aabab + 2aabba + 2abaab - 3ababa\\ &\qquad + 2abbaa - 3baaab + 2baaba + 2babaa - 3bbaaa),\\ \zeta^{bbaab} &= \tfrac{1}{30}(2aabbb - 3ababb + 2abbab + 2abbba - 3baabb,\\ &\qquad + 2babab + 2babba - 3bbaba - 3bbaab + 2bbbaa),\\ \zeta^{abbab} &= \tfrac{1}{30}(-aabbb - ababb + 4abbab - abbba - baabb,\\ &\qquad - babab + 4babba - bbaba - bbaab - bbbaa), \quad \text{and}\\ \zeta^{bbbab} &= \tfrac{1}{30}(abbbb + babbb - 4bbabb + bbbab + bbbba).\end{aligned}$$

The above system expresses the ζ^h for $h \in \mathcal{H}$ as linear combinations of the standard basis vectors $w \in Z^*$. For control applications and, potentially, for numerical integration algorithms, it is more useful to rewrite these formulas in a recursive format using the Zinbiel product, i.e. as a cascade nilpotent control system. The following may again be seen as an alternate normal form for a free nilpotent control system:

$$\begin{aligned}
\dot{z}^a &= u^a, \\
\dot{z}^b &= u^b, \\
\dot{z}^{ab} &= -\tfrac{1}{6} z^b u^a + \tfrac{1}{6} z^a u^b, \\
\dot{z}^{aab} &= \left(-\tfrac{1}{2} z^{ab} - \tfrac{1}{12} z^a z^b\right) u^a + \tfrac{1}{12} (z^a)^2 u^b, \\
\dot{z}^{bab} &= -\tfrac{1}{12} (z^b)^2 u^a + \left(-\tfrac{1}{2} z^{ab} + \tfrac{1}{12} z^a z^b\right) u^b, \\
\dot{z}^{aaab} &= \left(-\tfrac{1}{2} z^{aab} - \tfrac{1}{12} z^{ab} z^a\right) u^a, \\
\dot{z}^{baab} &= \left(-\tfrac{1}{2} z^{bab} - \tfrac{1}{12} z^{ab} z^b\right) u^a - \left(\tfrac{1}{2} z^{aab} + \tfrac{1}{12} z^{ab} z^a\right) u^b, \\
\dot{z}^{bbab} &= \left(-\tfrac{1}{2} z^{bab} - \tfrac{1}{12} z^{ab} z^b\right) u^b,
\end{aligned} \qquad [5.132]$$

$$\begin{aligned}
\dot{z}^{aaaab} &= \left(-\tfrac{1}{2} z^{aaab} - \tfrac{1}{12} z^a z^{aab} - \tfrac{1}{720} (z^a)^3 z^b\right) u^a, -\tfrac{1}{720} (z^a)^4 u^b \\
\dot{z}^{bbbab} &= \tfrac{1}{720} (z^b)^4 u^a - \left(\tfrac{1}{2} z^{bbab} + \tfrac{1}{12} z^b z^{bab} + \tfrac{1}{720} z^a (z^b)^3\right) u^b, \\
\dot{z}^{baaab} &= \left(\tfrac{1}{240} (z^a)^2 (z^b)^2 - \tfrac{1}{2} z^{baab} - \tfrac{1}{12} z^b z^{aab} - \tfrac{1}{12} z^a z^{bab}\right) u^a, \\
&\quad - \left(\tfrac{1}{2} z^{aaab} + \tfrac{1}{12} z^a z^{aab} + \tfrac{1}{240} (z^a)^3 z^b\right) u^b, \\
\dot{z}^{abaab} &= \left(-\tfrac{1}{2} z^{baab} + \tfrac{1}{12} z^b z^{aab} - \tfrac{1}{12} z^a z^{bab} - \tfrac{1}{12} (z^{ab})^2 + \tfrac{1}{360} (z^a)^2 (z^b)^2\right) u^a, \\
&\quad + \left(-\tfrac{1}{6} z^a z^{aab} - \tfrac{1}{360} (z^a)^3 z^b\right) u^b, \\
\dot{z}^{abbab} &= \left(-\tfrac{1}{2} z^{bbab} + \tfrac{1}{12} z^b z^{bab} + \tfrac{1}{720} z^a (z^b)^3\right) u^a, \\
&\quad - \left(\tfrac{1}{6} z^a z^{bab} + \tfrac{1}{12} (z^{ab})^2 + \tfrac{1}{720} (z^a)^2 (z^b)^2\right) u^b, \\
\dot{z}^{bbaab} &= \left(-\tfrac{1}{2} z^{bbab} - \tfrac{1}{12} z^b z^{bab} + \tfrac{1}{240} z^a (z^b)^3\right) u^a, \text{ and} \\
&\quad - \left(\tfrac{1}{2} z^{baab} + \tfrac{1}{12} z^b z^{aab} + \tfrac{1}{12} z^a z^{bab} + \tfrac{1}{240} (z^a)^2 (z^a)^2\right) u^b.
\end{aligned}$$

This system [5.132] for the ζ^h does not have the elegant algebraic form of the system [5.108], which immediately lends itself to studying geometric features, controllability and optimality properties. However, these formulas for coefficients are explicit, easily computable by machine and well suited for direct implementation when solving problems of tracking, path planning, and for geometric numerical integration.

5.5. References

Agračev, A.A. and Gamkrelidze, R.V. (1978). Exponential representation of flows and a chronological enumeration. *Mat. Sb. (N.S.)*, 107(149)(4), 467–532, 639.

Agračev, A.A. and Gamkrelidze, R.V. (1980). Chronological algebras and nonstationary vector fields. In *Problems in Geometry*, vol. 11, Akad. Nauk SSSR, Vsesoyuz. Inst. Nauchn. i Tekhn. Informatsii, Moscow, 135–176, 243 (in Russian).

Agrachev, A.A. and Sachkov, Y.L. (2004). *Control Theory from the Geometric Viewpoint.* Springer-Verlag, Berlin, Heidelberg.

Aguilar, C.O. and Lewis, A.D. (2012). Small-time local controllability for a class of homogeneous systems. *SIAM J. Control Optim.*, 50(3), 1502–1517 [Online]. Available at: https://doi-org.ezproxy1.lib.asu.edu/10.1137/100785892.

Beauchard, K. and Marbach, F. (2018). Quadratic obstructions to small-time local controllability for scalar-input systems. *J. Differ. Equations*, 264(5), 3704–3774 [Online]. Available at: https://doi-org.ezproxy1.lib.asu.edu/10.1016/j.jde.2017.11.028.

Bellaïche, A. (1996). The tangent space in sub-Riemannian geometry. In *Sub-Riemannian Geometry*, Bellaiche, A. and Risler, J.-J. (eds). Birkhäuser, Basel.

Bialynicki-Birula, I., Mielnik, B., Plebanski, J.F. (1969). Explicit solution of the continuous Baker–Campbell–Hausdorff problem and a new expression for the phase operator. *Ann. Phys.*, 51(1), 187–200 [Online]. Available at: http://www.sciencedirect.com/science/article/pii/0003491669903510.

Bianchini, R.-M. and Kawski, M. (2003). Needle variations that cannot be summed. *SIAM J. Control Optim.*, 42(1), 218–238 [Online]. Available at: http://dx.doi.org.ezproxy1.lib.asu.edu/10.1137/S0363012902402876.

Bianchini, R.M. and Stefani, G. (1984). Normal local controllability of order one. *Int. J. Control*, 39(4), 701–714 [Online]. Available at: http://dx.doi.org.ezproxy1.lib.asu.edu/10.1080/00207178408933198.

Bloch, A.M. (2003). *Nonholonomic Mechanics and Control.* Springer-Verlag, New York [Online]. Available at: http://dx.doi.org.ezproxy1.lib.asu.edu/10.1007/b97376.

Bourbaki, N. (1998). *Lie Groups and Lie Algebras.* Springer-Verlag, Berlin.

Bressan, A. and Piccoli, B. (2007). *Introduction to the Mathematical Theory of Control.* American Institute of Mathematical Sciences (AIMS), Springfield, MO.

Bullo, F. and Lewis, A.D. (2005). *Geometric Control of Mechanical Systems.* Springer-Verlag, New York.

Campbell, J.E. (1897). On a law of combination of operators bearing on the theory of continuous transformation groups. *Proc. London Math. Soc.*, 28, 381–390.

Chen, K.-T. (1954). Iterated integrals and exponential homomorphisms. *Proc. London Math. Soc.*, (3)4, 502–512.

Chen, K.-T. (1957). Integration of paths, geometric invariants and a generalized Baker–Hausdorff formula. *Ann. Math.*, 65(2), 163–178.

Chen, K.-T. (1958). Integration of paths – A faithful representation of paths by non-commutative formal power series. *Trans. Am. Math. Soc.*, 89, 395–407.

Clarke, F.H. (1990). *Optimization and Nonsmooth Analysis*, 2nd edition. Society for Industrial and Applied Mathematics (SIAM), Philadelphia, PA.

Coron, J.-M. (2007). *Control and Nonlinearity.* American Mathematical Society, Providence, RI.

Ebrahimi-Fard, K. and Manchon, D. (2009). A Magnus- and Fer-type formula in dendriform algebras. *Found. Comput. Math.*, 9(3), 295–316 [Online]. Available at: http://dx.doi.org.ezproxy1.lib.asu.edu/10.1007/s10208-008-9023-3.

Fer, F. (1958). Résolution de l'équation matricielle $dU/dt = pU$ par produit infini d'exponentielles matricielles. *Acad. Roy. Belg. Bull. Cl. Sci.*, 5(44), 818–829.

Fliess, M. (1980). Realizations of nonlinear systems and abstract transitive Lie algebras. *Bull. Amer. Math. Soc. (N.S.)*, 2(3), 444–446.

Fliess, M. (1981). Fonctionnelles causales non linéaires et indéterminées non commutatives. *Bull. Soc. Math. France*, 109(1), 3–40.

Frankowska, H. (1986). Théorème d'application ouverte pour les correspondances. *C. R. Acad. Sci. Paris Sér. I Math.*, 302(15), 559–562.

Gehrig, E. (2007). Hopf algebras, projections, and coordinates of the first kind in control theory. PhD Thesis, Arizona State University, Phoenix, AZ.

Gehrig, E. and Kawski, M. (2008). A Hopf-algebraic formula for compositions of noncommuting flows. *Proceedings IEEE Conference Decision and Control*, Cancun, 156–1574.

Ginzburg, V. and Kapranov, M. (1994). Koszul duality for operads. *Duke Math. J.*, 76(1), 203–272 [Online]. Available at: http://dx.doi.org.ezproxy1.lib.asu.edu/10.1215/S0012-7094-94-07608-4.

Gray, W.S. and Duffaut Espinosa, L.A. (2011). A Faà di Bruno Hopf algebra for a group of Fliess operators with applications to feedback. *Systems Control Lett.*, 60(7), 441–449 [Online]. Available at: http://dx.doi.org.ezproxy1.lib.asu.edu/10.1016/j.sysconle.2011.03.007.

Gray, W.S. and Li, Y. (2005). Generating series for interconnected analytic nonlinear systems. *SIAM J. Control Optim.*, 44(2), 646–672 [Online]. Available at: http://dx.doi.org.ezproxy1.lib.asu.edu/10.1137/S036301290343007X.

Gray, W.S. and Wang, Y. (2002). Fliess operators on L_p spaces: Convergence and continuity. *Syst. Control Lett.*, 46(2), 67–74.

Gray, W.S., Duffaut Espinosa, L.A., Ebrahimi-Fard, K. (2014). Faà di Bruno Hopf algebra of the output feedback group for multivariable Fliess operators. *Systems Control Lett.*, 74, 64–73 [Online]. Available at: https://doi-org.ezproxy1.lib.asu.edu/10.1016/j.sysconle.2014.10.005.

Grayson, M. and Grossman, R. (1990). Models for free nilpotent Lie algebras. *J. Algebr.*, 135(1), 177–191.

Grossman, R. and Larson, R.G. (1989). Hopf-algebraic structure of families of trees. *J. Algebr.*, 126(1), 184–210.

Grossman, R. and Larson, R.G. (1992). The realization of input–output maps using bialgebras. *Forum Math.*, 4(2), 109–121.

Grossman, R.L. and Larson, R.G. (2004). Bialgebras and realizations. In *Hopf Algebras*, vol. 237. Marcel Dekker, New York.

Hall Jr., M. (1950). A basis for free Lie rings and higher commutators in free groups. *Proc. Amer. Math. Soc.*, 1, 575–581.

Hambly, B. and Lyons, T. (2010). Uniqueness for the signature of a path of bounded variation and the reduced path group. *Ann. Math.*, 171(1), 109–167 [Online]. Available at: http://dx.doi.org.ezproxy1.lib.asu.edu/10.4007/annals.2010.171.109.

Hauser, J. (2010). On the controllability of the pendubot. *8th IFAC Symposium on Nonlinear Control Systems*, Bologna, September.

Haynes, G.W. and Hermes, H. (1970). Nonlinear controllability via Lie theory. *SIAM J. Control*, 8, 450–460.

Hermes, H.G. (1977). Controlled stability. *Ann. Mat. Pura Appl.* 4(114), 103–119.

Hermes, H.G. (1982). On local controllability. *SIAM J. Control Optim.*, 20(2), 211–220 [Online]. Available at: http://dx.doi.org.ezproxy1.lib.asu.edu/10.1137/0320017.

Hermes, H.G. (1986). Nilpotent approximations of control systems and distributions. *SIAM J. Control Optim.*, 24(4), 731–736 [Online]. Available at: http://dx.doi.org.ezproxy1.lib.asu.edu/10.1137/0324045.

Hermes, H.G. (1991). Nilpotent and high-order approximations of vector field systems. *SIAM Rev.*, 33(2), 238–264 [Online]. Available at: http://dx.doi.org.ezproxy1.lib.asu.edu/10.1137/1033050.

Jean, F. (2001). Uniform estimation of sub-Riemannian balls. *J. Dynam. Control Syst.*, 7(4), 473–500 [Online]. Available at: http://dx.doi.org.ezproxy1.lib.asu.edu/10.1023/A:1013154500463.

Jurdjevic, V. (1997). *Geometric Control Theory*. Cambridge University Press, Cambridge.

Kawski, M. (1987). A necessary condition for local controllability. In *Contemporary Mathematics 68: Differential Geometry: The Interface between Pure and Applied Mathematics*, Luksic, M., Martin, C., Shadwick, W. (eds). San Antonio, TX.

Kawski, M. (1988a). An angular open mapping theorem. In *Analysis and Optimization of Systems*, Bensoussan, A. and Lions, J.L. (eds). Springer, Berlin [Online]. Available at: http://dx.doi.org.ezproxy1.lib.asu.edu/10.1007/BFb0042228.

Kawski, M. (1988b). Control variations with an increasing number of switchings. *Bull. Amer. Math. Soc. (N.S.)*, 18(2), 149–152 [Online]. Available at: http://dx.doi.org.ezproxy1.lib.asu.edu/10.1090/S0273-0979-1988-15630-3.

Kawski, M. (1988c). Nilpotent Lie algebras of vectorfields. *J. Reine Angew. Math.*, 388, 1–17 [Online]. Available at: http://dx.doi.org.ezproxy1.lib.asu.edu/10.1515/crll.1988.388.1

Kawski, M. (1990). High-order small-time local controllability. *Nonlinear Controllability and Optimal Control*, Sussmann, H.J. (ed.). Marcel Dekker, New York.

Kawski, M. (2000). Calculating the logarithm of the Chen–Fliess series. *Proceedings MTNS*, Perpignan, 111–128.

Kawski, M. (2002). The combinatorics of nonlinear controllability and noncommuting flows. Lecture, Summer School on Mathematical Control Theory, Trieste, Italy, 3–28 September.

Kawski, M. and Sussmann, H.J. (1997). Noncommutative power series and formal Lie-algebraic techniques in nonlinear control theory. *Operators, Systems and Linear Algebra: Three Decades of Algebraic Systems Theory*, Helmke, U., Prätzel-Wolters, D., Zerz, E. (eds). Springer, Cham.

Knobloch, H.-W. (1981). *Higher Order Necessary Conditions in Optimal Control Theory*. Springer-Verlag, Berlin, Heidelberg.

Krener, A.J. (1977). The high order maximal principle and its application to singular extremals. *SIAM J. Control Optim.*, 15(2), 256–293.

Lazard, M. (1954). Sur les algèbres enveloppantes universelles de certaines algèbres de Lie. *Publ. Sci. Univ. Alger. Sér. A.*, 1, 281–294.

Lazard, M. (1960). Groupes, anneaux de Lie et problème de Burnside. Technical report, Centro Internazionale Matematico Estivo, Rome.

Li, Y. and Gray, W.S. (2006). The formal Laplace–Borel transform of Fliess operators and the composition product. *Int. J. Math. Math. Sci.*, Art. ID 34217, 14.

Loday, J.-L. (1993). Une version non commutative des algèbres de Lie : les algèbres de Leibniz. *Recherche Coopérative sur Programme*, 25(44), 127–151.

Luque, J.-G. and Thibon, J.-Y. (2007). Noncommutative symmetric functions associated with a code, Lazard elimination, and Witt vectors. *Discrete Math. Theor. Comput. Sci.*, 9(2), 59–72.

Lyons, T. and Qian, Z. (2002). *System Control and Rough Paths*. Oxford University Press, Oxford.

Magnus, W. (1954). On the exponential solution of differential equations for a linear operator. *Comm. Pure Appl. Math.*, 7, 649–673.

Melançon, G. (1992). Combinatorics of Hall trees and Hall words. *J. Combin. Theory Ser. A*, 59(2), 285–308.

Melançon, G. and Reutenauer, C. (1989). Lyndon words, free algebras and shuffles. *Can. J. Math.*, 41(4), 577–591.

Mielnik, B. and Plebański, J. (1970). Combinatorial approach to Baker–Campbell–Hausdorff exponents. *Ann. Inst. H. Poincaré Sect. A (N.S.)*, 12, 215–254.

Murua, A. (2006). The Hopf algebra of rooted trees, free Lie algebras, and Lie series. *Found. Comput. Math.*, 6(4), 387–426.

Nagano, T. (1966). Linear differential systems with singularities and an application to transitive Lie algebras. *J. Math. Soc. Jpn*, 18, 398–404.

Ree, R. (1958). Lie elements and an algebra associated with shuffles. *Ann. Math.*, 68(2), 210–220.

Reutenauer, C. (1993). *Free Lie Algebras*. The Clarendon Press, Oxford University Press, New York.

Sarychev, A.V. (2001). Lie- and chronologico-algebraic tools for studying stability of time-varying systems. *Syst. Control Lett.*, 43(1), 59–76 [Online]. Available at: http://dx.doi.org.ezproxy1.lib.asu.edu/10.1016/S0167-6911(01)00090-1.

Schützenberger, M.-P. (1958). Sur une propriété combinatoire des algèbres de Lie libres pouvant être utilisée dans un probléme de mathématiques appliquées. *Séminaire P. Dubreil, Algèbres et Théorie des Nombres*, 12(1).

Širšov, A.I. (1958). On free Lie rings. *Mat. Sb. N.S.*, 45(87), 113–122.

Širšov, A.I. (1962). On the bases of a free Lie algebra. *Algebra i Logika Sem.*, 1(1), 14–19.

Sklyar, G.M. and Ignatovich, S.Y. (2002). Representations of control systems in the Fliess algebra and in the algebra of nonlinear power moments. *Syst. Control Lett.*, 47(3), 227–235.

Sklyar, G.M. and Ignatovich, S.Y. (2007). Description of all privileged coordinates in the homogeneous approximation problem for nonlinear control systems. *C. R. Math. Acad. Sci. Paris*, 344(2), 109–114 [Online]. Available at: http://dx.doi.org.ezproxy1.lib.asu.edu/10.1016/j.crma.2006.11.016.

Sklyar, G.M. and Ignatovich, S. (2008). Fliess series, a generalization of the Ree's theorem, and an algebraic approach to a homogeneous approximation problem. *Int. J. Control*, 81(3), 369–378 [Online]. Available at: http://dx.doi.org.ezproxy1.lib.asu.edu/10.1080/00207170701561427.

Sontag, E.D. (1998). *Mathematical Control Theory*, 2nd edition. Springer-Verlag, New York.

Stefani, G. (1985). Local controllability of nonlinear systems: An example. *Syst. Control Lett.*, 6(2), 123–125.

Stefani, G. (1986). On the local controllability of a scalar-input control system. In *Theory and Applications of Nonlinear Control Systems*, Birnes, C. and Lindquist, A. (eds). North-Holland, Amsterdam.

Strichartz, R.S. (1987). The Campbell–Baker–Hausdorff–Dynkin formula and solutions of differential equations. *J. Funct. Anal.*, 72(2), 320–345 [Online]. Available at: http://dx.doi.org.ezproxy1.lib.asu.edu/10.1016/0022-1236(87)90091-7.

Sussmann, H.J. (1974). An extension of a theorem of Nagano on transitive Lie algebras. *Proc. Am. Math. Soc.*, 45, 349–356.

Sussmann, H.J. (1983). Lie brackets and local controllability: A sufficient condition for scalar-input systems. *SIAM J. Control Optim.*, 21(5), 686–713.

Sussmann, H.J. (1986). A product expansion for the Chen series. *Theory and Applications of Nonlinear Control Systems*, Byrnes, C.I. and Lindquist, A. (eds). North-Holland, Amsterdam.

Sussmann, H.J. (1987). A general theorem on local controllability. *SIAM J. Control Optim.*, 25(1), 158–194.

Sussmann, H.J. (2001). New theories of set-valued differentials and new versions of the maximum principle of optimal control theory. In *Nonlinear Control in the Year 2000*, Vol. 2, Françoise Lamnabhi-Lagarrigue, I. and Respondek, W. (ed.). Springer, London [Online]. Available at: http://dx.doi.org.ezproxy1.lib.asu.edu/10.1007/BFb0110322.

Sussmann, H.J. (2007). Set separation, approximating multicones, and the Lipschitz maximum principle. *J. Differential Equations*, 243(2), 448–488 [Online]. Available at: http://dx.doi.org.ezproxy1.lib.asu.edu/10.1016/j.jde.2007.05.041.

Thitsa, M. and Gray, W.S. (2012). On the radius of convergence of interconnected analytic nonlinear input–output systems. *SIAM J. Control Optim.*, 50(5), 2786–2813 [Online]. Available at: http://dx.doi.org.ezproxy1.lib.asu.edu/10.1137/110852760.

Viennot, G. (1973). Une généralisation des ensembles de Hall. *C. R. Acad. Sci. Paris Sér. A–B*, 276, A599–A602.

Viennot, G. (1978). *Algèbres de Lie libres et monoïdes libres*. Springer, Berlin.

6

Algebra is Geometry is Algebra – Interactions Between Hopf Algebras, Infinite Dimensional Geometry and Application

Alexander SCHMEDING

Technische Universität Berlin, Germany

In this chapter, we examine the interaction of algebra and (infinite-dimensional) geometry in the guise of Hopf algebras and certain associated character groups. We assume familiarity with Hopf algebras and refer to Chapter 1 of this book by D. Manchon for an overview (though we will fix some notation in the rest of this chapter). The title "Algebra is Geometry is Algebra"[1]) which the author enjoyed as an undergraduate alludes to the fact that to deal with applications, we will cross from algebra to geometry and back again. Furthermore, we wish to emphasize that by going back and forth we arrive, not at the same point, but at new and interesting structures.

1. The title is a reference to the famous Gertrude Stein quote "a rose is a rose is a rose". However, the author blatantly stole the title from a lecture of the same title (in German) by C.M. Ringel, Conference "*Representations of Algebras and their Geometry*", 2006 [Online]. Available at: https://www.math.uni-bielefeld.de/~ringel/lectures/lenzing/Welcome.html.

Algebra and Applications 2,
coordinated by Abdenacer MAKHLOUF.

6.1. The Butcher group and the Connes–Kreimer algebra

To motivate the approach outlined above, let us recall the correspondence between rooted trees, differential operators and the Butcher group, also known as the character group of the Butcher–Connes–Kreimer Hopf algebra.

6.1.1. *The Butcher group and B-series from numerical analysis*

In numerical analysis, we are interested in the differential equation

$$y' = f(y),\ y(0) = y_0 \qquad f\colon \mathbb{R}^n \supseteq U \to \mathbb{R}^n \text{ smooth} \tag{6.1}$$

Taking derivatives of y, we see that $y'' = (f(y))' = f'(y)y'$ (where the (iterated) derivatives act as (multi-)linear operators, i.e. $f'(y)x = J_f(y).x$, where J_f is the Jacobian, $f''(y).(x,y) := f''(y;x,y)$ and so forth). Thus, the Taylor expansion of the (unknown) solution y of equation [6.1] is determined by iterated derivatives of f. Going back to an idea by Cayley, these iterated derivatives can be encoded by rooted trees $\mathcal{T} = \{\bullet, \begin{smallmatrix}\bullet\\ |\\ \bullet\end{smallmatrix}, \cdots\}$. In more detail, we set

$$F_\bullet(f) := f \text{ and, recursively, } F_\tau(f) := f^{(m)}(F_{\tau_1}(f), \ldots, F_{\tau_m}(f)) \tag{6.2}$$

where $\tau = [\tau_1, \ldots, \tau_m]$ is the tree given by grafting the trees τ_i to a common root. To approximate the exact solution of equation [6.1], we can thus approximate the Taylor expansion of solution with a numerical method such as the well-known Runge–Kutta-type methods. Butcher (1972) discovered that the numerical methods and their convergence order can be studied by the so-called B-series (see also Chapter 4 of this book by A. Murua, where the important notion of order is explained). Moreover, the B-series form a group in which the multiplication of two elements yields the B-series associated with the composition of numerical methods. Later, in (Brouder 2004), Brouder discovered that the Butcher group can be identified with the character group of the Connes–Kreimer Hopf algebra. Taking this fruitful point of view, we will think of the Butcher group in the following as the character group $\mathcal{G}(\mathcal{H}_{CK}, \mathbb{R})$ of the Connes–Kreimer Hopf algebra of rooted trees. An element in the Butcher group (or, more generally, in the character groups we are about to consider) corresponds to a Taylor-like expansion of certain evolutions, induced, for example, by solutions generated by numerical schemes.

Summing up, for a smooth map f and a map which associates with every rooted tree a number (also known as a character $a \in \mathcal{G}(\mathcal{H}_{CK}, \mathbb{R})$ of the Connes–Kreimer Hopf algebra of rooted trees), we thus obtain a formal power series in a step-size parameter h, the so-called B-series

$$B_f(a, y, h) := y + \sum_{\tau \in \mathcal{T}} \frac{h^{|\tau|} a(\tau)}{\sigma(\tau)} F_f(\tau)(y), \tag{6.3}$$

Here, σ is a certain symmetry factor induced by the tree. Due to the Hairer and Wanner theorem (Hairer *et al.* 2006, Theorem III.1.10), taking the B-series of a product yields

$$B_f(\varphi \star \psi, y, h) = B_f(\psi, B_f(\varphi, y, h), h), \qquad \varphi, \psi \in \mathcal{G}(\mathcal{H}_{CK}, \mathbb{R}). \qquad [6.4]$$

Thus, for the group $\mathrm{Gf}_n(\mathbb{R})_1 = \{x + \sum_{|\mathbf{m}|>1} c_{\mathbf{m}} x^{\mathbf{m}}\}$[2] of pro-unipotent formal power series fixing $0 \in \mathbb{R}^n$, every smooth f and $y_0 = 0$ (chosen for simplicity) yield a group anti-homomorphism (due to the reversal of order in equation [6.4])

$$\mathrm{FB}_f \colon \mathcal{G}(\mathcal{H}_{\mathrm{CK}}, \mathbb{R}) \to \mathrm{Gf}_{n+1}(\mathbb{R})_1, \quad a \mapsto (B_f(a, y, h), h). \qquad [6.5]$$

Note that (for algebraic reasons), we have to consider h as an additional parameter such that the $(n+1)$st entry of all $c_{\mathbf{m}}$ vanishes (as suggested by the above pair notation).

To move on to geometry, we endow both groups with suitable (infinite-dimensional) Lie group structures (see (Bogfjellmo and Schmeding 2017, 2016)). The geometry then mirrors the algebra in that equation [6.5] becomes a Lie group anti-homomorphism. Furthermore, the geometric structure allows us to give intrinsic geometric meaning of certain constructions in numerical analysis, such as Lie derivatives and differential equations, on the groups. Note that these objects were originally studied in a purely formal algebraic way, see, for example, Hairer *et al.* (2006, Remark 9.4) and see the comments in Bogfjellmo and Schmeding (2017, Remark 2.2).

Our exposition has so far hinted at a beautiful and simple picture. As is often the case with applications, the truth is more complicated. Albeit (certain) numerical methods yield Taylor-like expansions and these expansions correspond to B-series, all sensible numerical schemes yield at least locally convergent Taylor series expansions. Due to the general setup of equation [6.5], algebra and topology on the character group do not provide control over the convergence behavior of the resulting B-series. In particular, we can show that every (topological) neighborhood of the identity element in the character group contains characters whose associated series expansion converges nowhere. We can even show that this can not be solved by choosing another "natural topology" (see Bogfjellmo *et al.* (2018, section 4, pp. 287–314) for a detailed discussion). One solution to eliminate the convergence problem is to truncate the B-series equation [6.3] above a certain size of the trees (i.e. truncate the Taylor expansion after finitely many orders of derivatives) and use approximation results. This is often sufficient for the numerical treatment of B-series and is indeed necessary to avoid differentiability and analyticity assumptions (see proposition 6.1). However,

2. Here, we use multi-index notation: $x^{\mathbf{m}} = x_1^{m_1} \cdots x_n^{m_n}$ for $\mathbf{m} = (m_1, \ldots m_n) \in \mathbb{N}_0^n$ and $c_{\mathbf{m}} \in \mathbb{R}^n$.

since all sane numerical schemes yield locally convergent B-series, this particular "solution" ignores the root of problem: the Butcher group is simply too large for the intended application.

To overcome this problem, we are therefore led to consider a subgroup of the Butcher group, which contains only elements leading to locally convergent B-series. To this end, we introduce growth bounds for a character $\varphi \in \mathcal{G}(\mathcal{H}_{\text{CK}}, \mathbb{R})$:

$$\exists C, K \in \mathbb{R} \text{ such that } |\varphi(\tau)| \leq CK^{|\tau|} \qquad \forall \tau \in \mathcal{T} \tag{6.6}$$

where $|\tau|$ is the order of the tree τ, i.e. the number of nodes in the tree. In view of the elementary differentials [6.2], the order of a tree is an upper bound for the order of iterated differential. If growth bounds for the derivatives of the vector field are available, equation [6.6] yields local convergence of equation [6.5]. For example, via Hairer and Lubich (1997), we deduce:

PROPOSITION 6.1 (Bogfjellmo and Schmeding 2016, Proposition 1.8).– *Let* $f\colon \mathbb{R}^n \supseteq U \to \mathbb{R}^n$ *be a real analytic vector field and* $\varphi \in \mathcal{G}(\mathcal{H}_{CK}, \mathbb{R})$*. If* φ *satisfies a bound equation* [6.6]*, then the B-series* $B_f(\varphi, y, h)$ *converges in a neighborhood of* y_0 *for all* h *small enough.*

On the algebraic level, estimates of the type equation [6.6] are stable under the group product of $\mathcal{G}(\mathcal{H}_{\text{CK}}, \mathbb{R})$ since equation [6.6] allows us to arbitrarily raise the constants. Thus, the growth restrictions lead to a subgroup $\mathcal{B}_{\text{TM}}$ of the character group, which is called the *tame Butcher group*. We can prove that the tame Butcher group is an infinite-dimensional Lie group (with a more detailed topology then that of the Butcher group) (Bogfjellmo and Schmeding 2016). This subgroup contains virtually all interesting elements of the Butcher group and is closely connected to a subgroup of bounded elements that already appears in Butcher's seminal work (Butcher 1972)[3]. Furthermore, for a real analytic f, the local convergence of the induced series allows us to write the correspondence as a Lie group antimorphism into a suitable group $\text{DiffGerm}_0(\mathbb{R}^n \times \mathbb{R})$ of germs of diffeomorphisms fixing 0. Summarizing these constructions, we obtain a commutative diagram of Lie group (anti-)morphisms

$$\begin{array}{ccc} \mathcal{B}_{\text{TM}} & \xrightarrow{a \mapsto \text{germ}_0(B_f(x,a,h),h)} & \text{DiffGerm}_0(\mathbb{R}^n \times \mathbb{R}) \\ \downarrow \subseteq & & \downarrow \text{Tay}_0 \\ \mathcal{G}(\mathcal{H}_{\text{CK}}, \mathbb{R}) & \xrightarrow{\text{FB}_f} & \text{Gf}_{n+1}(\mathbb{R}) \end{array} \tag{6.7}$$

3. The subgroup appearing in Butcher's work has a rather involved definition and therefore is currently unclear (though the author expects this to be the case) whether this subgroup coincides with the tame Butcher group or not.

Here, Tay_0 is the mapping taking the Taylor series at 0, see Bogfjellmo and Schmeding (2016, section 6). Hence, the analytic growth constraints lead to a (now more involved and possibly more interesting) algebro-geometric picture of the theory. We will develop this picture in greater generality in the rest of this chapter.

Before we continue, we wish to stress, however, that the analyticity requirement in proposition 6.1 is far stronger than the typical C^p requirements for convergence of numerical methods. Furthermore, in the setting of rough paths, the infinite Taylor-like series associated with the flow of a rough differential equation will, in general, not converge even under the above analyticity assumptions. That said, the aim of our exposition is certainly not to dismiss the usual analysis of numerical schemes or solutions to rough differential equations as superfluous. On the contrary, the aim is to offer a somewhat complementary point of view on Taylor-like series expansions, which adds new aspects to the analysis already available in the literature.

6.1.2. *Beyond the Butcher group*

The Butcher group is by no means the only interesting application of Hopf algebra characters. Different Hopf algebras and their characters are studied from the perspectives of numerical analysis (Munthe-Kaas and Wright 2008; Lundervold and Munthe-Kaas 2011, pp. 295–324), renormalization of quantum field theories (Manchon 2008, pp. 365–427; Connes and Marcolli 2008), the theory of rough paths (Hairer and Kelly 2015; Bruned *et al.* 2019) and control theory (Oberst 1985, pp. 214–269; Grossman and Larson 1992; Grünenfelder 1994; Gray *et al.* 2014; Foissy 2015). In all these contexts, the Hopf algebra encodes some combinatorial information, and the characters are related to certain Taylor-like series expansions relevant to applications (e.g. Chen–Fliess series in control theory (Duffaut Espinosa *et al.* 2016) or power series on spaces of coupling constants (Connes and Marcolli 2008, Chapter 1, 6.5)). Following in the spirit of the previous section on the Butcher group, our task is twofold:

1) to understand the geometry of the character groups for a large class of Hopf algebras and connect this to formal power series dictated by the application;

2) to find a suitable subgroup of characters, which satisfy sufficient growth bounds, which lead to local convergence of the series as required by the application.

Surely this program is too broad for us to cover it in the present chapter, as it requires in-depth knowledge of the application at hand (to single out suitable growth restrictions and the correct space of series). Thus, in this chapter we will just give an introduction to the geometry of groups of (suitably bounded) characters.

We will proceed in two steps: first, we will discuss the general theory for character groups of graded and connected Hopf algebras. Here, the emphasis will be that the geometry and topology mirror the algebra. Then, we will investigate ways to generalize the tame Butcher group. To this end, we have to restrict the choice of Hopf algebras to a class of Hopf algebras we call *combinatorial Hopf algebra* (following, in spirit, Loday and Ronco, see Loday and Ronco (2010, pp. 347–383)).

DEFINITION 6.1.– *A connected and graded Hopf algebra $\mathcal{H}$ is called* combinatorial *if it is isomorphic as an algebra to $\mathbb{R}[\Sigma]$ or to $\mathbb{R}\langle\Sigma\rangle$ (the algebra of commutative, respectively, noncommutative polynomials in Σ).*

The base Σ is part of the structure of a combinatorial Hopf algebra and in applications consists of combinatorial objects like trees, graphs, words or permutations. Using Σ, we can prescribe growth conditions, such as equation [6.6], referring to the combinatorial objects. Choosing suitable conditions, this leads to a subgroup of the character group called the group of "controlled characters". Again these groups turn out to be (infinite-dimensional) Lie groups (see Dahmen and Schmeding 2018).

What is not discussed in this chapter? To keep the exposition simple, we consider only characters of graded and connected Hopf algebras taking values that are real numbers. These conditions can be relaxed (while most results carry over) by allowing the following: characters may take values in a commutative Banach algebra (or even, more generally, in a commutative locally convex algebra) over real or complex numbers. Furthermore, some of the results also generalize to (a certain class of) non-connected Hopf algebras. We do not delve into these details and instead refer to Bogfjellmo *et al.* (2016), Bogfjellmo and Schmeding (2016, 2017) and Dahmen and Schmeding (2018).

6.2. Character groups of graded and connected Hopf algebras

In the rest of this section and the rest of the chapter, we fix a graded and connected Hopf algebra $(\mathcal{H}, m, 1_{\mathcal{H}}, \Delta, \epsilon, S)$ (we refer again to Chapter 1 for more information).

Denote by $\mathrm{Hom}(\mathcal{H}, \mathbb{R})$ the space of all linear mappings $\mathcal{H} \to \mathbb{R}$ and recall that the convolution product $\phi \star \psi := m_B \circ (\phi \otimes \psi) \circ \Delta$ turns $\mathrm{Hom}(\mathcal{H}, \mathbb{R})$ into a unital algebra (the unit being the counit ε of $\mathcal{H}$). We are then interested in the following subsets of the algebra of all linear maps:

– the *character group* $(G(\mathcal{H}, \mathbb{R}), \star)$ of $\mathcal{H}$ is defined as

$$G(\mathcal{H}, \mathbb{R}) := \left\{ \phi \in \mathrm{Hom}(\mathcal{H}, \mathbb{R}) \middle| {}^{\phi(ab)=\phi(a)\phi(b), \forall a,b \in \mathcal{H}}_{\text{and } \phi(1_{\mathcal{H}})=1} \right\}.$$

Inversion in $G(\mathcal{H},\mathbb{R})$ is induced by precomposition with the antipode, i.e. for a character $\iota(\varphi) = \varphi^{-1} = \varphi \circ S$. The counit ε of $\mathcal{H}$ thus becomes the unit of the character group;

– the *Lie algebra of infinitesimal characters* $(\mathfrak{g}(\mathcal{H},\mathbb{R}),[\cdot,\cdot])$ is given as

$$\mathfrak{g}(\mathcal{H},\mathbb{R}) := \{\psi \in \mathrm{Hom}(\mathcal{H},\mathbb{R}) \mid \psi(ab) = \psi(a)\varepsilon(b) + \varepsilon(a)\psi(b), \forall a,b \in \mathcal{H}\}$$

with the Lie bracket $[\psi,\kappa] := \psi \star \kappa - \kappa \star \psi$.

We now wish to add a topology to the algebra: choosing a vector space basis $\mathcal{B}$ for $\mathcal{H}$, we identify $\mathrm{Hom}(\mathcal{H},\mathbb{R}) \cong \mathbb{R}^{\mathcal{B}} := \prod_{b\in\mathcal{B}} \mathbb{R}$ as vector spaces. This suggests that we should topologize $\mathrm{Hom}(\mathcal{H},\mathbb{R})$ and the subsets $G(\mathcal{H},\mathbb{R})$ and $\mathfrak{g}(\mathcal{H},\mathbb{R})$ with the topology induced through the identification by the product topology on $\prod_{b\in\mathcal{B}} \mathbb{R}$. The resulting topology on the function spaces is also known as the *topology of pointwise convergence*. The reason for this is that a sequence $((x_b^n)_{b\in\mathcal{B}})_{n\in\mathbb{N}}$ converges in $\mathbb{R}^{\mathcal{B}}$ if and only if every component sequence $(x_b^n)_n$ converges in $\mathbb{R}$. Furthermore, this topology turns the point evaluation mappings $\mathrm{ev}_h \colon \mathrm{Hom}(\mathcal{H},\mathbb{R}) \to \mathbb{R}, \phi \mapsto \phi(h)$ into continuous maps, whence it is easy to see that $\mathfrak{g}(\mathcal{H},\mathbb{R})$ and $G(\mathcal{H},\mathbb{R})$ are closed subspaces. Moreover, $\mathrm{Hom}(\mathcal{H},\mathbb{R})$ becomes a locally convex topological vector space[4] such that the convolution $\star\colon \mathrm{Hom}(\mathcal{H},\mathbb{R}) \times \mathrm{Hom}(\mathcal{H},\mathbb{R}) \to \mathrm{Hom}(\mathcal{H},\mathbb{R})$ becomes a continuous bilinear map. In other words, $\mathrm{Hom}(\mathcal{H},\mathbb{R})$ is a locally convex algebra.

REMARK 6.1.– *In the algebraic treatment of characters, we often encounter another topology on the algebra* $\mathrm{Hom}(\mathcal{H},\mathbb{R})$ *(and on the (infinitesimal) characters). This topology is induced by the grading via the ultrametric*

$$d(x,y) = 2^{-ord(x-y)}, \quad ord(z) := \max\left\{ k\in\mathbb{N}_0 \cup \{\infty\} \,\middle|\, z(b) = 0 \quad \forall b \in \bigoplus_{n=1}^{k} \mathcal{H}_n \right\}.$$

The ultrametric topology measures the order in the grading up to which two elements agree. Its advantage is that it can be defined for any base field. The locally convex topology is restricted to $\mathbb{R}$ *(or* $\mathbb{C}$*) and is designed to enable differential calculus. To see that the ultrametric is unsuitable for this purpose, note that it induces the discrete topology on the subspace* $\mathrm{Hom}(\mathcal{H}_0,\mathbb{R}) \cong \mathbb{R}$.

Before we continue, we need a calculus which works beyond Banach spaces since, for an infinite-dimensional Hopf algebra, $\mathcal{H}$ will be at best a metrisable space (in case $\mathcal{B}$ is countable). To this end, we base our investigation on the so-called Bastiani calculus (requiring existence of directional derivatives). We urge the reader to consult the appendix, section 6.4 for a brief introduction. With respect to Bastiani calculus, the exponential and logarithm become smooth.

4. i.e. the vector space operations are continuous with respect to the topology and the topology is induced by a family of seminorms, see the appendix, section 6.4.

6.2.1. *The exponential and logarithm*

The formal exponential and logarithm series with respect to the convolution product converge and can be shown to yield continuous (and indeed smooth (Bogfjellmo *et al.* 2016, Lemma B.5, B.6)) mappings

$$\exp_\star\colon \mathfrak{g}(\mathcal{H},\mathbb{R}) \to G(\mathcal{H},\mathbb{R}), \psi \mapsto \sum_{n=0}^{\infty} \frac{\psi^{\star n}}{n!}$$

$$\log_\star\colon G(\mathcal{H},\mathbb{R}) \to \mathfrak{g}(\mathcal{H},\mathbb{R}), \varepsilon + \psi \mapsto \sum_{k=0}^{\infty} \frac{(-1)^k}{k} \varphi^{\star k}$$

In particular, $T_0 \exp_\star = \mathrm{id}_{\mathrm{Hom}\mathcal{H}\mathbb{R}}$ and this identifies the tangent space of the character group as $\mathfrak{g}(\mathcal{H},\mathbb{R})$. Moreover, from the algebraic identities, it is clear that $G(\mathcal{H},\mathbb{R}) = \varepsilon + \mathfrak{g}(\mathcal{H},\mathbb{R})$ is an affine subspace of $\mathrm{Hom}(\mathcal{H},\mathbb{R})$.

We can now establish the first main theorem on the Lie group structure of character groups.

THEOREM 6.1 (Bogfjellmo *et al.* 2016, Theorem 2.7).– *Let $\mathcal{H}$ be a graded and connected Hopf algebra. The topology of pointwise convergence then turns $(G(\mathcal{H},\mathbb{R}),\star)$ into an infinite-dimensional Lie group with associated Lie algebra $\mathfrak{g}(\mathcal{H},\mathbb{R})$. Moreover,* $\exp_\star$ *is the Lie group exponential of* $G(\mathcal{H},\mathbb{R})$.

PROOF.– Note that $G(\mathcal{H},\mathbb{R})$ is a closed affine subspace of $\mathrm{Hom}(\mathcal{H},\mathbb{R})$ (whence a smooth submanifold). Since $(\mathrm{Hom}(\mathcal{H},\mathbb{R}),\star)$ is a locally convex algebra, the map $\star$ is continuous bilinear, and therefore smooth. It restricts to a smooth map on $G(\mathcal{H},\mathbb{R})$. In the topology of pointwise convergence, the map $\mathrm{Hom}(\mathcal{H},\mathbb{R}) \to \mathrm{Hom}(\mathcal{H},\mathbb{R}), \psi \mapsto \psi \circ S$ is clearly continuous linear, and therefore smooth. It restricts to a smooth map on $G(\mathcal{H},\mathbb{R})$, whence the character group becomes a Lie group. As we have seen in the beginning of this section $\exp_\star$ is a diffeomorphism $\mathfrak{g}(\mathcal{H},\mathbb{R}) \to G(\mathcal{H},\mathbb{R})$, which shows that $\exp_\star$ is a global chart for the Lie group. As already discussed, the Lie algebra of $G(\mathcal{H},\mathbb{R})$ must be $\mathfrak{g}(\mathcal{H},\mathbb{R})$ (as a vector space). To identify the Lie bracket, we takes the derivative $\mathrm{Ad}(\phi) := T_\varepsilon c_\phi$ of the inner automorphism $c_\phi\colon G(\mathcal{H},\mathbb{R}) \to G(\mathcal{H},\mathbb{R}), \psi \mapsto \phi\star\psi\star\phi^{-1} = \phi\star\psi\star(\phi\circ S)$ at the unit ε. Since c_φ is the restriction of a continuous linear map to $G(\mathcal{H},\mathbb{R})$, we find for the Lie bracket (see (Neeb 2006, Example II.3.9) for the general method)

$$[\phi,\psi] = T_\varepsilon \mathrm{Ad}(\phi).\psi = \phi\star\psi + \psi\star(\phi\circ S) = \phi\star\psi - \psi\star\phi.$$

Here, we have used the fact that composition with the antipode S sends an infinitesimal character to its additive inverse (as S is an algebra antimorphism, we have $\exp_\star(-\phi) = \exp_\star(\phi)^{-1} = \exp_\star(\phi)\circ S = \exp_\star(\phi\circ S)$ for every infinitesimal character). Now, finally, as $\frac{d}{dt}\big|_{t=0} \exp_\star(t\phi) = \phi$, we see that $\gamma_\phi(t) := \exp_\star(t\phi)$ is the unique one-parameter group in $G(\mathcal{H},\mathbb{R})$, whose derivative at the unit is ϕ. We conclude that $\exp_\star$ is the Lie group exponential. □

The proof given for theorem 6.1 follows the one given in Bogfjellmo *et al.* (2016). Nevertheless, it is useful to recall these details since they will clarify an important algebraic property of the character group, which we discuss now. Namely, we will show that the character group $G(\mathcal{H}, \mathbb{R})$ is a pro-Lie group.

DEFINITION 6.2.– *A topological group G is called* pro-Lie group *if it is the projective limit of a directed system of finite-dimensional (real) Lie groups.*

For a pro-Lie group G, the space $\mathcal{L}(G)$ of continuous G-valued one parameter subgroups endowed with the compact-open topology is called the pro-Lie algebra *associated with G.*

Pro-Lie groups are a very well-studied class of topological groups (see the extensive monograph (Hofmann and Morris 2007; Hofmann and Neeb 2009) for their relation to infinite-dimensional Lie groups).

PROPOSITION 6.2.– *The group $G(\mathcal{H}, \mathbb{R})$ is a pro-Lie group and $\mathcal{L}(G(\mathcal{H}, \mathbb{R})) = \mathfrak{g}(\mathcal{H}, \mathbb{R})$.*

PROOF.– Note first that the algebra $(\mathrm{Hom}(\mathcal{H}, \mathbb{R}), \star)$ depends only on the coalgebra structure of $\mathcal{H}$. Now the fundamental theorem of coalgebras asserts that any coalgebra is an inductive limit of finite-dimensional coalgebras, i.e. $\mathcal{H} = \lim_{m\to\infty} \mathcal{C}_m$. Dualizing, we obtain a projective limit

$$(\mathrm{Hom}(\mathcal{H}, \mathbb{R}), \star) \cong \lim_{m\to\infty} (\mathrm{Hom}(\mathcal{C}_m, \mathbb{R}), \star)$$

in the category of topological algebras and on the level of unit groups $\mathrm{Hom}(\mathcal{H}, \mathbb{R})^\times = \lim_{m\to\infty} \mathrm{Hom}(\mathcal{C}_m, \mathbb{R})^\times$ (projective limit in the category of topological groups)[5]. Note that every $\mathrm{Hom}(\mathcal{C}_m, \mathbb{R})^\times$ is a finite-dimensional (linear) Lie group, whence we have an isomorphism (of Lie algebras) $(\mathrm{Hom}(\mathcal{C}_m, \mathbb{R}), [\cdot,\cdot]) \to \mathrm{Hom}(\mathcal{C}_m, \mathbb{R}), tx \mapsto (t \mapsto \exp(tx))$, where exp is the exponential series with respect to the algebra product and $[\cdot,\cdot]$ is the commutator bracket.

Now, $G(\mathcal{H}, \mathbb{R})$ is a closed topological subgroup of the pro-Lie group $\mathrm{Hom}(\mathcal{H}, \mathbb{R})^\times$, therefore a pro-Lie group (see Hofmann and Morris (2007, Theorem 3.39)). To identify the pro-Lie algebra, observe every continuous one-parameter subgroup of $G(\mathcal{H}, \mathbb{R})$ as a 1-parameter subgroup of $\mathrm{Hom}(\mathcal{H}, \mathbb{R})$, whence it is of the form $t \mapsto \exp_\star(t\phi)$ for some $\phi \in \mathrm{Hom}(\mathcal{H}, \mathbb{R})$. We already know that $\exp_\star$ is the Lie group exponential of $G(\mathcal{H}, \mathbb{R})$ and a bijection $\mathfrak{g}(\mathcal{H}, \mathbb{R}) \to C(\mathcal{H}, \mathbb{R})$. Thus, we conclude that $\mathcal{L}(G(\mathcal{H}, \mathbb{R})) = \mathfrak{g}(\mathcal{H}, \mathbb{R})$. □

5. On the algebraic level, the identity is clear. The topological level was clarified in Dahmen's fundamental theorem of weakly complete algebras (Bogfjellmo *et al.* 2016, Lemma 5.3).

In applications in numerical analysis and rough path theory, we often do not work with the character group but with a finite-dimensional (nilpotent) Lie group obtained by a "truncation procedure". Let us identify these truncated groups with the steps of a projective limit realizing $G(\mathcal{H}, \mathbb{R})$. Note first that due to the grading

$$\mathcal{C}_N := \bigoplus_{0 \leq n \leq N} \mathcal{H}_n$$

with the induced coproduct becomes a coalgebra. If we assume that the steps $\mathcal{H}_n$ are finite-dimensional (i.e. the Hopf algebra is of finite type, which is the case for the natural grading of almost all Hopf algebras appearing in these applications), then the $\mathcal{C}_N$ are finite-dimensional. Dualizing, the linear Lie groups $A_N := \mathrm{Hom}(\mathcal{H}, \mathbb{R})^\times$ contain

$$G_N := \{\phi \colon \mathcal{C}_N \to \mathbb{R} \mid \phi(ab) = \phi(a)\phi(b), \text{ if } a, b, ab \in \mathcal{C}_N\}$$

as a closed subgroup (since the topology on A_N is the topology of pointwise convergence). Closed subgroups of finite-dimensional Lie groups are Lie groups, whence G_N is a finite-dimensional nilpotent Lie group[6]. We now easily conclude, as in the proof of proposition 6.2, that $G(\mathcal{H}, \mathbb{R}) = \lim_{N \to \infty} G_N$ (projective limit in the category of topological groups). We observe that the elements of the group G_N are just characters truncated after the Nth level. Thus, the applications mentioned exploit the algebro-geometric fact that the full character group is a projective limit of the finite-dimensional truncated groups.

REMARK 6.2.–

1) Pro-Lie groups admit a remarkable structure theory along the lines of finite-dimensional Lie theory, see Hofmann and Morris (2007). For a character group, one is even in a more special situation: since theorem 6.1 implies that the Lie group exponential of a character group is a diffeomorphism, the Lie group is exponential, and this entails that the group $G(\mathcal{H}, \mathbb{R})$ is pro-solvable (Hofmann and Neeb 2009, S7), i.e. its Lie algebra is pro-solvable (Hofmann and Morris 2007, Theorem 10.18). Recall that a Lie algebra is pro-solvable if every finite-dimensional quotient of it is a solvable Lie algebra (meaning that its derived series terminates after finitely many steps).

2) As a result of these structural properties, the (unitary) representation theory of the character groups becomes manageable:[7] Somewhat similarly to the representation theory of compact groups, Neeb (2010, S12) shows that every unitary representation of $G(\mathcal{H}, \mathbb{R})$ is a direct sum of representations, which factor over finite-dimensional

6. It is not hard to see that these groups are actually even more special in that they are Carnot groups, i.e. they admit a Carnot–Caratheordory metric. This structure is used in the theory of rough paths. See Friz and Hairer (2014) for a friendly introduction.

7. The author wishes to thank K.–H. Neeb for pointing this out to him.

quotients. Thus, the classification of irreducible unitary representations reduces to finite-dimensional representations. Since $G(\mathcal{H}, \mathbb{R})$ is pro-solvable, the representation theory of solvable groups shows that equivalence classes of irreducibles are classified by coadjoint orbits of the character group (following Kirilov's "orbit philosophy", which becomes feasible since the dual space of the weakly complete algebra $\mathfrak{g}(\mathcal{H}, \mathbb{R})$ is "small").

Note that this has practical application: In Chevyrev and Lyons (2016), finite-dimensional unitary representations of certain character groups have been used to study moments of the signature of geometric rough paths.

This concludes our brief overview of the group of characters of a graded and connected Hopf algebra. We will now move on to certain subgroups of the character groups of combinatorial Hopf algebras.

6.3. Controlled groups of characters

In this section, we will discuss how subgroups of the group of characters can be constructed where the members satisfy certain growth bounds. Similar to the growth restrictions for the tame Butcher group (see section 6.1), we need a family of objects against which we can check the growth.

DEFINITION 6.3.– *Consider a graded and connected Hopf algebra $\mathcal{H}$ together with a subset $\Sigma \subseteq \mathcal{H}$. We call the pair $(\mathcal{H}, \Sigma)$* combinatorial Hopf algebra *if $\mathcal{H}$ is, as an associative algebra, isomorphic to $\mathbb{K}[\Sigma]$ (commutative polynomial algebra) or $\mathbb{K}\langle\Sigma\rangle$ (noncommutative polynomial algebra). A combinatorial Hopf algebra is of* finite type, *if, as graded Hopf algebras $\mathcal{H} = \bigoplus_{n\in\mathbb{N}} \mathcal{H}_n$ and every $\mathcal{H}_n$ is finite-dimensional*

We will think of Σ as the given combinatorial basis, i.e. in all concrete examples, Σ will be a set of combinatorial objects such as trees and words. For the combinatorial Hopf algebra $(\mathcal{H}, \Sigma)$, the basis (and thus the isomorphism with the polynomial algebra) is part of the structure of a combinatorial Hopf algebra.

6.3.1. *Conventions for this section*

For the rest of this section, we will only consider (combinatorial) Hopf algebras of finite type. To shorten the notation, we will drop "of finite type". We remark that the theory we are about to develop can also be generalized to Hopf algebras which are not of finite type, at the cost of several more technical assumptions and arguments.

Note that there are different ways to endow a given Hopf algebra with the structure of a combinatorial Hopf algebra. We will see two natural structures in the example of the Faá di Bruno algebra below.

EXAMPLE 6.1.–

1) The Butcher–Connes–Kreimer algebra $\mathcal{H}_{CK}$ *is a combinatorial Hopf algebra as it is isomorphic to the polynomial algebra* $\mathbb{R}[\mathcal{T}]$, *where* $\mathcal{T}$ *is the set of rooted trees with the grading by the number of nodes.*

2) Fix a set $\mathcal{A} \neq \emptyset$ *(called an alphabet). The shuffle algebra* $Sh(\mathcal{A})$ *(Reutenauer 1993) is the (noncommutative) polynomial algebra* $\mathbb{R}\langle\Sigma\rangle$[8]. *For* $a, b \in \mathcal{A}$ *("letters") and products ("words") of elements in* $\mathcal{A}$, u, w, *define recursively the* shuffle product

$$1 ⧢ w = w ⧢ 1 = w, \quad (au) ⧢ (bw) := a(u ⧢ (bw)) + b((au) ⧢ w)$$

The (bilinear extensions of the) shuffle product and the deconcatenation of words

$$\Delta(a_1 \cdot a_2 \cdot \dots \cdot a_n) = \sum_{i=0}^{n} a_1 \cdot a_2 \cdot \dots \cdot a_i \otimes a_{i+1} \cdot a_{i+2} \cdot \dots \cdot a_n$$

turn $\mathbb{R}\langle\mathcal{A}\rangle$ *with the grading by word length into a graded and connected Hopf algebra (see Chapter 1 on how to obtain an antipode). Given a total order on* $\mathcal{A}$, $Sh(\mathcal{A})$ *is isomorphic to the commutative polynomial algebra* $\mathbb{R}[\Sigma]$ *generated by the set of Lyndon words* Σ *due to Radford's theorem (Reutenauer 1993, section 6).*

The next example shows that a given graded and connected Hopf algebra can have several non-isomorphic[9] structures as a combinatorial Hopf algebra.

EXAMPLE 6.2 (The commutative Faà di Bruno algebra).– *The* Faà di Bruno algebra *is the combinatorial Hopf algebra* $\mathcal{H}_{FdB}$ *given by the following data:*

– fix a graded index set $\Sigma := \{a_1, a_2, \dots\}$ *with* $|a_n| := n$. *Conceptually, the variables correspond to the coordinate functions of formal diffeomorphisms, i.e.* $a_n(\phi) = \frac{1}{(n+1)!}\frac{\mathrm{d}^{n+1}}{\mathrm{d}x^{n+1}}\phi(0)$;

– as an algebra $\mathcal{H}_{FdB}$ *is the commutative polynomial algebra* $\mathbb{R}[\Sigma]$ *with the connected grading induced by* Σ. *The counit* $\epsilon\colon \mathcal{H}_{FdB} \to \mathbb{R}$ *is projection onto the* 0*th component (as in every connected Hopf algebra);*

– setting $a_0 := 0$, *the coproduct* $\Delta\colon \mathcal{H}_{FdB} \to \mathcal{H}_{FdB} \otimes \mathcal{H}_{FdB}$ *is given by the formula*

$$\Delta(a_n) := \sum_{r=0}^{n} \sum_{\substack{\beta_1+2\beta_2+\dots+n\beta_n=n-r \\ \beta_0+\beta_1+\dots+\beta_n=r+1}} \frac{(r+1)!}{\beta_0!\beta_1!\cdots\beta_n!} a_r \otimes a_1^{\beta_1}\cdots a_n^{\beta_n}. \qquad [6.8]$$

8. Note that this algebra is noncommutative with respect to the usual multiplication of polynomials, whereas the shuffle product defined now turns it into a commutative algebra.

9. Here, "isomorphic" means in the sense of the category of combinatorial Hopf algebras introduced in definition 6.7.

– *the antipode* $S\colon \mathcal{H}_{FdB} \to \mathcal{H}_{FdB}$ *is given (Brouder* et al. *2006) by the abelianization of*

$$S(a_n) := -a_n - \sum_{r=1}^{n-1}(-1)^r \sum_{\substack{n_1+\cdots+n_r+n_{r+1}=n \\ n_1,\ldots,n_{r+1}>0}} \lambda(n_1,\ldots,n_r)\, a_{n_1}\cdots a_{n_r} a_{n_{r+1}}, \qquad [6.9]$$

$$\textit{where } \lambda(n_1,\ldots,n_r) = \sum_{\substack{m_1+\cdots+m_r=r \\ m_1+\cdots+m_h\geq h \\ h=1,\ldots,r-1}} \binom{n_1+1}{m_1}\cdots\binom{n_r+1}{m_r}. \qquad [6.10]$$

The Faà di Bruno Hopf algebra encodes the combinatorial structure of the composition of formal diffeomorphisms on vector spaces, and it is well known that this composition can be described via the famous Faà di Bruno formula (see, for example (Frabetti and Manchon 2015, pp. 91–147)).

EXAMPLE 6.3 (Another combinatorial Hopf algebra structure on the Faà di Bruno algebra).– *Taking the Faà di Bruno algebra as in example 6.2, we construct another combinatorial basis by scaling the variables*

$$\Sigma' := \{X_n :- (n+1)!a_n \mid n \in \mathbb{N}\}$$

and consider $\mathcal{H}_{FdB}$ *as a combinatorial Hopf algebra with respect to* Σ'. *Introducing the coordinate transformation, the coproduct* [6.8] *changes to*

$$\Delta(X_n) = \sum_{k=0}^{n} X_k \otimes B_{n+1,k+1}(X_0, X_1, X_2, \ldots, X_{n-k+1}) \quad n \in \mathbb{N} \qquad [6.11]$$

where $B_{n+1,k+1}$ *is the partial Bell polynomial and* $X_0 = 1$. *For the antipode, an easy computation shows that equation* [6.9] *yields the following in the new variables.*

$$S(X_n) = -X_n - \sum_{r=1}^{n-1}(-1)^r \sum_{\substack{n_1+\cdots+n_r+n_{r+1}=n \\ n_1,\ldots,n_{r+1}>0}} \lambda(n_1,\ldots,n_r)\frac{(n+1)!X_{n_1}\cdots X_{n_r}X_{n_{r+1}}}{(n_1+1)!\cdots((n_{r+1}+1)!)} \qquad [6.12]$$

This yields another combinatorial Hopf algebra structure, which we will see is not isomorphic to the one from example 6.2.

Having fixed a combinatorial structure, our goal is to study functions which grow in a controlled way, i.e. satisfy certain growth restrictions with respect to the combinatorial basis. To make this precise, we fix a family of functions controlling this growth:

DEFINITION 6.4 (Growth family).– *A family* $(\omega_k)_{k\in\mathbb{N}}$ *of functions* $\omega_k : \mathbb{N}_0 \to \mathbb{N}$ *is called a* growth family *if it satisfies the following conditions. For all* $k \in \mathbb{N}$ *and* $n, m \in \mathbb{N}_0$

(W1) $\omega_k(0) = 1$ *and* $\omega_k(n) \leq \omega_{k+1}(n)$.

(W2) $\omega_k(n) \cdot \omega_k(m) \leq \omega_k(n+m)$.

(W3) For all $k_1 \in \mathbb{N}$, *there exists* $k_2 \geq k_1$ *such that* $\omega_{k_2}(n) \geq 2^n \cdot \omega_{k_1}(n)$.

Let us now collect some standard examples of growth families.

EXAMPLE 6.4.– *Hopf algebra and application refer to typical examples that are treated with these growth families.*

$\omega_k(n)$	Application	Hopf algebra
k^n	B-series, (Hairer and Lubich 1997)	(colored) Connes–Kreimer
2^{kn}	P-series, (Bogfjellmo and Schmeding 2016)	algebra of rooted trees
$k^n \cdot n!$	Chen–Fliess series, (Foissy 2015; Gray *et al.* 2014)	Faà di Bruno type
k^{n^2}	Formal power	Faà di Bruno
$k^n(n!)^k$	Series[a]	

[a]The growth bounds will be used to illustrate certain technical details of the theory.

Table 6.1. *Standard examples for growth families (Dahmen and Schmeding 2018, Proposition 1.4)*

A growth family will allow us to restrict the growth of a character on the combinatorial basis. To this end, we define the space of controlled linear maps.

DEFINITION 6.5 (Controlled functions).– *Let* $(\mathcal{H}, \Sigma)$ *be a combinatorial Hopf algebra and* $\{\omega_k\}_{k\in\mathbb{N}}$ *a growth family.*

1) Let $k \in \mathbb{N}$. *A map* $f : \Sigma \to \mathbb{R}$ *is called* k-controlled *if* $\sup_{\tau\in\Sigma} \frac{|f(\tau)|}{\omega_k(|\tau|)} < \infty$. *The space of all k-controlled functions* $\ell^\infty_k(\Sigma, \mathbb{R})$ *is a Banach space with the norm* $\|f\|_{\ell^\infty_k} := \sup_{\tau\in\Sigma} \frac{|f(\tau)|}{\omega_k(|\tau|)}$. *We call a map* $f : \Sigma \to \mathbb{R}$ controlled *if there is a* $k \in \mathbb{N}$ *such that* f *is* k*-controlled. The* space of all controlled functions *is denoted by* $\ell^\infty_{\to}(\mathcal{H}, \mathbb{R})$ *and we endow it with the structure of a locally convex space as the locally convex limit of the Banach spaces* $\ell^\infty_k(\mathcal{H}, \mathbb{R})$.

2) A (infinitesimal) character $\phi\colon \mathcal{H} \to \mathbb{R}$ *is called a* controlled (infinitesimal) character *if it restricts to a controlled function on* Σ. *We denote the set of controlled characters by* $G_{\mathrm{ctr}}(\mathcal{H}, \mathbb{R})$ *and the set of controlled infinitesimal characters by* $\mathfrak{g}_{\mathrm{ctr}}(\mathcal{H}, \mathbb{R})$.

Note that so far we do not know whether the controlled characters form a subgroup of the group of characters. Since multiplication and inversion in the character group is induced by the coproduct and the antipode of the Hopf algebra, growth estimates for these mappings are needed to ensure that the group structure of the character group restricts to the controlled characters. This leads to the notion of control pair, in which the coproduct and antipode are asked to satisfy an ℓ^1-type estimate.

DEFINITION 6.6.– *Let* $(\mathcal{H}, \Sigma)$ *be a combinatorial Hopf algebra and* $(\omega_n)_{n\in\mathbb{N}}$ *a growth family. We call* $((\mathcal{H}, \Sigma), \{\omega_n\}_{n\in\mathbb{N}})$ *a* control pair *if* $\Delta\colon \mathcal{H} \to \mathcal{H} \otimes \mathcal{H}$ *and* $S\colon \mathcal{H} \to \mathcal{H}$ *are* $\ell^1_{\leftarrow}$*-continuous with respect to the growth family* $(\omega_n)_{n\in\mathbb{N}}$.

Explicitly, this means for $M = \Sigma^*$ *or* $M = \Sigma^\dagger$ *(the (non-)commutative monoid generated by* Σ*)*[10] *and* $\tau \in M$ *the following: expanding the images* $S(\tau)$, $\Delta(\tau)$ *of the antipode and the coproduct with respect to the basis* M, *we ask for a constant* $C > 0$ *such that for each* $k_1 \in \mathbb{N}$, *there is a* $k_2 \in \mathbb{N}$ *such that*

$$\|S(\tau)\|_{\ell^1_{k_1}} := \sum_{\sigma\in M} |c_\sigma|\, \omega_{k_1}(|\tau|) \leq C\omega_{k_2}(|\tau|)$$

$$\|\Delta(\tau)\|_{\ell^1_{k_1}} := \sum_{\sigma_1,\sigma_2\in M} |c_{\sigma_1\sigma_2}|\, \omega_{k_1}(|\sigma_1| + |\sigma_2|) \leq C\omega_{k_2}(|\tau|)$$

EXAMPLE 6.5 (Control pairs for the Butcher–Connes–Kreimer algebra).– *Consider the algebra* $\mathcal{H}_{CK} := \mathbb{R}[\mathcal{T}]$ *of polynomials which is generated by the set of rooted trees* $\mathcal{T}$. $\mathcal{H}_{CK}$ *is graded by the number of nodes (in a tree or forest) grading.* $M = \mathcal{T}^\dagger$ *can then be identified with the set of forests of rooted trees. We recall from A. Murua's contribution to this volume that the coproduct and antipode are defined on forests as follows:*

$$\Delta\colon \mathcal{H}_{CK} \to \mathcal{H}_{CK}\mathcal{H}_{CK}, \quad u \mapsto \sum_{(V\succ W)\subset U} v \otimes \otimes w,$$

$$S\colon \mathcal{H}_{CK} \to \mathcal{H}_{CK}, \quad u \mapsto -u - \sum_{m\geq 2} (-1)^m \sum_{\substack{V_1\succ V_2 \succ V_m)\subset U \\ V_1,\dots V_m}} v_1 \otimes \cdots \otimes v_m$$

10. The noncommutative case deals with $\mathcal{H} \cong \mathbb{R}\langle\Sigma\rangle$, and the commutative case deals with $\mathcal{H} \cong \mathbb{R}[\Sigma]$. In both cases, we have $\mathcal{H} \cong \mathbb{R}^{(M)} := \{\sum_{\sigma\in M} c_\sigma \sigma \mid$ all but finitely many c_σ are zero$\}$.

It is not hard to see that the set of admissible cuts $(V \succ W) \subset U$ contains less then $2^{|u|}$ elements. Similarly, we can use an explicit formula for the antipode on trees and the fact that the antipode is an algebra morphism (see, for example (Bogfjellmo and Schmeding 2016, Appendix B)) to see that the formula for the antipode contains less than $2^{|u|}$ non-zero terms. Hence, $((\mathcal{H}_{CK}, \mathcal{T}), (\omega_n)_{n\in\mathbb{N}})$ is a control pair for all growth families in definition 6.4. In particular, we can consider the controlled characters with respect to the growth family $\omega_k(n) = k^n$. These coincide with the tame Butcher group discussed in section 6.1.

Having ℓ^1-type estimates, we can exploit the duality between ℓ^1-spaces and ℓ^∞-spaces to prove that the controlled characters form a subgroup of all characters:

PROPOSITION 6.3.– *Let $(\mathcal{H}, \Sigma), \{\omega_k\}_{k\in\mathbb{N}})$ be a control pair. Then, the controlled characters $G_{\mathrm{ctr}}(\mathcal{H}, \mathbb{R})$ form a subgroup of $G(\mathcal{H}, \mathbb{R})$ and $\mathfrak{g}_{\mathrm{ctr}}(\mathcal{H}, \mathbb{R})$ forms a Lie subalgebra of $\mathfrak{g}(\mathcal{H}, \mathbb{R})$.*

PROOF.– Note that the counit $\varepsilon\colon \mathcal{H} \to \mathbb{R}$ is automatically $\ell^1_{\leftarrow}$-continuous since it is just the projection on $\mathcal{H}_0$ for a graded and connected Hopf algebra. Hence, the unit $\mathbf{1} = \varepsilon$ is controlled, i.e. $\mathbf{1} \in G_{\mathrm{ctr}}(\mathcal{H}, \mathbb{R})$.

Now consider the set $\ell^\infty_{\rightarrow}(\mathcal{H}, \mathbb{R})$ of all linear mappings $f\colon \mathcal{H} \to \mathbb{R}$, which satisfy for some $k \in \mathbb{N}$ the estimate

$$\|f\|_{\ell^\infty_k} := \sup_{\tau\in M} \frac{|f(\tau)|}{\omega_k(|\tau|)} < K, \quad K \text{ constant}, \qquad [6.13]$$

where $M = \Sigma^*$ (noncommutative case) or $M = \Sigma^\dagger$ (commutative case). Note that $G_{\mathrm{ctr}}(\mathcal{H}, \mathbb{R})$ and $\mathfrak{g}_{\mathrm{ctr}}(\mathcal{H}, \mathbb{R})$ are subsets of $\ell^\infty_{\rightarrow}(\mathcal{H}, \mathbb{R})$. We will show that $\ell^\infty_{\rightarrow}(\mathcal{H}, \mathbb{R})$ is closed under convolution. To this end, fix $k_1 \in \mathbb{N}$ and consider $\phi, \psi \in \ell^\infty_{\rightarrow}(\mathcal{H}, \mathbb{R})$, which satisfy equation [6.13] with respect to k_1. We construct $k_2 \in \mathbb{N}$ such that $\phi \star \psi$ satisfies equation [6.13] with respect to k_2. Since Δ is $\ell^1_{\rightarrow}$-continuous, we can choose $k_2 > k_1$ and $C > 0$ such that for each $\tau \in M$ with $\Delta(\tau) = \sum_{\mu,\sigma\in M} c_{\mu,\sigma}\mu \otimes \sigma.$, we have

$$\sum_{\mu,\sigma} |c_{\mu,\sigma}|\, \omega_{k_1}(|\mu| + |\sigma|) \leq C\omega_{k_2}(|\tau|). \qquad [6.14]$$

This allows us the following computation:

$$\begin{aligned}
|\phi \star \psi(\tau)| = |(\phi \otimes \psi) \circ \Delta(\tau)| &= \left|\sum_{\mu,\sigma} c_{\mu,\sigma}\phi(\mu)\psi(\sigma)\right| \\
&\leq \sum_{\mu,\sigma} |c_{\mu,\sigma}|\, |\phi(\mu)|\, |\psi(\sigma)| \\
&\leq \sum_{\mu,\sigma} |c_{\mu,\sigma}|\, \|\phi|_M\|_{\ell^\infty_{k_1}}\, \omega_{k_1}(|\mu|)\, \|\psi|_M\|_{\ell^\infty_{k_1}}\, \omega_{k_1}(|\sigma|)
\end{aligned}$$

$$\overset{6.4}{\leq} \|\phi|_M\|_{\ell^\infty_{k_1}} \|\psi|_M\|_{\ell^\infty_{k_1}} \sum_{\mu,\sigma} |c_{\mu,\sigma}| \omega_{k_1}(|\mu| + |\sigma|)$$

$$\overset{[6.14]}{\leq} \|\phi|_M\|_{\ell^\infty_{k_1}} \|\psi|_M\|_{\ell^\infty_{k_1}} C\omega_{k_2}(|\tau|).$$

Divide both sides of the inequality by $\omega_{k_2}(|\tau|)$ and move to the supremum over $\tau \in M$ to see that $\phi \star \psi$ is contained in $\ell^\infty_{k_2}(\mathcal{H}, B)$. Thus, $\ell^\infty_{\rightarrow}(\mathcal{H}, \mathbb{R})$ is stable under convolution and we deduce that $\mathfrak{g}_{\mathrm{ctr}}(\mathcal{H}, \mathbb{R})$ becomes a Lie subalgebra of $\mathfrak{g}(\mathcal{H}, \mathbb{R})$. Furthermore, $G_{\mathrm{ctr}}(\mathcal{H}, \mathbb{R})$ is stable under convolution. To see that it is a subgroup, we recall that for a character $\varphi^{-1} = \varphi \circ S$. Since the antipode is $\ell^1_{\leftarrow}$-continuous, a similar calculation as for the coproduct shows that φ^{-1} is a controlled character if φ is a controlled character. □

We can even prove that for a control pair, the associated controlled characters form a manifold and this structure turns the convolution and inversion into smooth maps. We collect these results in the following theorem, for whose proof we refer to Dahmen and Schmeding (2018, Theorem 3.12 and Proposition 3.15).

THEOREM 6.2.– *Let* $(\mathcal{H}, \Sigma), \{\omega_k\}_{k\in\mathbb{N}})$ *be a control pair and* $G_{\mathrm{ctr}}(\mathcal{H}, \mathbb{R})$ *and* $\mathfrak{g}_{\mathrm{ctr}}(\mathcal{H}, \mathbb{R})$ *be the associated controlled (infinitesimal) characters. Then,* $(G_{\mathrm{ctr}}(\mathcal{H}, \mathbb{R}), \star)$ *is a Lie group modeled on the space* $\ell^\infty_{\rightarrow}(\Sigma, \mathbb{R})$. *Its Lie algebra is* $(\mathfrak{g}_{\mathrm{ctr}}(\mathcal{H}, \mathbb{R}), [\cdot, \cdot])$, *where* $[\cdot, \cdot]$ *is the commutator induced by the convolution product.*

Note that the topological and differentiable structure on the (infinitesimal) characters is not the one induced by the structures on the group $G(\mathcal{H}, \mathbb{R})$ or the Lie algebra $\mathfrak{g}(\mathcal{H}, \mathbb{R})$. Instead it is properly finer, turning the inclusion $\iota\colon G_{\mathrm{ctr}}(\mathcal{H}, \mathbb{R}) \to G(\mathcal{H}, \mathbb{R})$ into a morphism of Lie groups.

Finally, we construct a category of combinatorial Hopf algebras:

DEFINITION 6.7 (The category of combinatorial Hopf algebras).– *A Hopf algebra morphism between combinatorial Hopf algebras* $F\colon (\mathcal{H}, \Sigma) \to (\widetilde{\mathcal{H}}, \widetilde{\Sigma})$ *will be called a* morphism of combinatorial Hopf algebras *if*

– *it is of degree* 0, *i.e.* $F(\mathcal{H}_n) \subseteq \widetilde{\mathcal{H}}_n, \forall n \in \mathbb{N}_0$;

– *there is* $C > 0$ *constant, such that for every* $\tau \in \Sigma$ *with* $F(\tau) = \sum_{\sigma\in\widetilde{\Sigma}} c^\tau_\sigma \sigma$, *the estimate* $\sum_{\sigma\in\widetilde{\Sigma}} |c^\tau_\sigma| < C^{|\tau|}$ *holds.*

Combinatorial Hopf algebras and their morphisms form a category **CombHopf**.

Note that the category **CombHopf** is not a subcategory of the category of graded and connected Hopf algebras, as a Hopf algebra might support several combinatorial structures (see the two structures on the Faá di Bruno Hopf algebra).

The definition of morphism is geared towards preserving the controlled morphisms, i.e. we establish the following:

LEMMA 6.1.– *Let* $((\mathcal{H},\Sigma),\{\omega_k\}_{k\in\mathbb{N}})$ *and* $((\widetilde{\mathcal{H}},\widetilde{\Sigma}),\{\omega_k\}_{k\in\mathbb{N}})$ *be control pairs and* $F\colon(\mathcal{H},\Sigma)\to(\widetilde{\mathcal{H}},\widetilde{\Sigma})$ *be a morphism of combinatorial Hopf algebras, then for every controlled character* $\phi\colon\widetilde{\mathcal{H}}\to B$, $\phi\circ F$ *is also a controlled character.*

PROOF.– Assume that φ is a k-controlled character, then

$$\sup_{\tau\in\Sigma}\frac{|\phi(\sum_{\sigma\in\tilde{\sigma}}c_\sigma^\tau\sigma)|}{\omega_k(|\sigma|)}\leq\sup_{\tau\in\Sigma}\sum_{\sigma\in\widetilde{\Sigma}}|c_\sigma^\tau|\frac{|\phi(\sigma)|}{\omega_k(|\tau|)}\leq\sup_{\tau\in\Sigma}C^{|\tau|}\frac{|\phi(\sigma)|}{\omega_k(|\tau|)}. \quad [6.15]$$

The last inequality is due to $|\sigma|=|\tau|$ since F is of degree 0. Applying the definition of a growth family from definition 6.4, we see that by replacing k with some larger K, the right-hand side of equation [6.15] is bounded, and therefore $\phi\circ F$ is controlled. □

The idea in introducing the group of controlled characters was to have growth estimates available to guarantee local convergence of associated (formal) power series. However, looking at the Butcher group case in section 6.1, we see that the growth estimates themselves are not enough to guarantee convergence. Indeed, we needed to divide by a symmetry factor generated by the Taylor expansion. Thus, if we want to carry the convergence behavior to other types of controlled character groups, it seems to be indispensable to require such a renormalization by a certain factor depending on the elements in the combinatorial base (A. Murua, private communication). However, the symmetry factor from Taylor series expansions is related to the so-called tree factorial through the so-called Connes–Moscovici character (see Brouder (2004, section 4) for a detailed exposition of the symmetry factor and related notions). Recently, a generalization of the tree factorial (and thus also the symmetry factor)[11] has been proposed.

6.3.2. *Combinatorial Hopf algebras and the inverse factorial character*

In Curry *et al.* (2018, section 3), for a graded and connected Hopf algebra, *inverse factorial characters* were introduced: for a graded and connected Hopf algebra, $\mathcal{H}=\bigoplus_{n\in\mathbb{N}}\mathcal{H}_n$ and $\alpha\colon\mathcal{H}_1\to\mathbb{R}$ linear, we define the inverse factorial character $\varphi_\alpha\in G(\mathcal{H},\mathbb{R})$ as the unique character satisfying:

- $\varphi_\alpha(1)=1$;
- $\varphi_\alpha|_{\mathcal{H}_1}=\alpha$;
- $\varphi_\alpha\star\varphi_\alpha(x)=2^{|x|}\varphi_\alpha(x)$.

11. To carry the analogy further, a generalization of the Connes–Moscovici character would be needed. However, up until now, it is unclear how such a generalization could be achieved to carry the link beyond the case of the Butcher group.

In the case of $\mathcal{H} = \mathrm{Sh}(\mathcal{A})$ being the shuffle algebra, we can prove that $\varphi_\alpha(w) = \frac{1}{|w|!}$ for α being the linear map with $\alpha(a) = 1, \forall a \in \mathcal{A}$.

REMARK 6.3.– *In applications involving iterated integrals, the appearance of the inverse factorial factor is linked with the decay rate of iterated integrals over a fixed interval. Formulated differently, we can understand the tree factorial as (essentially) the volume of the iterated integral $\int_D \mathrm{d}g(t_1)\cdots \mathrm{d}g(t_n)$ with respect to the control path $g(t) = t$ over the domain D, which is generated by $t_1, \dots, t_n$ obeying the partial order induced by the given tree (of n-nodes). We refer to Curry* et al. *(2018, section 6) for a detailed exposition of the iterated integrals. Viewing an ordinary differential equation as a controlled differential equation with respect to the control $g(t)$, it becomes clear why tree factorials appear as coefficients in the Taylor expansion of the associated flow. Going back to the algebro-geometric level, these observations have a direct impact on what kind of growth family provides convergence within the associated character group.*

The inverse factorial characters seem to be the right generalization of the tree factorials. Moreover, it can be shown that under suitable assumptions on the Hopf algebra (Curry *et al.* 2018, section 3.2), inverse factorial characters are functorial and these Hopf algebras with their inverse factorial characters form a category called the category of combinatorial Hopf algebras in loc.cit. In general, their category of combinatorial Hopf algebras and our category **CombHopf** are incomparable, as the notion of combinatorial Hopf algebras is incomparable (though a key idea in both approaches is to single out a specific generating set). However, all the classical examples of combinatorial Hopf algebras considered in this section are contained in both categories of combinatorial Hopf algebras. This suggests that the intersection of both categories of combinatorial Hopf algebras will lead to a category of Hopf algebras for which the correspondence $G(\mathcal{H}, \mathbb{R}) \leftrightarrow \{\text{formal power series}\}$ factors through a correspondence $G_{\mathrm{ctr}}(\mathcal{H}, \mathbb{R}) \leftrightarrow \{\text{locally convergent power series}\}$. Thus, the intersection of the two categories should provide a natural framework for applications in which local convergence of these series is needed. This, however, is beyond the scope of the current chapter.

6.4. Appendix: Calculus in locally convex spaces

In this appendix, we recall several well-known facts on calculus in locally convex spaces.

DEFINITION 6.8.– *Let E be a vector space over the real numbers with a topology T.*

1) (E, T) is called topological vector space, *if the vector space operations are continuous with respect to T and the metric topology on $\mathbb{R}$.*

2) A topological vector space (E,T) is called locally convex space *if there is a family $\{p_i : E \to [0,\infty[\mid i \in I\}$ of continuous seminorms for some index set I, such that:*

- T is the initial topology with respect to the projections $\{q_i : E \to E/p_i^{-1}(0)\}_{i\in I}$ onto the normed spaces $E/p_i^{-1}(0)$;

- if $x \in E$ with $p_i(x) = 0$ for all $i \in I$, then $x = 0$ (the seminorms separate the points, i.e. T has the Hausdorff property).

The topology T is then generated by the family of seminorms $\{p_i\}_{i\in I}$. *Usually, we suppress T and write $(E, \{p_i\}_{i\in I})$ or simply E instead of (E,T).*

3) A locally convex space $(E, \{p_i\}_{i\in\mathbb{N}})$ with a countable systems of seminorms is metrisable, and if E is complete, it is called Fréchet space.

DEFINITION 6.9.– *Let $r \in \mathbb{N}_0 \cup \{\infty\}$ and E, F be locally convex vector spaces and $U \subseteq E$ open. We say a map $f : U \to F$ is a C^r-map if it is continuous and the iterated directional derivatives*

$$d^k f(x, y_1, \ldots, y_k) := (D_{y_k} \cdots D_{y_1} f)(x)$$

exist for all $k \in \mathbb{N}_0$ with $k \leq r$ and $y_1, \ldots, y_k \in E$ and $x \in U$, and the mappings $d^k f : U \times E^k \to F$ so obtained are continuous. If f is C^∞, we say that f is smooth.

The composition of C^r-maps (if possible) is again a $C^r_{\mathbb{K}}$-map (see Glöckner 2002, Propositions 2.7 and 2.9, pp. 43–59).

6.4.1. C^r-Manifolds and C^r-mappings between them

For $r \in \mathbb{N}_0 \cup \{\infty\}$, manifolds modeled on a fixed locally convex space can be defined as usual. We will always assume that the manifold as a topological space is Hausdorff. Direct products of locally convex manifolds, tangent spaces and tangent bundles as well as C^r-maps between manifolds may be defined as in the finite-dimensional setting. For C^r-manifolds M, N, we use the notation $C^r(M, N)$ for the set of all C^r-maps from M to N. We define a *(locally convex) Lie group* to be a group with a C^∞-manifold structure, which turns the group operations into smooth maps.

6.5. References

Bogfjellmo, G. and Schmeding, A. (2016). The tame Butcher group. *J. Lie Theory*, 26(4), 1107–1144.

Bogfjellmo, G. and Schmeding, A. (2017). The Lie group structure of the Butcher group. *Found. Comput. Math.*, 17(1), 127–159.

Bogfjellmo, G., Dahmen, R., Schmeding, A. (2016). Character groups of Hopf algebras as infinite-dimensional Lie groups. *Ann. Inst. Fourier (Grenoble)*, 66(5), 2101–2155.

Bogfjellmo, G., Dahmen, R., Schmeding, A. (2018). Overview of Lie group structures on Hopf algebra character groups. In *Discrete Mechanics, Geometric Integration and Lie-Butcher Series*, Barbero, M. and Ebrahimi-Fard, K. (eds). Springer, Cham.

Brouder, C. (2004). Trees, renormalization and differential equations. *BIT*, 44(3), 425–438.

Brouder, C., Frabetti, A., Krattenthaler, C. (2006). Non-commutative Hopf algebra of formal diffeomorphisms. *Adv. Math.*, 200(2), 479–524.

Bruned, Y., Chevyrev, I., Friz, P.K., Preiß, R. (2019). A rough path perspective on renormalization. *J. Funct. Anal.*, 277(11), 108283.

Butcher, J.C. (1972). An algebraic theory of integration methods. *Math. Comp.*, 26, 79–106.

Chevyrev, I. and Lyons, T. (2016). Characteristic functions of measures on geometric rough paths. *Ann. Probab.*, 44(6), 4049–4082.

Connes, A. and Marcolli, M. (2008). *Noncommutative Geometry, Quantum Fields and Motives*. American Mathematical Society, Providence, RI, and Hindustan Book Agency, New Delhi.

Curry, C., Ebrahimi-Fard, K., Manchon, D., Munthe-Kaas, H.Z. (2018). Planarly branched rough paths and rough differential equations on homogeneous spaces. *J. Differ. Equ.*, 269(2020), 9740–9782.

Dahmen, R. and Schmeding, A. (2020). Lie groups of controlled characters of combinatorial Hopf algebras. *Annales de l'institut Henri Poincaré D*, 7(3), 395–456.

Duffaut Espinosa, L.A., Ebrahimi-Fard, K., Gray, W.S. (2016). A combinatorial Hopf algebra for nonlinear output feedback control systems. *J. Algebr.*, 453, 609–643.

Foissy, L. (2015). The Hopf algebra of Fliess operators and its dual pre-Lie algebra. *Comm. Algebr.*, 43(10), 4528–4552.

Frabetti, A. and Manchon, D. (2015). Five interpretations of Faà di Bruno's formula. In *Faà di Bruno Hopf Algebras, Dyson-Schwinger Equations, and Lie-Butcher Series*, Ebrahimi-Fard, K. and Fauvet, F. (eds). European Mathematical Society, Zürich.

Friz, P.K. and Hairer, M. (2014). *A Course on Rough Paths*. Springer, Cham.

Glöckner, H. (2002). Infinite-dimensional Lie groups without completeness restrictions. In *Geometry and Analysis on Finite- and Infinite-dimensional Lie Groups*, Strasburger, A. (ed.)., Banach Center Publications, Polish Academy of Sciences, Warsaw.

Gray, W.S., Duffaut Espinosa, L.A., Ebrahimi-Fard, K. (2014). Faà di Bruno Hopf algebra of the output feedback group for multivariable Fliess operators. *Syst. Control Lett.*, 74, 64–73.

Grossman, R. and Larson, R.G. (1992). The realization of input–output maps using bialgebras. *Forum Math.*, 4(2), 109–121.

Grünenfelder, L. (1994). Algebraic aspects of control systems and realizations. *J. Algebr.*, 165(3), 446–464.

Hairer, M. and Kelly, D. (2015). Geometric versus non-geometric rough paths. *Ann. Inst. Henri Poincaré Probab. Stat.*, 51(1), 207–251.

Hairer, E. and Lubich, C. (1997). The life-span of backward error analysis for numerical integrators. *Numer. Math.*, 76(4), 441–462.

Hairer, E., Lubich, C., Wanner, G. (2006). *Geometric Numerical Integration*, 2nd edition. Springer-Verlag, Berlin.

Hofmann, K.H. and Morris, S.A. (2007). *The Lie Theory of Connected Pro-Lie Groups*. European Mathematical Society, Zürich.

Hofmann, K.H. and Neeb, K.-H. (2009). Pro-Lie groups which are infinite-dimensional Lie groups. *Math. Proc. Cambridge Philos. Soc.*, 146(2), 351–378.

Loday, J.-L. and Ronco, M. (2010). Combinatorial Hopf algebras. *Quanta of Maths, Clay Mathematics Proceedings*, Blanchard, É., Ellwood, D., Khalkhali, M., Marcolli, M., Moscovici, H., Popa, S. (eds). Clay Mathematics Institute, Cambridge, American Mathematical Society, Providence, RI.

Lundervold, A. and Munthe-Kaas, H.Z. (2011). Hopf algebras of formal diffeomorphisms and numerical integration on manifolds. In *Combinatorics and Physics*, Ebrahimi-Fard, K., Marcolli, M., van Suijlekom, W.D. (eds). American Mathematical Society, Providence, RI.

Manchon, D. (2008). Hopf algebras in renormalisation. In *Handbook of Algebra*, Hazewinkel, M. (ed.). Elsevier, Amsterdam.

Munthe-Kaas, H.Z. and Wright, W.M. (2008). On the Hopf algebraic structure of Lie group integrators. *Found. Comput. Math.*, 8(2), 227–257.

Neeb, K.-H. (2006). Towards a Lie theory of locally convex groups. *Jpn J. Math.*, 1(2), 291–468.

Neeb, K.-H. (2010). On differentiable vectors for representations of infinite dimensional Lie groups. *J. Funct. Anal.*, 259(11), 2814–2855.

Oberst, U. (1985). Actions of formal groups on formal schemes. Applications to control theory and combinatorics. In *Séminaire d'algèbre Paul Dubreil et Marie-Paule Malliavin : Procédés Paris 1983–1984 (36ème Année)*, Malliavin, M.-P. (ed.). Springer, Berlin.

Reutenauer, C. (1993). *Free Lie Algebras*. The Clarendon Press, Oxford University Press, New York.

List of Authors

Kurusch EBRAHIMI-FARD
Norwegian University of Science and Technology
Trondheim
Norway

Matthias KAWSKI
Arizona State University
Tempe
USA

Abdenacer MAKHLOUF
IRIMAS-Department of Mathematics
University of Haute Alsace
Mulhouse
France

Dominique MANCHON
University of Clermont Auvergne
Clermont-Ferrand
France

Ander MURUA
University of the Basque Country
UPV/EHU
Donostia
Spain

Frédéric PATRAS
CNRS
Côte d'Azur University
Nice
France

Alexander SCHMEDING
Technische Universität Berlin
Germany

Jean-Yves THIBON
Gustave Eiffel University
Marne-la-Vallée
France

Index

Printed and bound by CPI Group (UK) Ltd, Croydon, CR0 4YY
13/09/2026
14961044-0005